Selected Papers from Experimental Stress Analysis 2020

Selected Papers from Experimental Stress Analysis 2020

Editors

Radim Kocich
Lenka Kunčická

MDPI • Basel • Beijing • Wuhan • Barcelona • Belgrade • Manchester • Tokyo • Cluj • Tianjin

Editors
Radim Kocich
VŠB—Technical University of Ostrava
Czech Republic

Lenka Kunčická
Czech Academy of Sciences
Czech Republic

Editorial Office
MDPI
St. Alban-Anlage 66
4052 Basel, Switzerland

This is a reprint of articles from the Special Issue published online in the open access journal *Materials* (ISSN 1996-1944) (available at: https://www.mdpi.com/journal/materials/special_issues/EAN2020).

For citation purposes, cite each article independently as indicated on the article page online and as indicated below:

LastName, A.A.; LastName, B.B.; LastName, C.C. Article Title. *Journal Name* **Year**, *Volume Number*, Page Range.

ISBN 978-3-0365-3455-8 (Hbk)
ISBN 978-3-0365-3456-5 (PDF)

Cover image courtesy of Lenka Kunčická

© 2022 by the authors. Articles in this book are Open Access and distributed under the Creative Commons Attribution (CC BY) license, which allows users to download, copy and build upon published articles, as long as the author and publisher are properly credited, which ensures maximum dissemination and a wider impact of our publications.
The book as a whole is distributed by MDPI under the terms and conditions of the Creative Commons license CC BY-NC-ND.

Contents

About the Editors

Radim Kocich, Ph.D. Prof. Kocich has been studying plastic deformation of non-ferrous materials for about 25 years. His specialization is on methods of severe plastic deformation, one of which he invented (twist channel angular pressing, TCAP), and the design and optimization of thermomechanical processing. He also has extensive experience with numerical modeling of deformation processes via the finite element method and experimental evaluation of the mechanical properties of materials. He participated in numerous scientific projects both national and international, authored or co-authored more than 260 papers, 3 books, and 4 book chapters, acquired 3 patents, and developed 2 utility designs and 7 software programs.

Lenka Kunčická, Ph.D. Dr. Kunčická has extensive experience in the field of materials forming, including thermomechanical processing and severe plastic deformation processes. She also focuses on powder metallurgy and on the subsequent deformation of compacted specimens. In addition, she has experience with numerical modeling of plastic deformation tasks. She specializes in structure analyses via electron microscopy. She has authored or co-authored more than 60 publications, most of which are in high-impact journals.

Editorial

Special Issue: Selected Papers from Experimental Stress Analysis 2020

Lenka Kunčická [1], Radim Halama [2,*] and Martin Fusek [2]

[1] Institute of Physics of Materials, Czech Academy of Sciences, 61662 Brno, Czech Republic; kuncicka@ipm.cz
[2] Faculty of Mechanical Engineering, VŠB-Technical University of Ostrava, 70800 Ostrava, Czech Republic; martin.fusek@vsb.cz
* Correspondence: radim.halama@vsb.cz

Citation: Kunčická, L.; Halama, R.; Fusek, M. Special Issue: Selected Papers from Experimental Stress Analysis 2020. *Materials* **2021**, *14*, 1136. https://doi.org/10.3390/ma14051136

Received: 23 February 2021
Accepted: 24 February 2021
Published: 28 February 2021

Publisher's Note: MDPI stays neutral with regard to jurisdictional claims in published maps and institutional affiliations.

Copyright: © 2021 by the authors. Licensee MDPI, Basel, Switzerland. This article is an open access article distributed under the terms and conditions of the Creative Commons Attribution (CC BY) license (https://creativecommons.org/licenses/by/4.0/).

The contemporary way of living brings about increasing demands on materials used in our everyday lives. Challenging requirements are put on the performance of modern metallic, as well as non-metallic, materials throughout commerce and industry. These requirements are not only from the viewpoint of their lifetime and durability, but also in the context of their user friendliness and safety. A great example of the importance of human safety in everyday life is presented by the published paper studying the effects of a collision involving a pedestrian with a tramway vehicle [1].

A great method that is of help during the research and development of innovative materials and components is the finite element method (FEM). Based on numerical computations performed with reliable models of real assemblies with optimized boundary conditions [2], FEM is not only useful for the thorough testing of high-level properties of durable metallic materials and components [3,4], but is also helpful in numerous other applications. Such applications include the abovementioned investigation of the impacts of a pedestrian/tram collision [1], simulations of the performance of bio-applicable components and materials [5–7], and in solving challenging industrial tasks [8,9].

However, the greatest impact on the successful and satisfactory performance of innovative components has been on the type of the material used, and the applied production and processing technologies. In addition to conventional alloys [10] and commercially pure elements [11], modern materials also include light and durable metallic composites [12], carbon tube reinforced polymers (CTRP) [13], and thermoplastic composites [14]. Additive manufacturing (or 3D printing) of powders, including methods such as selective laser melting (SLM) [3], has recently gained a great deal of attention. However, conventional production technologies, such as casting and welding [10,15], are still popular, especially for specific applications (large components or semi-products). Nevertheless, conventional (rolling, forging, etc. [16]), as well as unconventional (severe plastic deformation (SPD) methods, rotary swaging, etc. [17,18]) processing methods are applicable for both types of semi-products, i.e., conventionally produced and 3D printed. Optimized selection of the methods and conditions of processing can greatly enhance the performance and properties of virtually any semi-product.

Last but not least, research of innovative materials and components goes hand in hand with the development of modern testing methods, advantageously non-destructive ones, some of which are also of interest to the presented Special Issue. Popular methods are those based on optical systems, such as the PhotoStress and digital image correlation (DIC) methods [19], or the acoustic emission method based on transferring acoustic signals through the tested material [20]. Neutron diffraction is also among the top-level non-destructive testing methods [21,22].

Conflicts of Interest: The authors declare no conflict of interest.

References

1. Špirk, S.; Špička, J.; Vychytil, J.; Křížek, M.; Stehlík, A. Utilization of the Validated Windshield Material Model in Simulation of Tram to Pedestrian Collision. *Materials* **2021**, *14*, 265. [CrossRef] [PubMed]
2. Kocich, R.; Greger, M.; Macháčková, A. Finite element investigation of influence of selected factors on ECAP process. In Proceedings of the Metal 2010, Rožnov pod Radhoštěm, Czech Republic, 18–20 May 2010; pp. 166–171.
3. Kořínek, M.; Halama, R.; Fojtík, F.; Pagáč, M.; Krček, J.; Krzikalla, D.; Kocich, R.; Kunčická, L. Monotonic Tension-Torsion Experiments and FE Modeling on Notched Specimens Produced by SLM Technology from SS316L. *Materials* **2021**, *14*, 33. [CrossRef] [PubMed]
4. Kocich, R.; Kunčická, L.; Davis, C.F.; Lowe, T.C.; Szurman, I.; Macháčková, A. Deformation behavior of multilayered Al-Cu clad composite during cold-swaging. *Mater. Des.* **2016**, *90*, 379–388. [CrossRef]
5. Kunčická, L.; Kocich, R.; Drápala, J.; Andreyachshenko, V. FEM simulations and comparison of the ecap and ECAP-PBP influence on Ti6Al4V alloy's deformation behaviour. In Proceedings of the Metal 2013, Brno, Czech Republic, 15–17 May 2013; pp. 391–396.
6. Marsalek, P.; Sotola, M.; Rybansky, D.; Repa, V.; Halama, R.; Fusek, M.; Prokop, J. Modeling and Testing of Flexible Structures with Selected Planar Patterns Used in Biomedical Applications. *Materials* **2021**, *14*, 140. [CrossRef] [PubMed]
7. Zach, L.; Kunčická, L.; Růžička, P.; Kocich, R. Design, analysis and verification of a knee joint oncological prosthesis finite element model. *Comput. Biol. Med.* **2014**, *54*, 53–60. [CrossRef] [PubMed]
8. Kocich, R. Design and optimization of induction heating for tungsten heavy alloy prior to rotary swaging. *Int. J. Refract. Met. Hard Mater.* **2020**, *93*, 105353. [CrossRef]
9. Macháčková, A.; Kocich, R.; Bojko, M.; Kunčická, L.; Polko, K. Numerical and experimental investigation of flue gases heat recovery via condensing heat exchanger. *Int. J. Heat Mass Transf.* **2018**, *124*, 1321–1333. [CrossRef]
10. Čapek, J.; Trojan, K.; Kec, J.; Černý, I.; Ganev, N.; Němeček, S. On the Weldability of Thick P355NL1 Pressure Vessel Steel Plates Using Laser Welding. *Materials* **2021**, *14*, 131. [CrossRef] [PubMed]
11. Kunčická, L.; Kocich, R.; Král, P.; Pohludka, M.; Marek, M. Effect of strain path on severely deformed aluminium. *Mater. Lett.* **2016**, *180*, 280–283. [CrossRef]
12. Kunčická, L.; Kocich, R.; Strunz, P.; Macháčková, A. Texture and residual stress within rotary swaged Cu/Al clad composites. *Mater. Lett.* **2018**, *230*, 88–91. [CrossRef]
13. Kostroun, T.; Dvořák, M. Application of the Pulse Infrared Thermography Method for Nondestructive Evaluation of Composite Aircraft Adhesive Joints. *Materials* **2021**, *14*, 533. [CrossRef]
14. Hron, R.; Kadlec, M.; Růžek, R. Effect of the Test Procedure and Thermoplastic Composite Resin Type on the Curved Beam Strength. *Materials* **2021**, *14*, 352. [CrossRef]
15. Thomasová, M.; Seiner, H.; Sedlák, P.; Frost, M.; Ševčík, M.; Szurman, I.; Kocich, R.; Drahokoupil, J.; Šittner, P.; Landa, M. Evolution of macroscopic elastic moduli of martensitic polycrystalline NiTi and NiTiCu shape memory alloys with pseudoplastic straining. *Acta Mater.* **2017**, *123*, 146–156. [CrossRef]
16. Khalaj, O.; Jirková, H.; Burdová, K.; Stehlík, A.; Kučerová, L.; Vrtáček, J.; Svoboda, J. Hot Rolling vs. Forging: Newly Developed Fe-Al-O Based OPH Alloy. *Metals* **2021**, *11*, 228. [CrossRef]
17. Kunčická, L.; Kocich, R. Deformation behaviour of Cu-Al clad composites produced by rotary swaging. In *IOP Conference Series: Materials Science and Engineering*; IOP publishing: Bristol, UK, 2018; Volume 369, p. 012029.
18. Kocich, R.; Kunčická, L.; Král, P.; Macháčková, A. Sub-structure and mechanical properties of twist channel angular pressed aluminium. *Mater. Charact.* **2016**, *119*, 75–83. [CrossRef]
19. Pástor, M.; Hagara, M.; Virgala, I.; Kal'avský, A.; Sapietová, A.; Hagarová, L. Design of a Unique Device for Residual Stresses Quantification by the Drilling Method Combining the PhotoStress and Digital Image Correlation. *Materials* **2021**, *14*, 314. [CrossRef] [PubMed]
20. Šofer, M.; Cienciala, J.; Fusek, M.; Pavlíček, P.; Moravec, R. Damage Analysis of Composite CFRP Tubes Using Acoustic Emission Monitoring and Pattern Recognition Approach. *Materials* **2021**, *14*, 786. [CrossRef] [PubMed]
21. Wang, Z.; Chen, J.; Besnard, C.; Kunčická, L.; Kocich, R.; Korsunsky, A.M. In situ neutron diffraction investigation of texture-dependent Shape Memory Effect in a near equiatomic NiTi alloy. *Acta Mater.* **2021**, *202*, 135–148. [CrossRef]
22. Mikula, P.; Ryukhtin, V.; Šaroun, J.; Strunz, P. High-Resolution Strain/Stress Measurements by Three-Axis Neutron Diffractometer. *Materials* **2020**, *13*, 5449. [CrossRef] [PubMed]

 materials

Article

Damage Analysis of Composite CFRP Tubes Using Acoustic Emission Monitoring and Pattern Recognition Approach

Michal Šofer [1,*], Jakub Cienciala [1], Martin Fusek [1], Pavel Pavlíček [1] and Richard Moravec [2]

1 Department of Applied Mechanics, Faculty of Mechanical Engineering, VŠB—Technical University of Ostrava, 17. listopadu 2172/15, 708 00 Ostrava, Czech Republic; jakub.cienciala@vsb.cz (J.C.); martin.fusek@vsb.cz (M.F.); pavel.pavlicek@vsb.cz (P.P.)

2 Havel Composites CZ s.r.o., Svésedlice 67, 783 54 Přáslavice, Czech Republic; moravecr@havel-composites.com

* Correspondence: michal.sofer@vsb.cz; Tel.: +420-731-664-248

Abstract: The acoustic emission method has been adopted for detection of damage mechanisms in carbon-fiber-reinforced polymer composite tubes during the three-point bending test. The damage evolution process of the individual samples has been monitored using the acoustic emission method, which is one of the non-destructive methods. The obtained data were then subjected to a two-step technique, which combines the unsupervised pattern recognition approach utilizing the short-time frequency spectra with the boundary curve enabling the already clustered data to be additionally filtered. The boundary curve identification has been carried out on the basis of preliminary tensile tests of the carbon fiber sheafs, where, by overlapping the force versus time dependency by the acoustic emission activity versus time dependency, it was possible to identify the boundary which will separate the signals originating from the fiber break from unwanted secondary sources. The application of the presented two-step method resulted in the identification of the failure mechanisms such as matrix cracking, fiber break, decohesion, and debonding. Besides the comparison of the results with already published research papers, the study presents the comprehensive parametric acoustic emission signal analysis of the individual clusters.

Keywords: acoustic emission; CFRP composite tube; unsupervised learning approach; failure mechanism

Citation: Šofer, M.; Cienciala, J.; Fusek, M.; Pavlíček, P.; Moravec, R. Damage Analysis of Composite CFRP Tubes Using Acoustic Emission Monitoring and Pattern Recognition Approach. *Materials* **2021**, *14*, 786. https://doi.org/10.3390/ma14040786

Academic Editor: Michele Bacciocchi

Received: 12 November 2020

Accepted: 28 January 2021

Published: 7 February 2021

Publisher's Note: MDPI stays neutral with regard to jurisdictional claims in published maps and institutional affiliations.

Copyright: © 2021 by the authors. Licensee MDPI, Basel, Switzerland. This article is an open access article distributed under the terms and conditions of the Creative Commons Attribution (CC BY) license (https://creativecommons.org/licenses/by/4.0/).

1. Introduction

Over the past decades, carbon-fiber-reinforced polymer (CFRP) composites have shown a constant increase in a variety of applications such as car or aircraft components, sports and medical equipment [1], and recently also additive manufacturing [2–4]. Their main benefit lies primarily in the relatively high strength/weight ratio or the ability to customize the material properties for dedicated purposes by changing the stacking sequence and related fiber orientation. A relative drawback of CRFP composites is the lack of ductile-like behavior and the corresponding absence of pre-warning phase before the structural collapse [5,6] leading to the brittle failure. CRFP composites are also characterized by the accumulation of damage inside the structure without any evidence on the structure surface [6] thus leading to a relatively challenging damage assessment. There are many non-destructive testing approaches, which can be applied on composite structures, namely infrared tomography [7], eddy current testing [8], ultrasonic testing [9], and X-ray tomography [10].

One of the most promising approaches, especially coupled with other methods [11] such as Scanning Electron Microscopy (SEM), is the acoustic emission (AE) method, which is also used in various applications as a real time monitoring tool [6]. The AE method exhibits great sensitivity including considerable reliability of active cracks detection [12], even in the case of initiation phase [13]. The AE technique is even capable of detecting the onset of plastic deformation [14], which has the character of white noise with low energy [15]. For gaining a more detailed insight into the damage monitoring process within the meaning of AE source characterization, it is favorable to incorporate an adequate signal analysis tool. The supervised/unsupervised pattern recognition (UPR) approach [16] has become a very suitable and promising approach to tackle a wide variety of problems such as fatigue tests [17], structure health monitoring [18], and condition assessment of pressure vessels [19] and pressure components in operation [20]. Numerous studies [21–26] have been conducted in order to assess characteristic features of the AE transients originating from various failure mechanism in the CFRP composites such as matrix cracking, delamination, fiber break, and debonding (see Table 1 for further explanation).

Table 1. Basic characterization of damage mechanisms occurring in carbon-fiber-reinforced polymer (CFRP) composites.

Damage Mechanism	Characterization
Fiber break	Disintegration of single and/or multiple carbon fibers
Delamination	Separation of two adjacent plies (Interface failure)
Debonding	Integrity failure between fiber and matrix (Interface failure)
Matrix cracking	Nucleation and further propagating of (micro)cracks in the matrix

Although Chou [21] points to a discrepancy concerning, in particular, the signal amplitude, duration as well as frequency spectra of the individual damage mechanisms, it was possible to compile a general overview, which is given in the following table (Table 2).

Table 2. Summary of the acoustic emission (AE) signal characteristics for given damage mechanisms in CFRP composites

Damage Mechanism	AE Signal Characteristics *
Fiber break	A: 50–100 dB_{AE}, D: 100–10,000 μs, f = 300–700 kHz
Matrix micro cracks	A: 30–40 dB_{AE}, D: <1000 μs, f = 100–250 kHz
Matrix micro cracks (propagation)	A: 40–80 dB_{AE}, D: 1000–10,000 μs, f = 100–250 kHz
Delamination	A: >70 dB_{AE}, D: 1000–10,000 μs, f = 250–300 kHz
Debonding	A: <60 dB_{AE}, f $\cong$ 300 kHz

* A—amplitude, D—duration.

In the last decade, several/numerous studies utilizing advanced techniques for classification of failure modes, such as the use of statistical analysis of wavelet coefficients [27] or infrared thermography (IT) [28], have been conducted. Another interesting approach can be found in the work published by Munoz et al. [29], who identified and further characterized the damage mechanisms in the unidirectional CFRP composites subjected to axis and off-axis static tensile tests using the acoustic emission method and infrared thermography. Further utilization of unsupervised pattern recognition technique together with the IT method resulted in the identification of the failure mechanisms such as matrix cracking, fiber breakage, and interface failure, for which the characterization in terms of the signal amplitude or energy has been performed. In 2011, Gutkin et al. published an extensive research [30], in which the AE signal data from various test configurations were analyzed by three different pattern recognition approaches. The analysis resulted in characteristic frequency spectra for matrix cracking, delamination, debonding, fiber pull-out, and fiber failure. It has to be noted that the given findings in terms of the frequency spectra are to some extent similar to the results summarized in Table 2 and therefore confirms the factual accuracy of the study [30].

The main objective of this study is to investigate and comprehensively describe the AE signal characteristics of the damage mechanisms in three different types of CFRP composite tubes using a two-step method combining the unsupervised pattern recognition approach with the utilization of the boundary curve. The construction of the boundary curve has been conducted on the data from the preliminary carbon fiber sheaf tensile tests. The already identified boundary curve has then been used for further refinement of the data across individual clusters. Using the presented approach, it was possible to identify a total of four damage mechanisms presented in Table 1 with subsequent comparison of the obtained results with the already published research papers. The part of the study is also the comprehensive AE waveform analysis of the representative signals belonging to the individual clusters.

2. Experimental Procedure

2.1. Test Sample Characterization

The experiments were carried out on three types of CFRP tubes with a different number of layers, their orientation, and woven fiber density of the used material (see Figure 1), where each type of CFRP tube has been represented by three test samples. The samples labeled "A" were manufactured using four layers of unidirectional carbon woven fabric with density of 200 g/m^2 and one layer of aramid/carbon woven fabric (0–90°) with density of 175 g/m^2 and average wall thickness of 1.45 mm. The production of samples labeled "B" included the use of two layers of unidirectional carbon woven fabric with density 300 g/m^2 and one layer of carbon woven fabric (0–90°) with density 280 g/m^2 with average wall thickness of 0.9 mm, while samples labeled "C" were manufactured using solely four layers of unidirectional carbon woven fabric with density of 300 g/m^2 with average wall thickness of 1.42 mm. Table 3 summarizes the specification of the tested CFRP tubes.

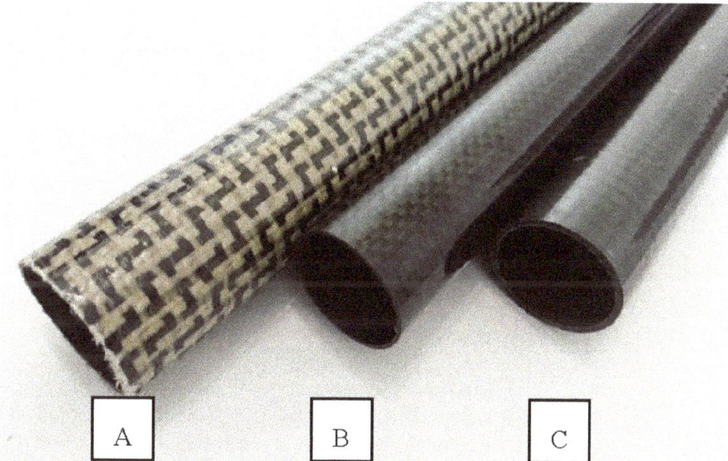

Figure 1. CFRP composite tubes under test; A—A series sample, B—B series sample, C—C series sample.

Table 3. Specification of the tested CFRP tubes

Property	A Series	B Series	C Series
Wall th. (mm)/Diameter (mm)	1.45/32	0.9/32	1.42/32
Fabrication	4 layers of 200 g/m^2 unidir. carbon fabric1 layer of 175 g/m^2 aramid/carbon fabric (0°–90°)	2 layers of 300 g/m^2 unidir. carbon fabric1 layer of 280 g/m^2 carbon fabric (0°–90°)	4 layers of 300 g/m^2 unidir. carbon fabric

2.2. Three-Point Bending Test

The three considered types of CFRP composite tubes under test are being used for paddle production; therefore, the three-point bending test has been selected in order to simulate as much as possible the real nature of the loading process during the use of the given sports equipment. The experiments were carried out on the universal Testometric M500-50CT testing machine (The Testometric Company Ltd., Rochdale, UK) with dedicated weldment, which enables its geometry to be modified with its moving parts for a wide variety of such experiments. The distance between the supports was equal to 1040 mm with the force acting point in a distance of 440 mm from AE sensor #1 (see Figure 2). The tested CFRP tube with attached AE sensors was additionally placed in a plastic pipe to prevent damage to the AE sensors and other equipment due to sudden structural integrity violation. The supports were covered by thin felt to allow free movement of the tube during its bending. The test has been deformation-controlled with the upper anvil speed equal to 10 mm/min.

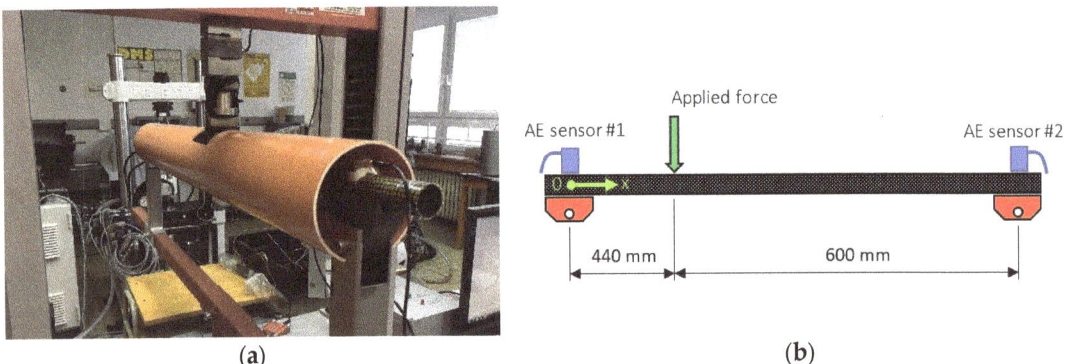

Figure 2. (**a**) In-Situ photograph of the test rig including the specimen equipped with AE sensors; (**b**) Schematic representation of the three-point bending test setup.

2.3. Acoustic Emission Monitoring

The acoustic emission activity has been monitored using the Vallen AMSY-6 AE system (Vallen Systeme GmbH, Icking, Germany) with two utilized measuring channels (ASIP-2A dual channel signal processor card), equipped with the AEP5H 34 dB preamplifiers and the broad-band Vallen VS-900 AE sensors. The sensors were attached onto the tube with the use of oil-based plasticine. The sampling frequency of the AE data was set to 10 MHz while the transient data (wave transients) were sampled with 20 MHz in the frequency range between 50 and 1100 kHz. The detection threshold has been set to 32 dB owing to a relatively greater distance of both sensors from the area in which the breach will most likely occur and related higher attenuation of the AE signal in composites. Only localized AE events, which fall within the ⟨250,650⟩ (mm) (see Figure 2b) interval of the x coordinate will be included for further data processing.

The filtered data was followingly analyzed with the Vallen VisualClass software package (Vallen Systeme GmbH, Icking, Germany), which uses the pattern recognition method [15] to associate similar waveform types into separate groups. Due to the nature of the task, an unsupervised learning approach was chosen. The procedure starts with loading the selected database of AE transients into Vallen VisualClass software, where the number of time windows including their span and the starting point of the segmentation analysis in the time domain must be specified (see Figure 3).

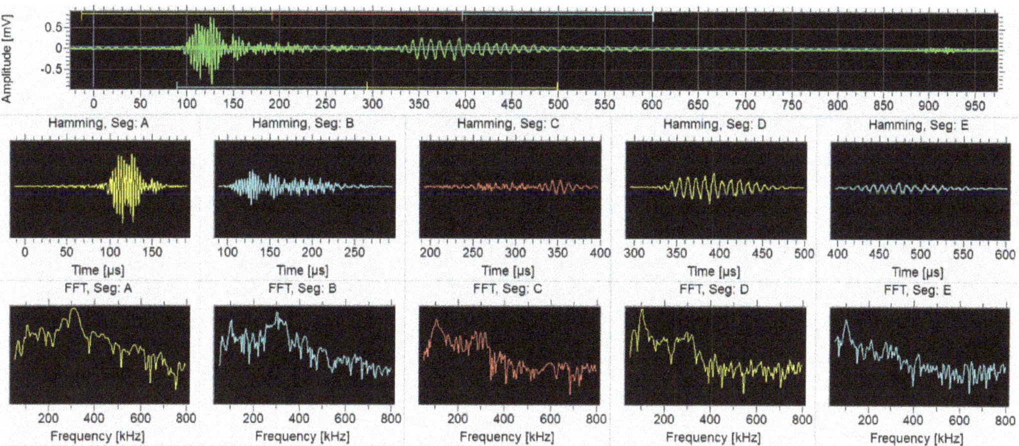

Figure 3. Setting up the Hamming windowed time segments with corresponding results in the frequency domain.

The current analysis uses the following settings for the AE transients with relation to the VisualClass software package: Number of time segments: 5; Size of single time segment in terms of points: 4096; Rel. trigger offset: −256 points; min/max frequency limit: 50/800 kHz. The software then performs the assembly of multidimensional feature vector, the size of which depends on the chosen number of time segments including their size. The pattern recognition analysis will result in the basic feature space identification, which is then linearly transformed for maximizing inter-class distance and minimizing the intra-class extension, at the same time (see Figure 4).

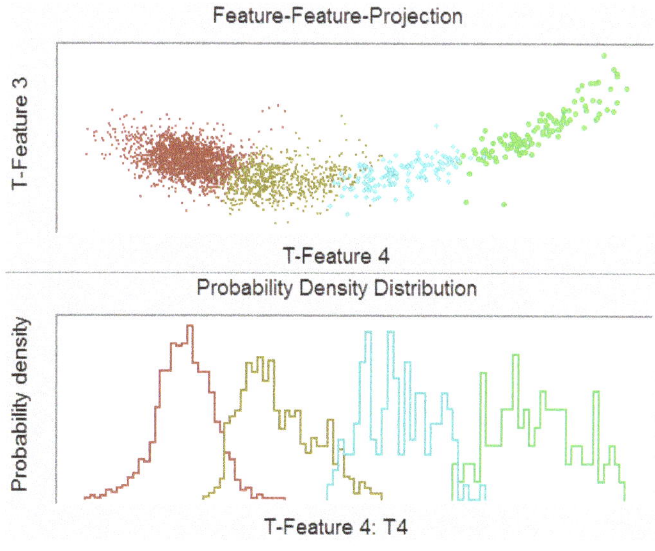

Figure 4. Transformed features projection—results for four selected clusters.

The results are then transferred into VisualAE software for further postprocessing. Four clusters were chosen for subsequent analysis, since the additional increase of the number of clusters did not lead to better differentiation of individual transients. One sample from each series has been subjected to the attenuation measurement of the AE signal using Hsu-Nielsen source [31] (pencil lead diameter: 0.35 mm; hardness: 2H) in order to properly evaluate the real AE signal amplitude in subsequent data analysis. The results of the attenuation measurements are shown in Table 4. The propagation velocity has been determined experimentally using Hsu-Nielsen source with value varying between 3200 and 3300 m/s across A/B/C series samples.

Table 4. Attenuation measurement on A/B/C series of CFRP composite tubes

Sample Series	Near Field Attenuation (dB/m)	Far field Attenuation (dB/m)
A	90	33.3
B	66.6	33.2
C	222.2	36.2

The preliminary measurements also included a series of six tensile tests of carbon fiber sheafs (see Figure 5) in order to construct the above-mentioned boundary curve, which will be further used for the detection of the carbon fiber breaks across the identified clusters. The deformation-controlled tests (upper anvil speed equal to 0.5 mm/min) were carried out on the universal Testometric M500-50CT testing machine equipped with 100 N load cell. The relatively small scale of the load cell enabled us to detect in time the events corresponding to the failure of certain number of fibers thanks to the registered force drop. The AE activity has been monitored using the Vallen AMSY-6 AE system with three utilized measuring channels (ASIP-2S dual channel signal processor card), equipped with the AEP5H 34 dB preamplifiers and the broad-band DAKEL MIDI AE sensors, where the top and bottom sensor acted as guard elements, while the middle sensor has been used for the data acquisition. The fiber sheafs were glued on their ends thus providing a clamping support for the attachment into the jaws.

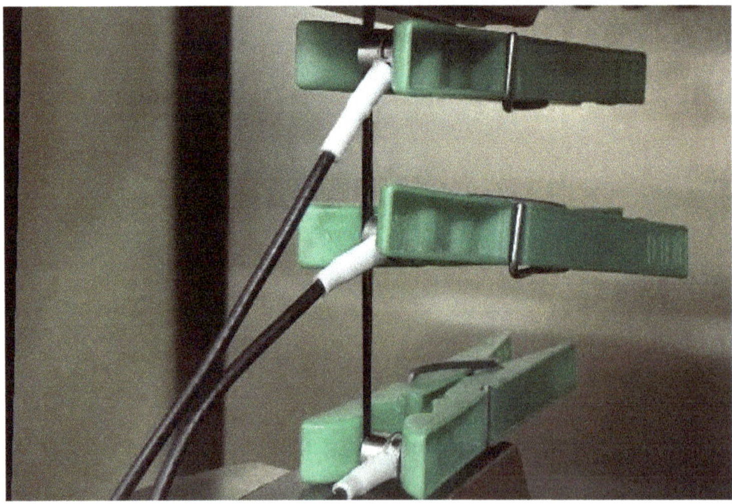

Figure 5. Experimental setup for tensile test of the carbon fibers.

The basic frequency analysis of the AE signal is, besides the exploited pattern recognition approach, relatively efficient and powerful tool for filtering the AE signal, which can be then affiliated to different failure mechanisms [32]. Chou in his work [21] states that fiber breakages in the case of carbon fiber/glass fiber composite systems produce extensional wave signals with frequencies between 350–700 kHz, while matrix cracks generate flexural wave modes with frequencies up to 350 kHz. It has to be noted that the given finding has been verified on the preliminary tensile tests of the carbon fibers, where the AE signals originating from fiber failure exhibit a higher power fraction in the 300(350)–600 kHz frequency interval. Based on this consideration, there will be defined a variable denoted as p_f factor, which will relate the power fraction of the AE signal in a certain frequency band to the AE signal power in the entire considered frequency range:

$$P_f = \frac{P_{(350\text{–}800)\text{kHz}}}{P_{(50\text{–}800)\text{kHz}}} 100(\%) \quad (1)$$

where $P_{(350\text{–}800)\text{kHz}}$ represents the power fraction of the AE signal in the 350–800 kHz frequency range and $P_{(50\text{–}800)\text{kHz}}$ is the power of the AE signal in the entire considered frequency range, i.e., 50–800 kHz. Note that the power of the AE signal in the given frequency interval has been calculated using Parseval's theorem. The boundary curve then states the p_f factor and the AE signal amplitude in relation with subsequent intention to filter the AE signal originating from the fiber failure from the other failure mechanisms or the secondary AE signals, which figure as noise (interaction of the individual fibers between each other (rubbing) and/or AE activity arising from the sample attachment points). The identification procedure of the boundary curve is based on the overlapping the force versus time dependency by the acoustic emission activity versus time dependency, where the detected force drops caused by the failure of the individual/multiple fibers can be directly matched to the emerged AE signals.

The following figures display the dependency between the force and amplitude of individual AE hits on time (Figure 6) and the relation between the p_f factor and the AE signal amplitude for the entire series (Figure 7). Note that the AE signal amplitude is being referred to dB_{AE} unit, thus expressing the voltage amplitude of non-amplified signal as a gain related to 1 μV. The blue line in Figure 7 represents the boundary line separating AE signals belonging to the fiber break from signals originating from the interaction between the individual fibers and the other interfering AE sources. At this point, it has to be noted that the high-frequency range appearing in numerator in Equation (1) has been for further application of the CFRP tubes extended to 300–800 kHz range due to a frequency attenuation arising from the geometric dimensions of the CFRP tubes and the mutual distance between the AE source and the AE sensors.

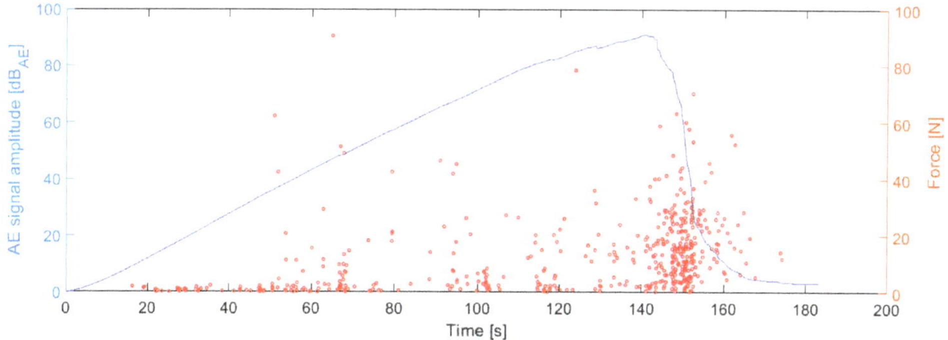

Figure 6. Force and AE activity as the function of time for the selected sample.

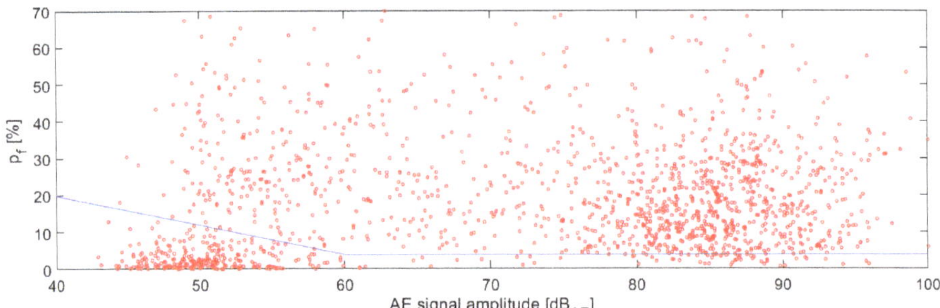

Figure 7. Dependency between p_f factor and AE signal amplitude for the entire series.

3. Results and Discussion

3.1. Mechanical Properties and Basic AE Signal Analysis

Figure 8 shows relation between force and displacement of the anvil for individual A/B/C series production samples. As expected, the highest stiffness is being reached by the C series samples, followed by A and B series samples. All three manufacturing modifications show within their group very similar trend in terms of the force-displacement course except the A series samples, namely A2 sample, which exhibits marginally lower stiffness, most likely due to the fabrication process, which is not in the form of the automated production. The above-mentioned consistency in terms of the sample stiffness is, however, not valid for the maximum force across individual series, where differences from 10 to 28 percent related to the maximum achieved force in each production series can be observed. Again, the reason for such results variation can be found in the production form itself. A somewhat similar trend can be registered in the case of the number of located AE events across the 0–F_{max} range for the individual samples (see Figure 9), where a relatively large variation has been registered. However, even despite this fact, A/B series report considerably higher level of the located AE events, which is most likely caused due to the presence of (0°–90°) fabric figuring as a top layer. Note that the given statement is currently a hypothesis, which needs to be verified in the future.

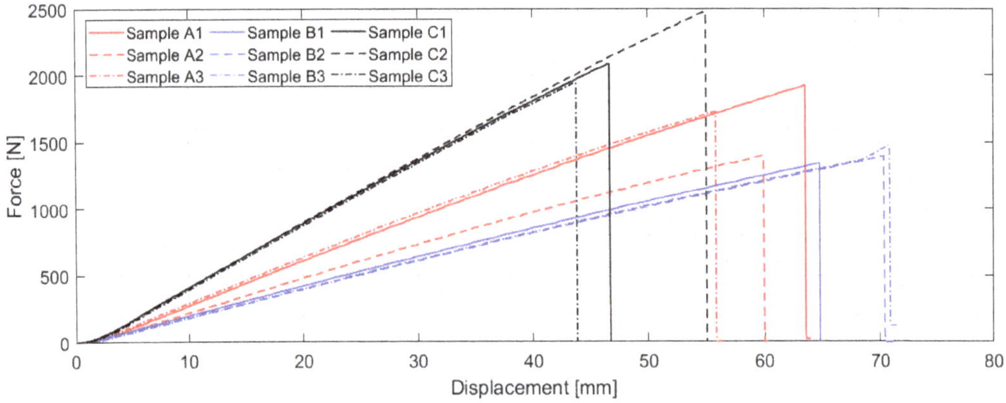

Figure 8. Force as the function of displacement for individual A/B/C series production samples

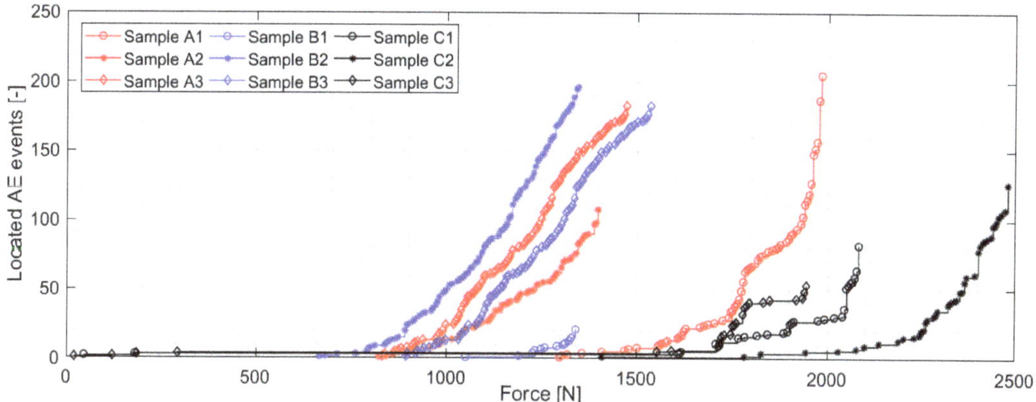

Figure 9. Cumulative number of located events as the function of force for individual A/B/C series samples

The energy of accumulated AE events versus force (Figure 10) is another important dependency, which can bring us closer to the overall structure behavior. The maximum value of the released AE energy is for all samples between 5×10^8 and 10^9 aJ. The difference, however, lies in the character how the energy is being released during the loading process. The A and B series specimen exhibit almost gradual AE energy release, with the difference in the final loading stage. While the A series specimen tend to gradually continue with the cumulation of the AE events and gradual release of the AE energy, the B series samples tend to suddenly lose integrity without any significant warning phase. A completely different behavior can be found in the case of the C series samples, which have considerably larger energy per event ratio with a lack of any warning phase before the integrity lose.

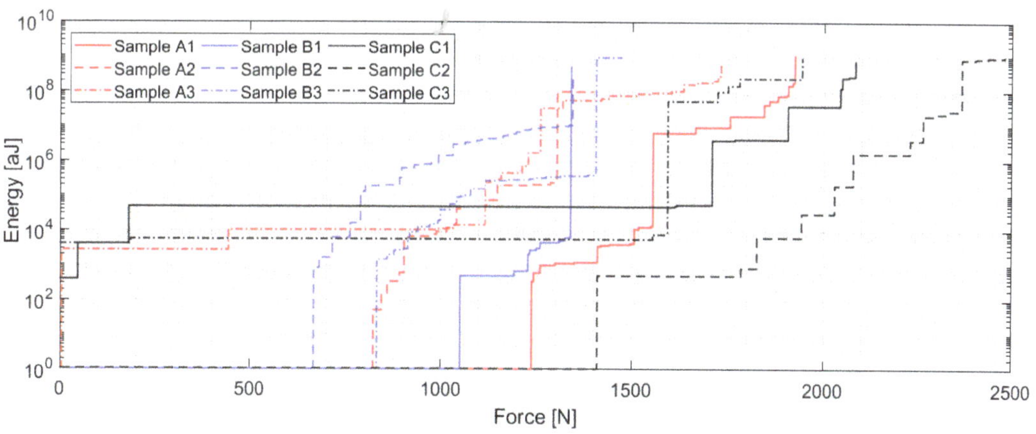

Figure 10. Energy as the function of force for individual A/B/C series samples

3.2. AE Signal Analysis Using Pattern Recognition Approach

The utilized unsupervised pattern recognition analysis resulted in identification of four clusters of AE signals with the following features. The signals affiliated to the first cluster are characterized by high amplitude, in most cases exceeding 90 dB_{AE} with energy value usually above 1×10^6 aJ and frequencies in the span from 50 kHz to 150 kHz (see Figure 11a), whereas this cluster also partially contains signals with frequency content above 300 kHz. The second cluster is characterized by the amplitudes mostly below

65 dB_{AE} with AE energy in the order of hundreds to the tens of thousands of aJ and the frequency in the 50–450 kHz range (see Figure 11b). The third cluster is represented by the signals with the amplitude in the 60–80 dB_{AE} range with AE energy in the order of ten thousand aJ and the frequency in the 50–300 kHz interval (see Figure 11c). The fourth cluster is characterized by the signals in the 90–110 dB_{AE} amplitude range, the AE energy of 1×10^5–1×10^6 aJ and the frequency in the 50–200 kHz range (see Figure 11d). For better clarity, the results are summarized in the following table and Figure 12, respectively.

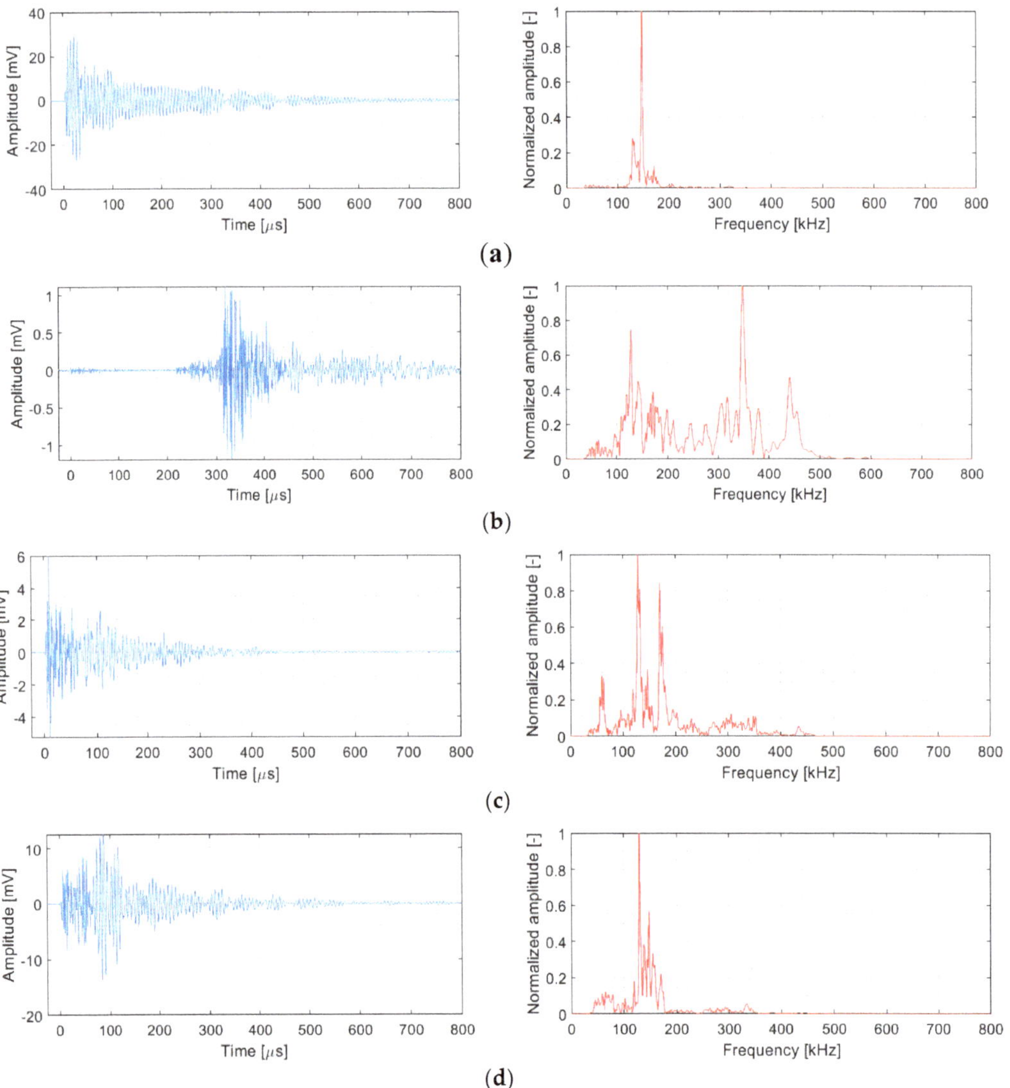

Figure 11. Examples of characteristic signal belonging to individual clusters: (**a**) Cluster 1; (**b**) Cluster 2; (**c**) Cluster 3; (**d**) Cluster 4.

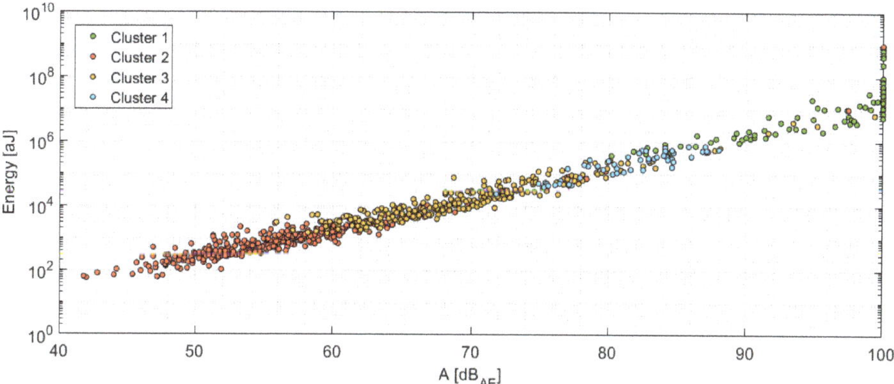

Figure 12. Amplitude versus Energy for individual clusters.

Due to the fact that there was no 100% separation of individual clusters, it was necessary to subject the obtained cluster data to additional analysis, in which the p_f factor for each clustered AE event has been calculated. The subsequent separation of the events affiliated to the fiber break has been performed using the already introduced boundary curve. However, care has to be taken whether the boundary curve relates to the AE hits of the already localized AE events, where in the latter case there must be considered an additional amplitude shift, which depends on the attenuation curve for the individual sample series.

Figure 13 displays the resulting $p_f = f(A_{corr}$,cluster) dependency for individual sample series including the drawn boundary curve, which has already been shifted using the identified attenuation curve for the given sample series and the reference distance (440 mm). Figure 14 shows the $A = f(F/F_{max})$ dependency for the localized AE events, which were detected for all three-sample series. The filled symbols represent the fiber break while empty ones represent failure mechanism affiliated to the given cluster. For the straightforward comparison the original vertical axis, showing the distance corrected amplitude A_{corr}, has been replaced by the AE signal amplitude—A in dB_{AE}. Figure 15 displays the relation between the p_f factor and the F/F_{max} quantity for all considered series.

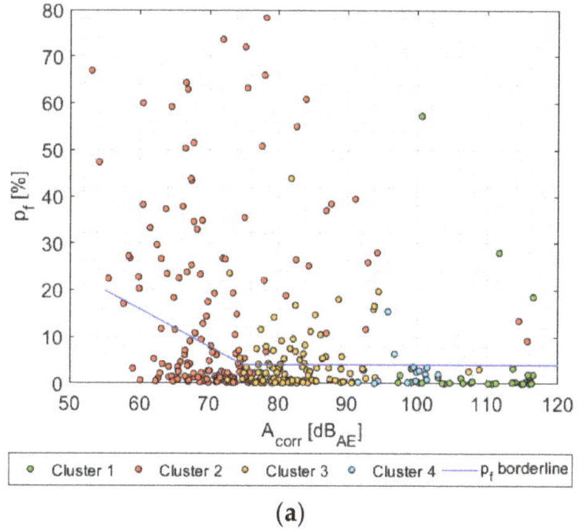

(**a**)

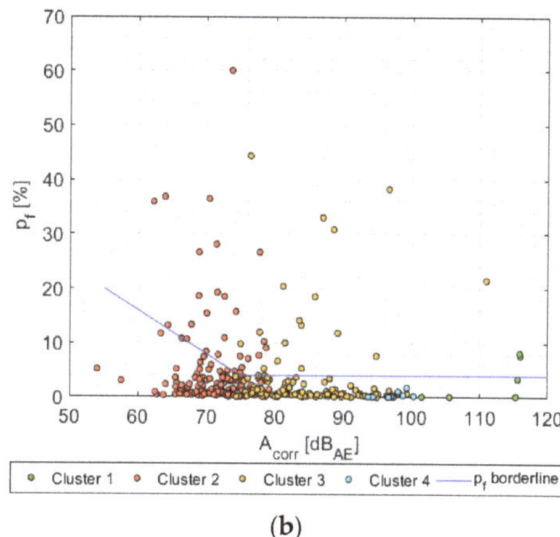

(**b**)

Figure 13. *Cont.*

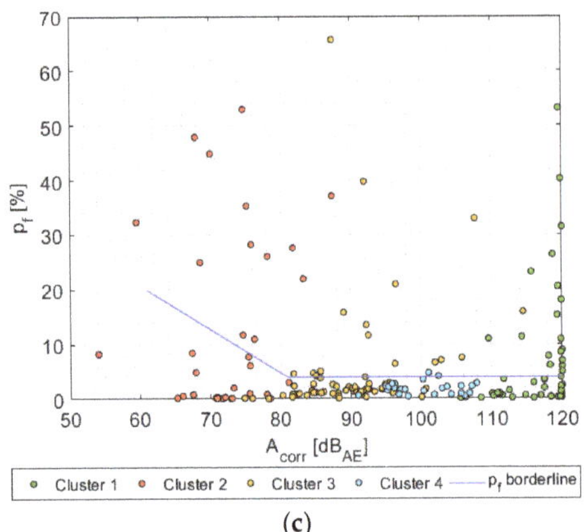

(c)

Figure 13. p_f factor as the function of distance corrected amplitude A_{corr}—A series samples (**a**); B series samples (**b**); C series samples (**c**).

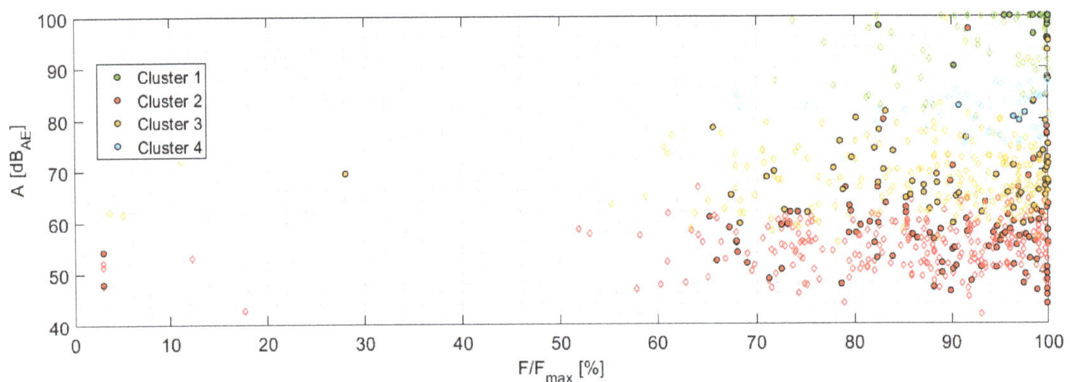

Figure 14. Amplitude as the function of F/F_{max}—A,B,C series.

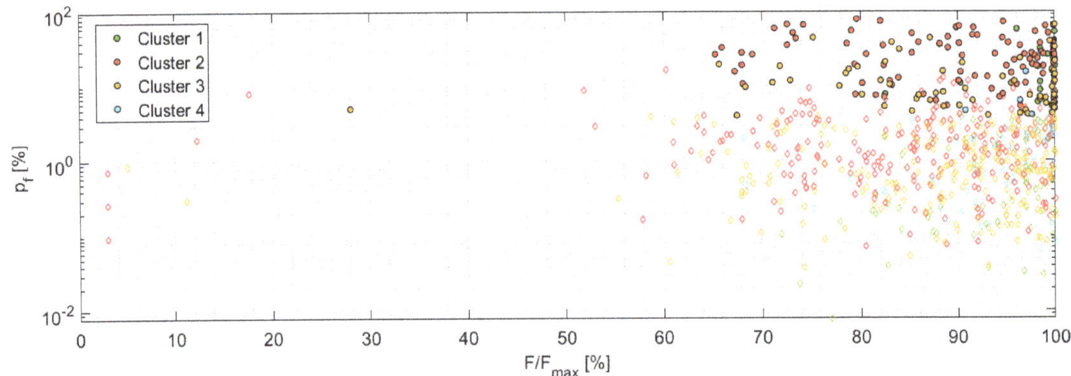

Figure 15. p_f factor as the function of F/F_{max}—A,B,C series.

It is obvious that the damage is being initiated with the AE signals assigned to the cluster 2 with low values of the p_f factor at the same time at the level of 50% of the maximum force, while the first indication of the fiber break appears at 65% of the maximum force. The AE signals affiliated to the cluster 2 can be therefore assigned to the matrix cracking in the initiation phase, including the high-frequency AE signals evaluated as the fiber breakage (29.5% incidence). Classification of the cluster 2, at least its low-frequency content group, is fully in accordance with literature [25], where the amplitudes corresponding to this failure mechanism were below 70 dB. The same applies to the frequency content (see Table 5), especially its low-frequency part, which is identical to results published in [24] or [32]. The formation of the two separate groups, i.e., low- and high-frequency content, within the cluster 2 becomes more evident if the p_f = f(duration) dependency is displayed, see Figure 16, where the same applies also to the other clusters. Gutkin et al. [30], on the other hand, reports a lower frequency band compared to the above-mentioned research papers.

Table 5. Characterization of the individual clusters.

Cluster No.	Frequency Range (kHz)	Amplitude Range (dB_{AE})	Energy Range (aJ)
1	50–150 (>300, minor cases)	>90	$>10^6$
2	50–450	<65	10^2–10^4
3	50–300	60–80	10^3–10^5
4	50–200	75–90	10^4–10^6

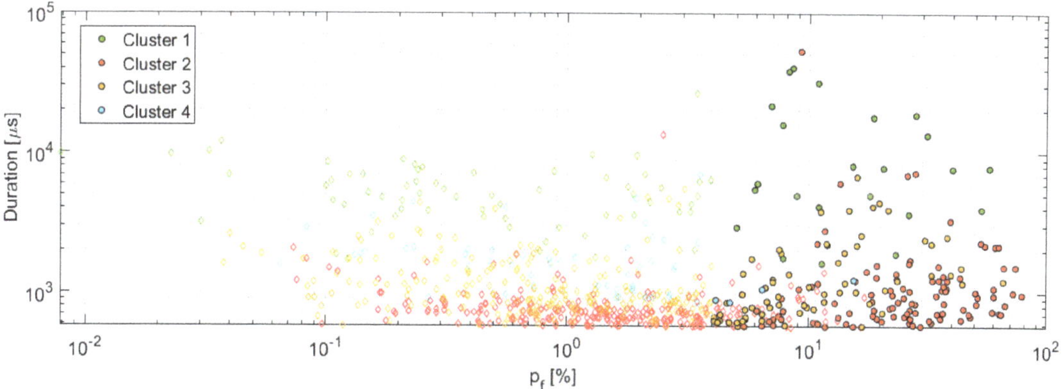

Figure 16. p_f factor versus duration—A,B,C series.

With a gradually increasing value of the F/F_{max} variable, we can register the emergence of the AE signal classified to the cluster 3 and the cluster 4, which contain 22.8% and 7.4% of the AE hits classified as the fiber break, respectively. Cluster 3 is likely to represent the debonding, while the cluster 4 affiliates to the delamination. Both these classifications are in accordance with [32] including the duration of the AE signal, which does not exceed 10 ms.

While the already obtained results indicate relatively similar frequency bands of the delamination and the debonding failure mechanism, where the delamination exhibits higher limit equal to 300 kHz (see Table 5), the results provided in [30] report a considerably lower frequency band for the delamination, which is contrary to [32] or [33]. Cluster 1, on the other hand, reflects cluster 2 to a certain extent, while having much higher levels of the AE signal energy including the amplitude. Its low- and high-frequency sections are well separated as can be seen in Figure 16. The F/F_{max} values of the AE signal belonging to this cluster are higher than 80%, thus pointing to the emergence of dominant events with loss of the structural integrity, which are represented even by propagation of the interlaminar matrix cracks or the failure of larger amounts of the carbon fibers [34]. A somewhat specific is the fiber break failure mechanism, which occurs across all clusters as a high-frequency content AE signal with amplitudes ranging from 45 to 100 dB including wide AE energy span (10^2–10^9 aJ), see Figure 17, and high-frequency content above 350 kHz. This finding has been supported by various research papers [22,23,30,34].

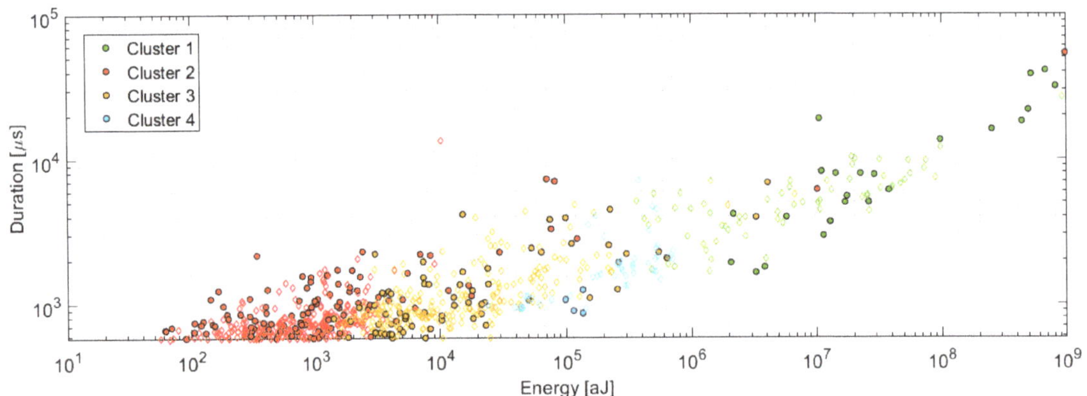

Figure 17. Energy versus duration—A,B,C series.

4. Concluding Remarks

The AE monitoring technique together with an unsupervised pattern recognition technique and the introduced boundary curve have been used to analyze the various damage mechanisms in three types of the CFRP composite tubes, which were subjected to the three-point bending test. The utilized two-step approach involved the preliminary tensile tests of the carbon fiber sheafs, which enabled the boundary curve to be designed for further identification of the AE signal originating from the fiber break. The boundary curve was then adopted on the clustered data, which has been obtained using the unsupervised pattern recognition approach, thus enabling the AE events originating from the fiber break to be additionally filtered across all the identified clusters. The conclusions obtained within the framework of this research study are summarized below.

1. Four damage mechanisms have been identified using the above-mentioned techniques, namely the fiber break, delamination, debonding, and matrix cracking. The application of the boundary curve appeared to be an effective tool for the further refinement of the results across the individual clusters.
2. The fiber break failure mechanism has been identified across all clusters resulting in the wide amplitude as well as energy span. This finding has been supported by various research projects/studies/papers.
3. It was found that the matrix cracking failure mechanism generates AE signals with the frequency band between 50 and 200 kHz. The result is in accordance with most studies; however, even in this matter a certain contradiction can be found [30].
4. The distinction between delamination/debonding failure mode seems to be a relatively challenging, since both failure mechanisms report very similar frequency spectra [21]. However, according to the presented study, we can find the difference in the energy as well as amplitude values of both mechanisms.

Author Contributions: M.Š. wrote the manuscript and performed the experiments and data analysis. J.C. designed the dedicated weldment. M.F. performed data analysis and reviewed the manuscript. P.P. performed data analysis and reviewed the manuscript. R.M. provided consultation on the strategy. All authors have read and agreed to the published version of the manuscript.

Funding: This work was supported by Specific Research (SP2020/23) and by project Innovative and additive manufacturing technology—new technological solutions for 3D printing of metals and composite materials—CZ.02.1.01/0.0/0.0/17_049/0008407 and by the Czech Science Foundation (GACR) project No. 19-03282S. The authors would also like to thank the Havel Composites Ltd. company for providing the test samples and consulting during the elaboration of the paper.

Institutional Review Board Statement: Not applicable.

Informed Consent Statement: Not applicable.

Data Availability Statement: Data can be provided upon request from the correspondent author.

Conflicts of Interest: The authors declare no conflict of interest.

References

1. Unnthorsson, R.; Runarsson, T.P.; Jonsson, M.T. Acoustic emission based fatigue failure criterion for CFRP. *Int. J. Fatigue* **2008**, *30*, 11–20. [CrossRef]
2. Majko, J.; Handrik, M.; Vaško, M.; Sága, M. Influence of Fiber Deposition and Orientation on Stress Distribution in Specimens Produced Using 3D Printing. *J. Mech. Eng.* **2019**, *69*, 81–88.
3. Dorčiak, F.; Vaško, M.; Handrik, M.; Bárnik, F.; Majko, J. Tensile test for specimen with different size and shape of inner structures created by 3D printing. *Transp. Res. Procedia* **2019**, *40*, 671–677. [CrossRef]
4. Majko, J.; Sága, M.; Vaško, M.; Handrik, M.; Barnik, F.; Dorčiak, F. FEM analysis of long-fibre composite structures created by 3D printing. *Transp. Res. Procedia* **2019**, *40*, 792–799. [CrossRef]
5. Awad, Z.K.; Aravinthan, T.; Zhuge, Y.; Gonzalez, F. A review of optimization techniques used in the design of fibre composite structures for civil engineering applications. *Mater. Des.* **2012**, *33*, 534–544. [CrossRef]
6. Xu, J.; Wang, W.; Han, Q.; Liu, X. Damage pattern recognition and damage evolution analysis of unidirectional CFRP tendons under tensile loading using acoustic emission technology. *Comput. Struct.* **2020**, *238*, 1–10. [CrossRef]
7. Libonati, F.; Vergani, L. Damage assessment of composite materials by means of thermographic analyses. *Compos. Part B* **2013**, *50*, 82–90. [CrossRef]
8. Heuer, H.; Schulze, M.; Pooch, M.; Gäbler, S.; Nocke, A.; Bardl, G. Review on quality assurance along the CFRP value chain—Nondestructive testing of fabrics, preforms and CFRP by HF radio wave techniques. *Compos. Part B* **2015**, *77*, 494–501. [CrossRef]
9. Dong, J.; Kim, B.; Locquet, A.; McKeon, P.; Declercq, N.; Citrin, D.S. Nondestructive evaluation of forced delamination in glass fiber reinforced composites by terahertz and ultrasonic waves. *Compos. Part B* **2015**, *79*, 667–675. [CrossRef]
10. Garcea, S.C.; Wang, Y.; Withers, P.J. X-ray computed tomography of polymer composites. *Compos. Sci. Technol.* **2018**, *156*, 305–319. [CrossRef]
11. Dahmene, F.; Yaacoubi, S.; Mountassir, M.E.L. Acoustic Emission of Composites Structures: Story, Success, and Challenges. *Phys. Procedia* **2015**, *70*, 599–603. [CrossRef]
12. Saeedifar, M.; Fotouhi, M.; Najafabadi, M.A.; Toudeshky, H.H.; Minak, G. Prediction of quasi-static delamination onset and growth in laminatedcomposites by acoustic emission. *Compos. Part B* **2016**, *85*, 113–122. [CrossRef]
13. Crivelli, D.; Guagliano, M.; Eaton, M.; Pearson, M.; Al-Jumaili, S.; Holford, K.; Pullin, R. Localization and identification of fatigue matrix cracking and delamination in a carbon fibre panel by acoustic emission. *Compos. Part B* **2015**, *74*, 1–12. [CrossRef]
14. Kocich, R.; Cagala, M.; Crha, J.; Kozelsky, P. Character of acoustic emission signal generated during plastic deformation. In Proceedings of the 30th European Conference on Acoustic Emission Testing & 7th International Conference on Acoustic Emission, University of Granada, Granada, Spain, 12–15 September 2012.
15. Tang, J.; Soua, S.; Mares, C.; Gan, T.-H. Pattern Recognition Approach to acoustic emission data originating from fatigue of wind turbine blades. *Sensors* **2017**, *17*, 2507. [CrossRef]
16. Assi, L.; Soltangharaei, V.; Anay, R.; Ziehl, P.; Matta, F. Unsupervised and supervised pattern recognition of acoustic emission signals during early hydration of Portland cement paste. *Cem. Concr. Res.* **2018**, *103*, 216–225. [CrossRef]
17. Doan, D.D.; Ramasso, E.; Placet, V.; Zhang, S.; Boubakar, L.; Zerhouni, N. An unsupervised pattern recognition approach for AE data originating from fatigue tests on polymer–composite materials. *Mech. Syst. Signal Process.* **2015**, *64*, 465–478. [CrossRef]
18. Chen, B.; Wang, Y.; Zhaoli, Y. Use of Acoustic Emission and Pattern Recognition for Crack Detection of a Large Carbide Anvil. *Sensors* **2018**, *18*, 386. [CrossRef]
19. Anastasopoulos, A. Pattern recognition techniques for acoustic emission based condition assessment of unfired pressure vessels. *J. Acoust. Emiss.* **2005**, *23*, 318–330.
20. Baran, I.; Nowak, M.; Ono, K. Acoustic emission pattern recognition analysis applied to the over-strained pipes in a polyethylene reactor. *J. Acoust. Emiss.* **2006**, *24*, 44–51.

21. Chou, H.-Y. Damage Analysis of Composite Pressure Vessels Using Acoustic Emission Monitoring. Ph.D. Thesis, School of Aerospace, Mechanical & Manufacturing Engineering College of Science Engineering and Health, RMIT University, Melbourne, Australia, 2011.
22. Ono, K.; Kawamoto, K. Digital signal analysis of acoustic emission from carbon fiber/epoxy composites. *J. Acoust. Emiss.* **1990**, *9*, 109–116.
23. Ono, K. Acoustic emission behavior of flawed unidirectional carbon fiber-epoxy composites. *J. Reinf. Plast. Compos.* **1988**, *7*, 90–105. [CrossRef]
24. Bohse, J. Acoustic emission characteristics of micro-failure processes in polymer blends and composites. *Compos. Sci. Technol.* **2000**, *60*, 1213–1226. [CrossRef]
25. Komai, K.; Minoshima, K.; Shibutani, T. Investigations of the fracture mechanism of carbon/epoxy composites by AE signal analyses. *JSME Int. J. Ser. 1 Solid Mech. Strength Mater.* **1991**, *34*, 381–388. [CrossRef]
26. Godin, N.; Huguet, S.; Gaertner, R.; Salmon, L. Clustering of acoustic emission signals collected during tensile tests on unidirectional glass/polyester composite using supervised and unsupervised classifiers. *NDT&E Int.* **2004**, *37*, 253–264.
27. Baccar, D.; Söffker, D. Identification and classification of failure modes in laminated composites by using a multivariable statistical analysis of wavelet coefficients. *Mech. Syst. Signal Process.* **2017**, *96*, 77–87. [CrossRef]
28. Prakash, R.V.; Maharana, M. Damage detection using infrared thermographz in a carbon-flax fiber hybrid composite. *Procedia Struct. Integr.* **2017**, *7*, 283–290. [CrossRef]
29. Munoz, V.; Valés, B.; Perrin, M.; Pastor, M.L.; Welemane, H.; Cantarel, A. Damage detection in CFRP by coupling acoustic emission and infrared thermography. *Compos. Part B* **2016**, *85*, 68–75. [CrossRef]
30. Gutkin, R.; Green, C.J.; Vangrattanachai, S.; Pinho, S.T.; Robinson, P.; Curtis, P.T. On acoustic emission for failure investigation in CFRP: Pattern recognition and peak frequency analyses. *Mech. Syst. Signal Process.* **2011**, *25*, 1393–1407. [CrossRef]
31. Sause, M.G.R. Investigation of pencil lead breaks as acoustic emission sources. *J. Acoust. Emiss.* **2011**, *29*, 184–196.
32. Krietsch, T.; Bohse, J. Selection of acoustic emissions and classification of damage mechanisms in fiber composite materials. *J. Acoust. Emiss.* **1998**, *16*, 233–242.
33. Nam, K.-W.; Ahn, S.-H.; Moon, C.-K. Fracture behavior of carbon fiber reinforced plastics determined by the time-frequency analysis method. *J. Appl. Polym. Sci.* **2003**, *88*, 1659–1664. [CrossRef]
34. Siron, O.; Tsuda, H. Acoustic emission in carbon fibre-reinforced plastic materials. *Ann. Chim. Sci. Mater.* **2000**, *25*, 533–537. [CrossRef]

 materials

Article

Application of the Pulse Infrared Thermography Method for Nondestructive Evaluation of Composite Aircraft Adhesive Joints

Tomáš Kostroun [1,*] and Milan Dvořák [2]

[1] Department of Aerospace Engineering, Faculty of Mechanical Engineering, Czech Technical University in Prague, Technická 4, 160 00 Praha 6, Czech Republic

[2] Department of Mechanics, Biomechanics and Mechatronics, Faculty of Mechanical Engineering, Czech Technical University in Prague, Technická 4, 160 00 Praha 6, Czech Republic; milan.dvorak@fs.cvut.cz

* Correspondence: tomas.kostroun@fs.cvut.cz

Abstract: In this article, we examine the possibility of using active infrared thermography as a nontraditional, nondestructive evaluation method (NDE) for the testing of adhesive joints. Attention was focused on the load-bearing wing structure and related structural joints, specifically the adhesive joints of the wing spar caps and the skins on the wing demonstrator of a small sport aircraft made mainly of a carbon composite. The Pulse Thermography (PT) method, using flash lamps for optical excitation, was tested. The Modified Differential Absolute Contrast (MDAC) method was used to process the measured data to reduce the effect of the heat source's inhomogeneity and surface emissivity. This method demonstrated a very high ability to detect defects in the adhesive joints. The achieved results are easy to interpret and use for both qualitative and quantitative evaluation of the adhesive joints of thin composite parts.

Keywords: NDE; infrared thermography; Infrared Nondestructive Testing; composite; CFRP

Citation: Kostroun, T.; Dvořák, M. Application of the Pulse Infrared Thermography Method for Nondestructive Evaluation of Composite Aircraft Adhesive Joints. *Materials* **2021**, *14*, 533. https://doi.org/10.3390/ma14030533

Academic Editors: Radim Kocich and Lenka Kunčická

Received: 30 November 2020
Accepted: 19 January 2021
Published: 22 January 2021

Publisher's Note: MDPI stays neutral with regard to jurisdictional claims in published maps and institutional affiliations.

Copyright: © 2021 by the authors. Licensee MDPI, Basel, Switzerland. This article is an open access article distributed under the terms and conditions of the Creative Commons Attribution (CC BY) license (https://creativecommons.org/licenses/by/4.0/).

1. Introduction

This article describes our research into the viability of the quality control of a strength-critical adhesive joint on an all-composite aircraft and follows up on the work described in [1]. At present, the market for small high-performance sports aircraft, which are mostly built of composite materials, is developing rapidly. These aircraft are designed and operated in the Light Sport Aircraft (LSA) category. The LSA regulations require a certificated strength test of the prototype. Other aircraft in the production series are then operated using periodic inspections, but these are only done by sight. None of the known Nondestructive Evaluation (NDE) methods are mandatory and are also not used for financial reasons. The variance in production quality, which is typical for composite structures, is covered by the special safety factor of 1.20–1.50 [2]. The basic safety factor of 1.5 is then multiplied by this special factor. It is clear that the use of a simple and fast control method would make it possible to safely use the lower limit of the increasing coefficient and thus safely operate the optimized lightweight aircraft structure. Another factor is that LSA-category planes are not monitored while in service from the point of view of their fatigue life; the general assumption is that the composite aircraft is effectively obsolete before it reaches the technical end of its life. Due to the nature of LSA production, which is small-scale or a piece type with a wide range of designs, the pressure on the price of such NDE solutions is high, as it is inevitably passed on to the sale price. As previously mentioned, the current trend is for all-composite aircraft design. Typically, these aircrafts' constructions are based on precured composite parts. They are made by a contact hand lamination method, using vacuum-assisted resin transfer methods or by means of prepreg technology using an autoclave [3,4]. These assemblies, for example the fuselage halves, the wing skins, wing spars, etc., are typically connected using adhesive joints. Both the composites and the adhesive joints bring new issues and problems to the subject of NDE testing.

The aim of this experimental work was to evaluate the bonding quality of small composite aircraft wing adhesive joints. The all-composite wing was assembled from precured carbon-fiber-reinforced plastic (CFRP) composite parts, using a two-component epoxy adhesive system. This work was focused on the critical adhesive joints of the wing skin to the spars and ribs.

The composite design of LSA-category aircraft creates new demands for the detection capabilities of NDE methods. Although in general aviation, the thickness of glued composite parts can be within a range of millimeters to tens of millimeters [5], in the LSA category, composite parts are often very thin, with thicknesses in the tenths of millimeters [3,4].

The problem is an interesting and challenging one due to the combined requirements involved in evaluating the quality of the adhesive joints of the composite parts. In addition to the currently developed methods, such as Fiber optic Bragg Grating (FBG) sensors integrated directly in the adhesive joint layer [6], only some Nondestructive Evaluation methods are usable. NDE methods are generally used to detect hidden defects in a material, such as air bubbles, poorly impregnated composite fabric or defects caused by mechanical impact or excessive load, without damaging the part being tested or affecting its working properties. Traditional NDE methods include eddy current testing, acoustic emission, bond testing, X-ray and ultrasonic testing, infrared testing and others [7–9].

One of the methods that can potentially provide a solution for NDE testing of the glued joints of small sports aircraft structures is a group of methods using active infrared thermography. Active infrared thermography is an inspection technique that requires an external energy source to create a temperature difference between defective and nondefective areas of the specimen during testing. There are several methods of providing a heat source from which to choose, such as excitation by light, laser, hot air, ultrasonic or microwave radiation [9–11].

According to available sources, the Pulse Active Infrared Thermography Nondestructive Testing (PT IRNDT) method seems to be a suitable solution for solving this problem. This method uses a short light pulse generated by a flash lamp or a halogen light. The PT IRNDT method allows for the rapid inspection of specific areas of the tested object with a direct 2D graphical output. This provides relatively easy-to-interpret results. It is suitable for the finding of flaws and voids located close to the surface [12–24].

In the case of the tests performed on the composite wing, defects in the adhesive joints were expected to be found at a depth of 0.5–1.8 mm below the tested wing skin surface. These would be difficult to detect using different standard NDE methods, such as ultrasonic testing or X-rays.

The first method described assumes the use of either an attenuation (through- transmission) method or a reflection (pulse-echo) method. The attenuation method cannot be used here because adhesive joints are not accessible along their entire length from both sides. Due to the small sizes of the measured thicknesses, the use of the reflection method is only possible when using an immersion technique, which requires immersion of the part in a water bath, or the use of a water-squirt configuration (Water Jet Device), which is more suitable for the pass-through method. The immersion technique requires the use of a scanning device, or even a large vessel, into which the test part must be immersed. With the size of the construction being tested (a wingspan larger than 7 m), such a device is not commonly available. In addition, even when using a water-squirt configuration instead of the immersion of the complete part, the resulting moisture is generally problematic for composites [10,11].

The second potential method is with X-ray, but, again, the major problem is the size of the part and accessibility to the tested area from both sides of an adhesive joint. In addition, this method evaluates the change of the thickness of the inspected part of the structure so it will not detect the disbonding of the joint if the adherends still overlap.

2. Methods

2.1. Experimental Composite Wing Specimen

The composite wing being tested came from a two-seater aircraft designed for the LSA category. The maximum take-off mass of the aircraft is 600 kg, with a wingspan of about 7.2 m and a length of 7.5 m. The aircraft is powered by a piston engine. This category of aircraft is characterized by the use of extremely thin-walled structures. Commonly used NDE methods are not suitable for this task because of their lower sensitivity.

2.1.1. Composite Wing Description

The test sample was the right half of the wing of a small composite aircraft demonstrator. The wing was used for the development of static structural strength testing. Its structure consisted of spars and ribs made from CFRP. The wing skins were made of a sandwich structure consisting of a carbon sheet and a foam core. The outer and inner skins were made from one layer of CFRP fabric, using TeXtreme® 100 material (Oxeon AB, Boras, Sweden), with a nominal layer thickness of 0.1 mm. Wing skins were glued to the load-bearing structure using the HexBond® EA9394 epoxy adhesive (Hexcel Corporation, Stamford, CT, USA). Bonding was done by applying a thin layer of adhesive to the wing skin and a thicker layer of glue in the shape of a "snake" on the wing spar caps and ribs. Subsequent curing was done in the assembly jig. The maximum thickness of the adhesive layer should be up to 1.5 mm. Figure 1 shows a system drawing of the upper half of the wing (the lower half is similar), with the areas of interest marked for testing. The scheme of the configuration of the bonded joints in these areas is pictured in Figure 2.

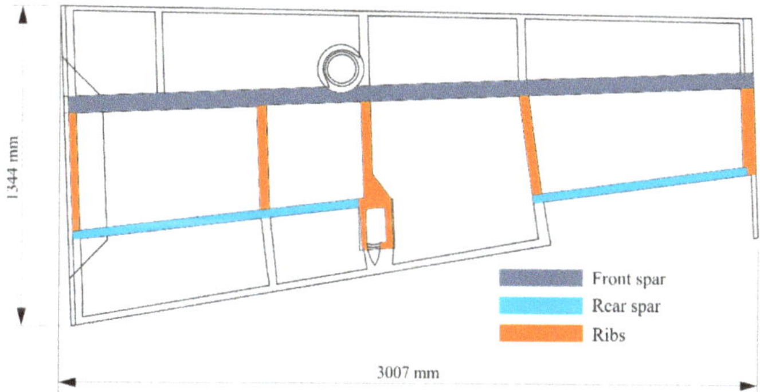

Figure 1. Composite wing configuration with the area of interest.

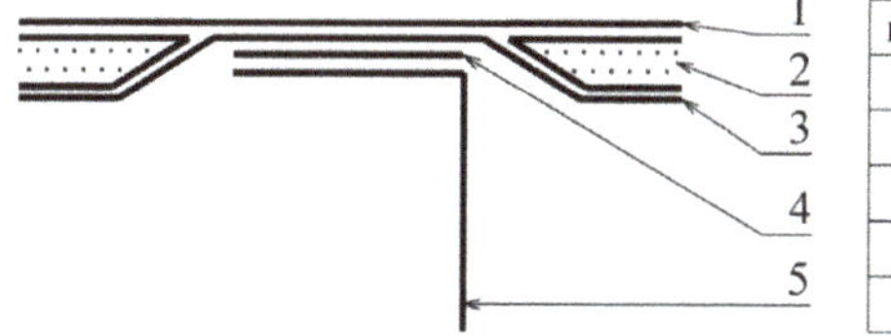

no.	description	material
1	outer skin	TeXtreme 100 (CFRP)
2	foam	Rohacel
3	inner skin	TeXtreme 100 (CFRP)
4	adhesive	EA9394
5	spar (rib)	CFRP

Figure 2. Bonded joint configuration.

2.1.2. Material Properties

Within this work, a determination of the basic thermal properties of both HexBond® EA9394 and TeXtreme® 100, in the direction perpendicular to their surfaces, was performed. The density ρ, thermal diffusivity α and specific heat capacity c_p were determined experimentally. From the data obtained, the values of thermal conductivity λ and thermal effusivity e were subsequently determined by calculation.

Experimental specimens from the examined materials were made with an outer diameter of D = 48 mm. The thicknesses of the specimens were z = 1.745 mm for the EA9394 material and z = 2.510 mm for the TeXtreme® 100 material. The composite lay-up of the second specimen was as follows: $[0°, 90°]_{6S}$. The thickness of the reference specimen, which was made of aluminum alloy 6061 T6, was z = 1.240 mm.

The densities of the materials being examined were determined by measuring the difference between the weights of the specimens in air and the weights of the same specimens when immersed in distilled water according to Archimedes' law. The configuration of the test is shown in Figure 3. The resulting densities were calculated according to Equation (1)

$$\rho = \frac{m}{\Delta m_w}(\rho_w - \rho_a) + \rho_a \tag{1}$$

where m is the weight of the sample weighed in air, Δm_w is the difference between the weights of the sample measured in air and the samples immersed in water, ρ_w is the density of water (ρ_w = 998 kg·m^{-3} at 20 °C) and ρ_a is the density of air (ρ_a = 1.2 kg·m^{-3}). The resulting density values for the individual materials are given in Table 1.

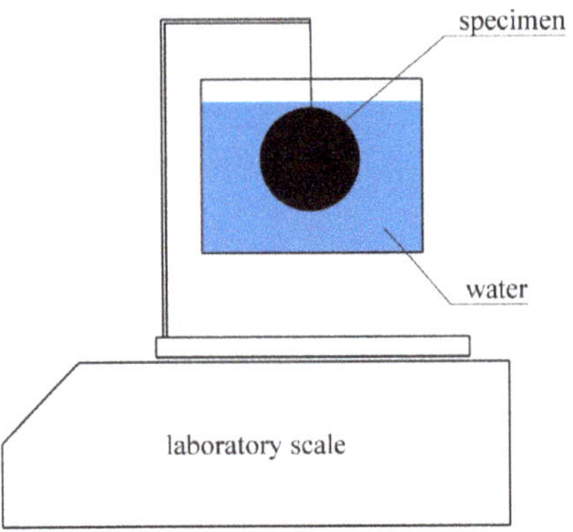

Figure 3. Density measurement configuration.

Table 1. Used materials' densities.

Material	m (g)	Δm_w (g)	ρ (kg·m^{-3})
TeXtreme® 100	7.235	4.650	1552
EA9394	4.187	3.000	1392
6061 T6	6.440	2.390	2687

The thermal diffusivity, α, of these materials was measured on the basis of the thermal curve of the surface of a thin specimen, after excitation by a short thermal pulse. Excitation was performed from the other side of the specimen using a flash lamp. This measuring technique is based on the ASTM E 1461 standard [10]. The value of thermal diffusivity is calculated from the thickness of the sample z and the time t (measured from the excitation moment), when the temperature increase, ΔT, on the measured surface reaches a certain percentage of the maximum surface temperature increase ΔT_{max}. The surface temperature of the specimen was measured using an Infrared (IR) camera and determined as the average value from a circular area in the center of the specimen, which had a diameter of 10 mm. To ensure the same surface emissivity for all specimens, a thin layer of paint with a defined emissivity ε (Therma Spray 800 with ε = 0.96) was applied to both sides of the specimen.

For each specimen, 5 measurements were performed and evaluated, and the resulting value of the thermal diffusivity α is their average.

Equation (2) [25] was used to determine the thermal diffusivity $\alpha_{0.5}$ in time $t_{0.5}$:

$$\alpha = 0.13879\frac{z^2}{t_{0.5}} \tag{2}$$

where z is the specimen thickness, and $t_{0.5}$ is the time taken for the temperature to rise to 50% ΔT_{max}.

Since the above equation is based on the simplified assumption that no heat losses occur during the test, a correction for these losses needs to be done. One possibility is to use the correction reported by Clark and Taylor [25,26], which is based on the ratio of times $t_{0.25}$ and $t_{0.75}$, when the temperature rise reaches 25% and 75% of the maximum temperature rise, respectively. The correction factor K_R is then calculated according to Equation (3):

$$K_R = -0.3461467 + 0.361578\frac{t_{0.75}}{t_{0.25}} - 0.06520543\left(\frac{t_{0.75}}{t_{0.25}}\right)^2 \tag{3}$$

The corrected value of the thermal diffusivity α_{corr} is then (4):

$$\alpha_{corr} = \frac{\alpha_{0.5}\,K_R}{0.13885} \tag{4}$$

A measuring device was assembled for the purpose of this experimental work. Its scheme can be seen in Figure 4. The device consisted of a flash lamp with a reflector, a specimen holder and an IR camera to record the temperatures.

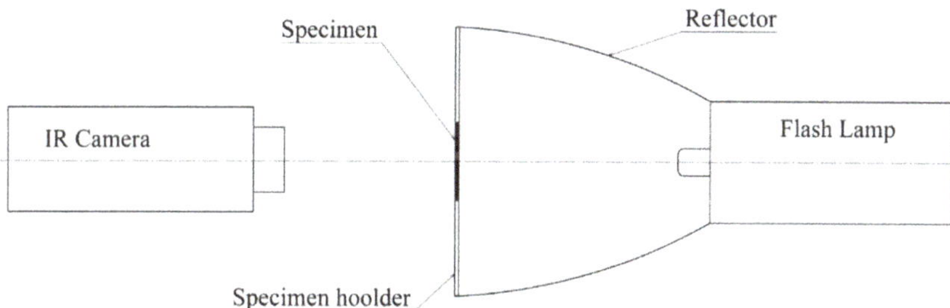

Figure 4. Thermal diffusivity and specific heat capacity measurement configuration.

The resulting values of thermal diffusivity are given in Tables 2 and 3. Figure 5a,b shows the cooling curves for the EA9394 material specimen (left) and the TeXtreme® 100 material specimen (right).

Table 2. Thermal diffusivity—EA9394.

Measurement No.	z (mm)	$t_{0.25}$ (s)	$t_{0.50}$ (s)	$t_{0.75}$ (s)	$\alpha_{0.5}$ (m^{2}s^{-1})	α_{corr} (m^{2}s^{-1})
1	1.745	1.253	1.797	2.622	2.352×10^{-7}	2.118×10^{-7}
2	1.745	1.254	1.853	2.789	2.281×10^{-7}	2.227×10^{-7}
3	1.745	1.289	1.885	2.816	2.242×10^{-7}	2.140×10^{-7}
4	1.745	1.246	1.848	2.760	2.287×10^{-7}	2.220×10^{-7}
5	1.745	1.245	1.821	2.755	2.320×10^{-7}	2.250×10^{-7}
Average	-	-	-	-	2.297×10^{-7}	2.191×10^{-7}
Stand. deviation	-	-	-	-	4.169×10^{-9}	5.831×10^{-9}

Table 3. Thermal diffusivity—TeXtreme® 100.

Measurement No.	z (mm)	$t_{0.25}$ (s)	$t_{0.50}$ (s)	$t_{0.75}$ (s)	$\alpha_{0.5}$ (m^2s^{-1})	α_{corr} (m^2s^{-1})
1	2.510	1.522	2.278	3.318	3.839×10^{-7}	3.657×10^{-7}
2	2.510	1.581	2.330	3.365	3.753×10^{-7}	3.460×10^{-7}
3	2.510	1.574	2.302	3.441	3.799×10^{-7}	3.629×10^{-7}
4	2.510	1.521	2.203	3.266	3.969×10^{-7}	3.704×10^{-7}
5	2.510	1.525	2.297	3.402	3.807×10^{-7}	3.728×10^{-7}
Average	-	-	-	-	3.833×10^{-7}	3.636×10^{-7}
Stand. deviation	-	-	-	-	8.195×10^{-9}	1.054×10^{-8}

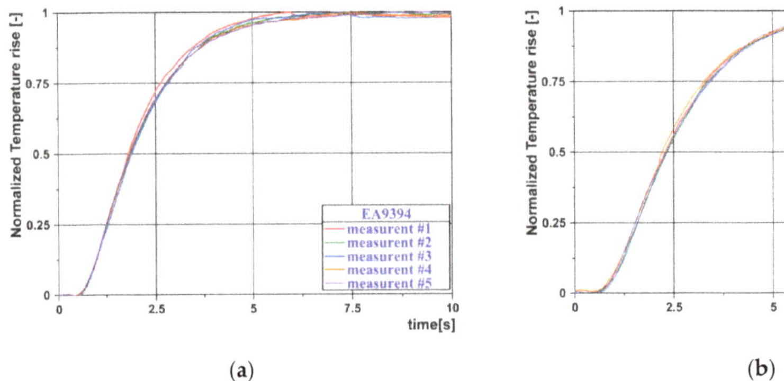

Figure 5. Measurement of thermal diffusivity—heating curves: (**a**) EA9394; (**b**) TeXtreme® 100.

The standard deviation value given in Tables 2 and 3 is the sample standard deviation from the measured (calculated) values defined by Equation (5):

$$s = \sqrt{\frac{\Sigma(x_i - x_a)^2}{n-1}} \tag{5}$$

where x_i is the measured (calculated) value from the i-th measurement, x_a is the arithmetic mean of the values from the measured (calculated) data and n is the number of measurements.

The determination of the specific heat capacity of the material is based on the assumption that the supplied heat Q per unit area A is manifested by an increase in temperature depending on the density of the material ρ, the specific heat capacity c_p and the specimen thickness z, as shown in Equation (6).

$$\frac{Q}{A} = \Delta T \times c_p \times z \times \rho \tag{6}$$

If we perform these measurements under the same conditions (temperature, heat input) for different specimens, where for one reference specimen (index R), all parameters are known, and for the others, the only unknown value is the specific heat capacity, it is possible to determine unknown specific heat capacity c_p according to Equation (7) [27]:

$$c_p = \frac{\Delta T_R}{\Delta T} \frac{z_R \times \rho_R}{z \times \rho} c_{pR} \tag{7}$$

The specimen was made of the aluminum alloy 6061 T6 with a thickness of $z_R = 1.240$ mm, density of $\rho_R = 2687$ kg·m^{-3} and specific heat capacity of $c_p = 896$ J·kg^{-1}·K^{-1} was used as a reference standard. This measurement was performed on the same equipment, with the same specimens and under the same conditions as the thermal diffusivity measurement. The individual values of the temperature increases are given in Table 4. The specific heat capacity values were subsequently determined from the average values of the temperature increase parameter.

Table 4. Values of the temperature increase.

Measurement No.	EA9394	ΔT (K) TeXtreme 100	6061 T6
1	7.809	6.660	8.867
2	7.744	6.655	8.874
3	7.785	6.738	8.874
4	7.832	6.597	8.930
5	7.861	6.547	8.959
average	7.806	6.639	8.901

The values of thermal conductivity λ and thermal effusivity e were calculated from the determined thermal properties according to Equations (8) and (9):

$$\lambda = \alpha \times \rho \times c_p \tag{8}$$

$$e = \sqrt{\lambda \times \rho \times c_p} \tag{9}$$

A summary of the determined and calculated thermal properties is given in Table 5.

Table 5. Determined thermal properties.

Material	ρ (kg·m^{-3})	α_{corr} (m^{2}s^{-1})	c_p (J·kg^{-1}K^{-1})	λ (W·m^{-1}K^{-1})	e (J·s$^{-0.5}$m^{-2}K^{-1})
EA9394	1392	2.191×10^{-7}	1394	0.425	908.4
TeXtreme	1552	3.636×10^{-7}	1022	0.557	957.0

2.2. PT Experimental Method Description

2.2.1. PT Theory

The PT method is based on the principle of heating a sample from one side with a short thermal pulse (for example, a halogen light or a flash lamp) and the subsequent monitoring of the cooling curve at each point of the surface using an IR thermal camera. By sending a pulse, the heat wave begins to propagate through the material. The surface cools due to heat wave propagation (conduction) into the depth of the material as well as due to convection and radiation losses. If beneath the surface there is a defect with a different thermal effusivity to that of the base material (delamination, cavity or void in an adhesive joint), the heat wave will be reflected back to the surface and the cooling process will change at this point. This behavior of the surface cooling curves is demonstrated in Figure 6. Defects that occur at a greater depth will appear on the thermogram with a time delay [12,16]. The time t required to manifest the temperature deviation is a function of the depth of the defect z and the thermal diffusivity α according to the relation (10) [12]

$$t \propto \frac{z^2}{\alpha} \tag{10}$$

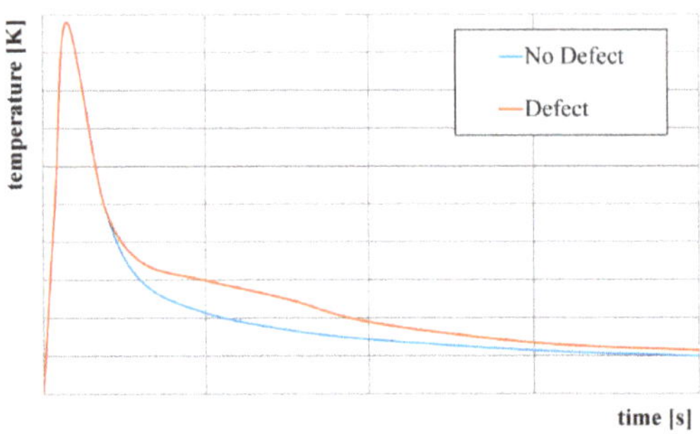

Figure 6. Surface cooling curves.

Figure 7 shows an example of the time evolution of thermograms for the different moments after the excitation pulse. The figure shows a time-sequential drawing of the deeper layers of the adhesive joint. At time t = 5 s, the poor quality of the joint can be seen as resulting from inadequate technology (the adhesive bead was not compressed and spread sufficiently). At the same time, it can be seen that due to lateral diffusion, thermograms lose their sharpness with increasing time.

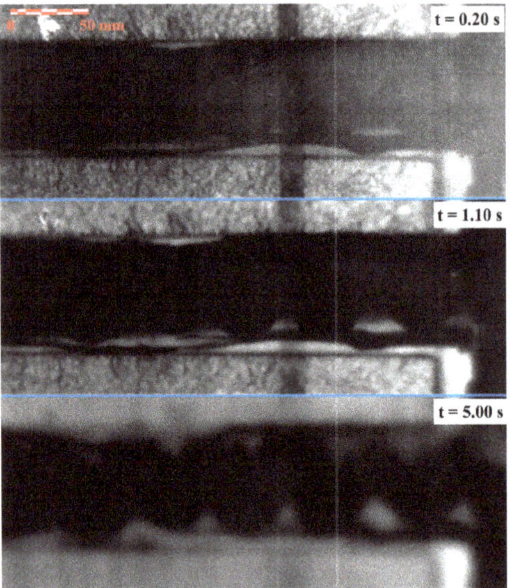

Figure 7. Sequence of the thermograms of the bonded joint.

The disadvantage of this method is a sensitivity to the unevenness of the heat source and the distribution of emissivity on the surface. This can be partially eliminated by subsequent postprocessing.

2.2.2. PT Method Verification

For the purpose of verifying the adhesive joints testing method, and for the setting up of the measuring device, a reference gauge was produced. This gauge corresponds in its composition to the point of the adhesive joints on the wing (Layers 1, 3, 4 in Figure 2). The individual thicknesses represent the depths of the occurrence of the defects of the adhesive joint formed within the adhesive-air interface. The total thickness of the gauge is cut in a range of z = 0.25 mm (skin alone) to z = 2.3 mm (skin + adhesive). With the assumed maximum thickness of the adhesive layer in the adhesive joint of approx. 1.5 mm, this larger gauge thickness should represent a correctly glued joint. Due to the homogenization of the emissivity of the surface, the gauge was sprayed with paint with a defined emissivity of ε = 0.96. The dimensions of the reference gauge are shown in Figure 8.

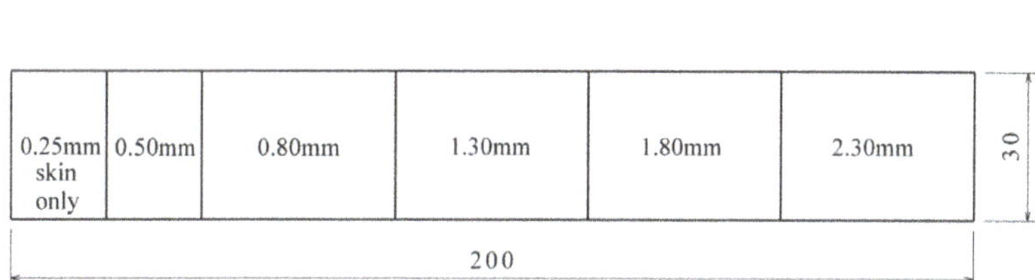

Figure 8. Dimensions of the reference gauge.

Figure 9 shows the sequence of thermograms at five different time points t after excitation. It represents the temperature distribution on the surface of the gauge. The warmest spot is represented by a white color; the coldest by black, with each image in the sequence being normalized to achieve the maximum dynamic range. This figure clearly shows the gradual delineation of individual thicknesses and the gradual blurring of the boundaries between the individual steps of the gauge due to lateral diffusion. Local deviations in temperature distribution within the individual steps of the gauge are caused by imperfections in its production (deviations from the optimal thickness and the occurrence of air voids in the adhesive).

Figure 9. Thermogram sequence for reference gauge.

The following graph in Figure 10 shows the course of temperature distribution along the longitudinal axis of the gauge for the same time steps as in the previous thermograms. The temperature curves are again normalized separately for each curve. From the above thermograms and temperature distribution curves, it is clear that the performed measurement confirms an ability to detect defects in the adhesive joint within the entire range of expected depths.

Figure 11a,b shows the time course of the cooling curves measured at the centers of the individual steps of the reference gauge. The first image has a linear timeline, while the second image is logarithmic. The second figure clearly shows a turning point in the temperature decrease at time t = 0.05–0.10 s after excitation, which is caused by the different thermal diffusivity of the skin and of the adhesive material. Both graphs show a time-varying deviation of the individual cooling curves from the curve, representing a correctly made joint (z = 2.3 mm).

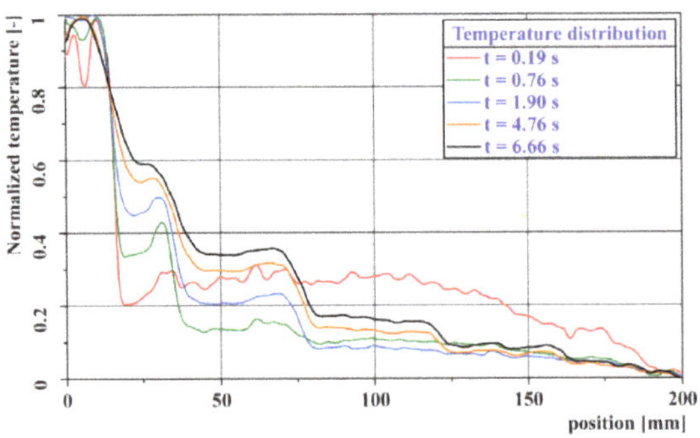

Figure 10. Course of temperature distribution along the longitudinal axis of the gauge.

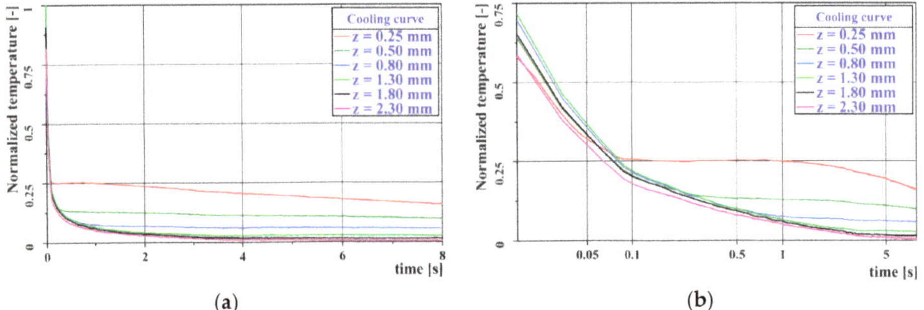

Figure 11. Cooling curves for the reference gauge: (**a**) linear time axis; (**b**) logarithmic time axis.

Since the resulting thermograms are normalized to the temperature range for a given time, the following graph in Figure 12 shows the temperature profile normalized to the temperature range from the highest temperature (curve for z = 0.25 mm) to the lowest temperature (curve for z = 2.3 mm) at each measurement point. This graph shows the dimensionless contrasts within the data obtained at a given time.

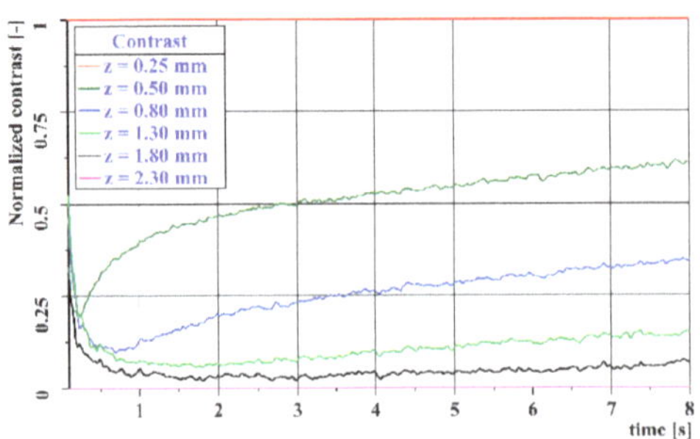

Figure 12. Image contrast curves for the reference gauge.

2.2.3. Differential Absolute Contrast (DAC) Evaluation

A Modified Differential Absolute Contrast (DAC) method was used to process the measured data. This allows partial elimination of heat source unevenness, such as reflections from the surroundings (e.g., IR camera, flash lamps) and emissivity distribution on the surface. The DAC method compares the temperature of the tested place containing the defect, with the theoretical value of the temperature if there were no defect in the place being tested. This theoretical temperature is calculated on the basis of the 1D form of the Fourier equation of heat conduction in a semi-infinite medium from the measured temperature, at a point in time where the temperature defect does not manifest itself [12]. The standard DAC method works with the temperature at that point in time just before the manifestation of the defect, which, however, usually requires the manual intervention of the test operator. This thermal contrast is calculated according to Equation (11) below, which describes the relation of temperature $T(t)$ at the observed time t and the temperature $T(t')$ at the reference time t' for each individual pixel of the record [28].

$$\Delta T_{DAC}(t) = T(t) - \sqrt[b]{\frac{t'}{t}}T(t') \quad (11)$$

Since the theoretical temperature decrease using the standard DAC method does not involve heat transfer by radiation, but only by conduction, this decrease is significantly lower than in reality. For this reason, the square root in Equation (11) was replaced by a power with the general parameter b representing the slope of the temperature decrease (12).

$$\Delta T_{DAC}(t) = T(t) - \left(\frac{t'}{t}\right)^b T(t') \quad (12)$$

This parameter was determined from experimental data based on the approximation of the cooling curve on the reference gauge for the thickness $z = 2.30$ mm, which represents the adhesive joint without any defect. The reference time for calculating the contrast is the time $t' = 0.15$ s, which corresponds to the thickness just behind the interface between the skin and the adhesive, and the value of the parameter $b = 0.65$. The whole calculation is performed for temperatures normalized to the maximum and minimum temperatures reached during the measurement from excitation to a steady state at the end of the measurement, over the entire measured area. Figure 13 shows the actual and theoretical cooling curves plotted using Equation (11) for three different thicknesses on the reference gauge (Figure 13a) and the DAC curves for all the thicknesses on the reference gauge (Figure 13b).

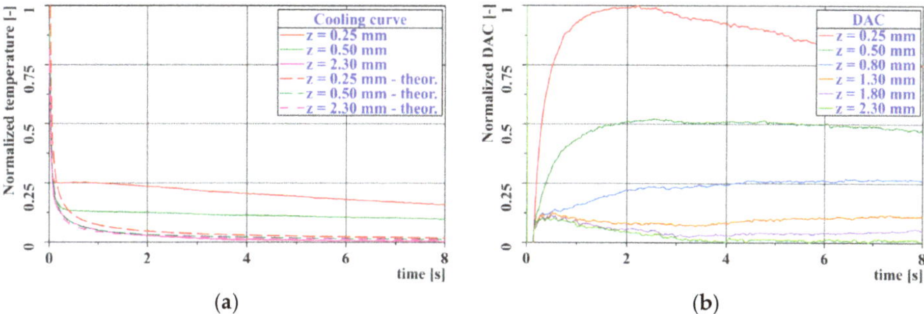

Figure 13. Cooling curves for the reference gauge: (**a**) linear time axis; (**b**) Differential Absolute Contrast (DAC) curves.

As in Figure 12, the following graph in Figure 14 shows the DAC profile normalized to the temperature range from the highest temperature (curve for $z = 0.25$ mm) to the lowest temperature (curve for $z = 2.3$ mm) at each measurement point. This graph shows the dimensionless contrast within the data obtained at a given time. A comparison of the two graphs shows that the DAC contrast calculation does not have a significant effect on the resulting image contrast.

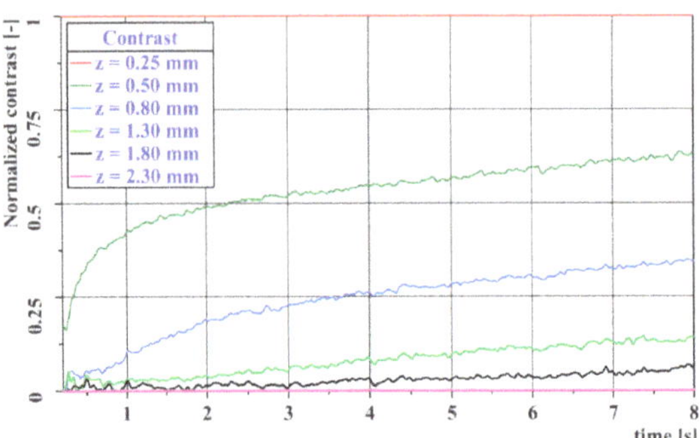

Figure 14. DAC image contrast curves for the reference gauge.

Figure 15 shows a comparison between the raw thermogram image and the modified DAC-processed image. These images represent the same test area (end of the front wing spar) at the same time after excitation. In the first image, the reflection of the IR camera is visible in the lighter parts (the foam sandwich area, blue circle), and in the dark area (the area of the adhesive joint), a significant unevenness in the emissivity of the surface can be observed (green marking). The effect of the heating unevenness is marked by a red circle. These imperfections have been largely eliminated by the application of the DAC method.

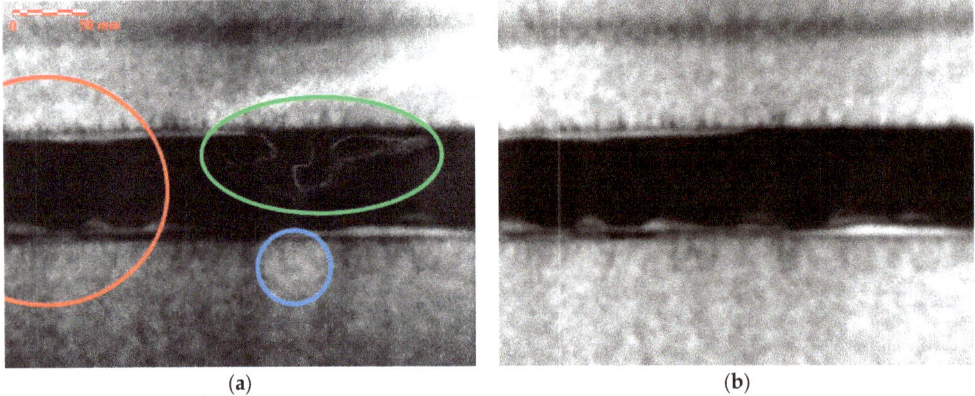

(**a**) (**b**)

Figure 15. Comparison of the resulting images: (**a**) RAW thermal data; (**b**) modified DAC-processed data.

2.3. Experimental System Description

A modular test system was designed and employed for the PT NDE method. The experimental setup can be seen in Figure 16. The basic hardware elements of this system consist of a FLIR A325SC bolometric uncooled IR camera (resolution 320 × 240 pixels; NETD < 50 mK; maximum scanning frequency, 60 Hz), an instrument unit equipped with a PC for test control and data recording and two flash lamps (2 × 1200 Ws).

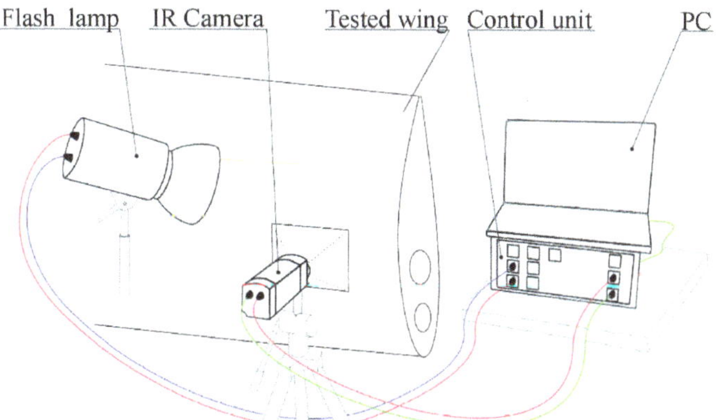

Figure 16. Pulse Thermography (PT) method experimental configuration.

The instrument unit works as the communication interface between the PC, IR camera and excitation lamps. It consists mainly of the cDAQ measuring and control system from National Instruments, equipped with analog output cards (for excitation lamp control) and digital input/output as well as other necessary auxiliary electronics.

A program was specifically created for this purpose in the LabView environment from National Instruments, and this was used to control the test's processes and record the measured data. The processing and evaluation of the obtained data were performed in the MATLAB environment from MathWorks [29].

3. Results and Discussion

The following figures represent examples of the test results. Each individual image covers the tested area, which has a size of approx. 320 × 240 mm and, at a given resolution of the IR camera, represents a resolution of 1 × 1 mm for each pixel of the image. The final images were adjusted so that the grayscale range covers the entire range of the evaluated data (from white to black). The maximum temperature increase at the moment after excitation is up to 10 K. The temperature is not measured during the NDE process. Figures 17 and 18 show a representative selection of the images on which the test results are demonstrated, with an explanation of the individual indications.

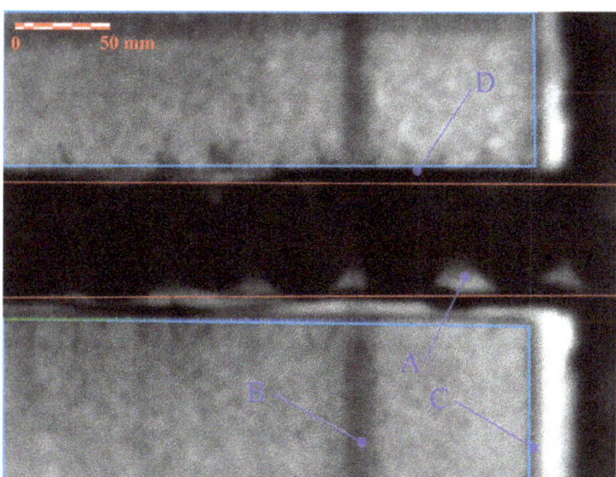

Figure 17. Results of the PT Nondestructive Evaluation (NDE) method: adhesive joint of the front wing spar and the wing tip rib on the lower wing side.

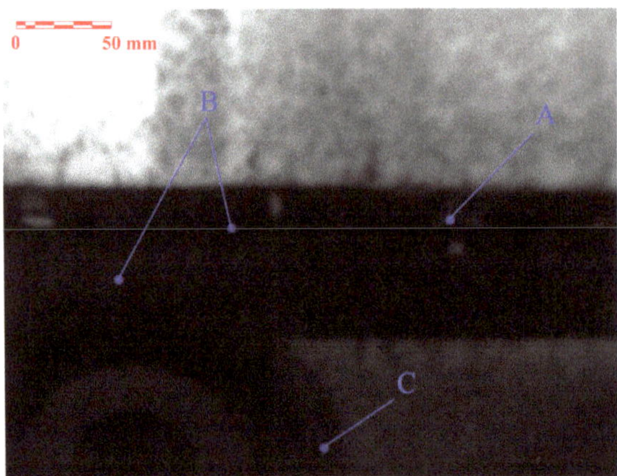

Figure 18. Results of the PT NDE method: adhesive joint of the front wing spar and the upper wing side in the fuel tank area.

The area of the adhesive joint of the wing tip rib (right) and of the front wing spar cap on the lower wing skin side is pictured in Figure 17. The red lines mark the area of the front wing spar for the PT NDT evaluation process. The blue lines mark the area of the sandwich foam core reinforcements of the wing skin. From the point of view of the adhesive joint evaluation, the critical places are represented by the lighter shades of the corresponding color (Figure 17A), which can be interpreted as the voids in the adhesive joint. In Figure 17D, the indication marked represents an overflow of excessive adhesive outside the joint area. In Figure 17B, the indication marked is caused by the overlapping of the top layers of the skin and thus its localized doubling. In Figure 17C, the indication marked represents the region of resin accumulation at the point where foam core ends and sandwich skins are joined together.

Figure 18 represents the area of the front wing spar at the location of the fuel tank on the upper wing skin side. Again, the places with the voids in the adhesive layer are clearly visible (Figure 18A). The use of two different types of adhesives is visible in the bonded area (Figure 18B). This is due to the need for increased resistance to the influence of fuel in the fuel tank area (use of C-resin type). In Figure 18C, the indication marked represents the local reinforcement in the area of the fuel tank lid by the addition of one layer of fabric to the outer skin lay-up.

Due to the fact that no etalons with artificial defects were available for testing, an additional comparison of the NDE findings, with the actual condition of the adhesive joint at the failure area, was performed following the static strength test of the wing demonstrator. In the main wing spar area, the CFRP wing skin was removed to the depth of the adhesive joint. Figure 19 shows a comparison of the NDE findings with the actual condition of the adhesive joint. The upper part of the image contains an evaluated picture of the adhesive joint before the structural strength test. The lower part shows the condition of the adhesive joint after the strength test and the resulting wing damage. The individual colored circles mark the corresponding defects of the joint. The NDE measurement shows a good match with the actual condition of the joint. The smallest detected defect (marked in red) is about 3 mm in diameter. Due to the resolution of the IR camera, this dimension can be considered as the smallest detectable defect in the adhesive joint for a given test configuration. Except for defects in the adhesive layer, the use of two adhesives (EA9394 and C-resin) is clearly visible in the picture. In the lower half of the picture, two vertical cracks caused by a failure during the strength test of the wing demonstrator can be also seen.

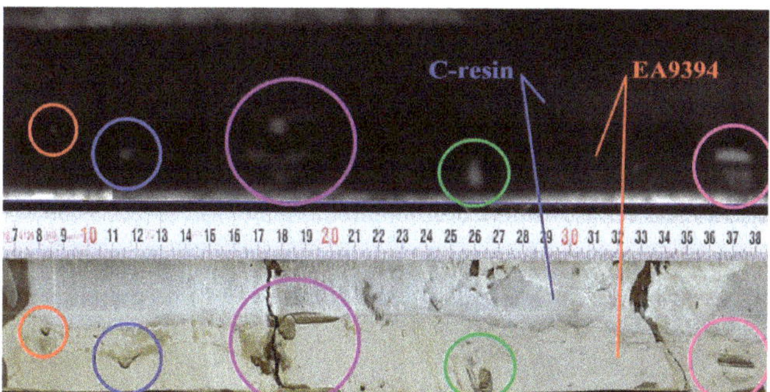

Figure 19. Comparison of NDE findings (**top**) with the actual condition of the adhesive joint (**bottom**).

Figure 20 compares the result of the NDE test with the actual condition of the adhesive joint in the end of the wing spar area. Areas without adhesive are clearly visible. This was caused by insufficient squeezing of the adhesive onto the whole area of the joint. Figure 21 shows a similar comparison of the adhesive joint of the skin and the rib at the location of the flap lever seating. The comparison shows a good match of the NDE with the actual condition of the adhesive joint. It is only in the upper left part of the image that nonglued areas indicated by the NDE method cannot be seen, which is probably due to the removal of the nonglued layer during the removal of the wing skin.

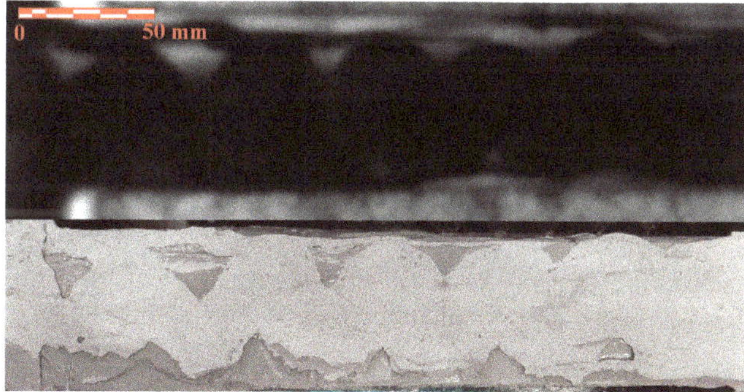

Figure 20. Comparison of NDE findings in the end of the wing spar area (**top**) with the actual condition of the adhesive joint (**bottom**).

Figure 22 shows an overview image (a top and bottom view of the right wing) resulting from the test of the entire wing, composed of individual images. The tested area is limited to the adhesive joints only. The dark vertical areas represent the adhesive joints of the spar caps to the wing skin. The horizontal areas represent the adhesive joints of the ribs to the wing skin. Areas with insufficient adhesive coverage (brighter areas in the adhesive joints) can be seen along almost the entire length of the adhesive joints of the spar caps to the wing skin. Furthermore, the areas of overlap of the outer layers of the wing skin, or the reinforcements of the structural openings in the skin, are clearly visible. To give the reader an idea of the speed of the described method, we present an overview of the time required. The measurement itself lasted about 20 h, which could be shortened when using custom tooling for setting up the camera and flash lamps. Data processing took about 10 h. In total, there are 26 measurements on the top of the wing and 21 measurements on the bottom of the wing.

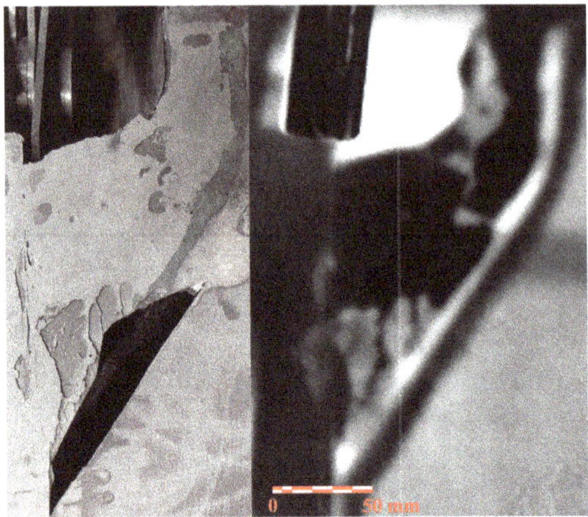

Figure 21. Comparison of NDE findings at the location of the flap lever seating (**right**) with the actual condition of the adhesive joint (**left**).

(a)

(b)

Figure 22. Results of the PT NDE method—overview image of the tested wing area: (**a**) lower wing side on the left; (**b**) upper wing side on the right.

4. Conclusions

This article presents a possible method for testing the quality of adhesive joints in the composite thin-walled structures of light sports aircraft. It also presents a simplified procedure for determining the thermal properties of the materials used, which are not usually reported by the manufacturers of basic materials, using an IR camera. This measurement procedure is not intended to be an alternative to accurate laboratory methods. Its purpose is to give an idea of the properties that depend, among other things, on the production technology used by the manufacturer of the specific composite part under investigation. Prior to the actual testing of the adhesive joints of the wing, the procedure for the testing and evaluation of the measured data was set and verified on a reference gauge, which simulated the adhesive joint. These tests demonstrated a sufficient depth resolution of the test method. The generally achieved sensitivity of the method was confirmed, where it was possible to detect a defect of a size twice its depth below the surface of the inspected part [12,30,31]. The anticipated method sensitivity, which is sufficient to detect a defect of at least 3 mm in size, was confirmed. Although no verification was performed on the samples with artificial defects, the sensitivity was found to be quite sufficient when compared to known cases of catastrophic failure of adhesive joints in this category of aircraft. The authors of [32] describe the case of a nonglued area with a size of 500 mm leading to a catastrophic failure of the composite wing of the LSA-category aircraft. According to the authors of [33], the failure of an all-composite sailplane wing due to a 200 mm long, nonglued area is described. In addition, this technique makes it possible to assess the nature of defects in the adhesive and improve the production process. A comparison of the results of measurements performed on the adhesive joints of the tested wing showed a sufficient matching of the measured results with the actual state of the adhesive joint.

The NDE testing method described in this article demonstrated very good usability for the detection of the flows in the adhesive joints of the wing skins, ribs and the load-bearing spar structure made from thin CFRP. The achieved results are clearly interpretable and usable for both the qualitative and quantitative evaluation of adhesive joints. As well as defects in the adhesive layer, manufacturing technology defects and inaccuracies such as adhesive overflow, foam insert misalignments or composite layer overlaps are detectable.

Of course, it would be possible to improve the readability of the resulting images in particular by: (1) increasing the image sharpness and thus more accurately determining the shape of the defect, which can be achieved by using a higher-resolution IR camera; (2) reducing noise in images, which can be achieved by using an IR camera with greater sensitivity and stability or, alternatively, due to the relatively low initial heating, by the use of more powerful flash lamps. This would increase the signal-to-noise ratio.

Author Contributions: Conceptualization, T.K.; methodology, T.K. and M.D.; thermal properties measurement, T.K.; measurement of wing adhesive joints, T.K. and M.D.; DAC data evaluation, T.K.; writing—original draft preparation, T.K.; writing—review and editing, T.K. and M.D.; project administration, T.K. All authors have read and agreed to the published version of the manuscript.

Funding: This work has been supported by Project No. SGS20/162/OHK2/3T/12 of the Grant Agency of the Czech Technical University in Prague.

Data Availability Statement: Data sharing is not applicable to this article.

Conflicts of Interest: The authors declare no conflict of interest.

References

1. Kostroun, T.; Dvořák, M. Non-Destructive Evaluation of Small Aircraft Wing Adhesive Joints using Pulse Infrared Thermography Method. In Proceedings of the 58th International Scientific Conference Experimental Stress Analysis, Sobotín, Czech Republic, 19–22 October 2020; VSB-TUO: Ostrava, Czech Republic, 2020; pp. 225–232.
2. AMC CS-VLA. *Acceptable Means of Compliance-Certification Specifications for Very Light Aeroplanes*; EASA: Colagne, Germany, 2009. Available online: www.easa.eu (accessed on 22 January 2021).

3. Torres, M.; Piedra, S.; Ledesma, S.; Velázquez, A.E.C.; Angelucci, G. Manufacturing Process of High Performance–Low Cost Composite Structures for Light Sport Aircrafts. *Aerospace* **2019**, *6*, 11. [CrossRef]
4. Li, R.; Lu, Z.H. Analysis of Manufacturing Technology of Composite Materials in Light Sport Aircraft. *Adv. Mater. Sci. Technol.* **2019**, *1*, 7–12. [CrossRef]
5. Pora, J. Composite materials in the airbus A380—From history to future. In Proceedings of the ICCM13, Beijing, China, 25–29 June 2001; Plenary Lecture: Beijing, China, 2001.
6. Dvořák, M.; Růžička, M.; Kulíšek, V.; Běhal, J.; Kafka, V. Damage Detection of the Adhesive Layer of Skin Doubler Specimens Using SHM System Based on Fibre Bragg Gratings. In Proceedings of the 5th European Workshop on Structural Health Monitoring, Sorento, Italy, 28 June–4 July 2010; DEStech Publications, Inc.: Lancaster, PA, USA, 2010; pp. 70–75.
7. Heslehurst, R.B. *Defects and Damage in Composite Materials and Structures*; CRC Press: Boca Raton, FL, USA, 2014; pp. 59–77. [CrossRef]
8. Beine, C.; Boller, C.; Netzelmann, U.; Porsch, F.; Venkat, R.; Schulze, M.; Bulavinov, A.; Heuer, H. NDT for CFRP Aeronautical Components A Comparative Study. In Proceedings of the 2nd International Symposium on NDT in Aerospace 2010, DGZfP, Hamburg, Germany, 22–24 November 2010.
9. Ehrhart, B.; Valeske, B.; Muller, C.E.; Bockenheimer, C. Methods for the Quality Assessment of Adhesive Bonded CFRP Structures—A Resumé. In Proceedings of the 2nd International Symposium on NDT in Aerospace 2010, DGZfP, Hamburg, Germany, 22–24 November 2010.
10. Yılmaz, B.; Jasiūnienė, E. Advanced ultrasonic NDT for weak bond detection in composite-adhesive bonded structures. *Int. J. Adhes. Adhes.* **2020**, 102. [CrossRef]
11. Workman, G.L.; Kishoni, D.; Moore, P.O. *Nondestructive Testing Handbook, Volume 7–Ultrasonic Testing*, 3rd ed.; ASNT: Columbus, OH, USA, 2007; pp. 267–299.
12. Meola, C.; Boccardi, S.; Carlomagno, G.M. *Infrared Thermography in Evaluation of Aerospace Composite Materials*; Woodhead Publishing: Cambridge, UK, 2017; pp. 25–56, 98–103.
13. Ciampa, F.; Mahmoodi, P.; Pinto, F.; Meo, M. Recent Advances in Active Infrared Thermography for Non-Destructive Testing of Aerospace Components. *Sensors* **2018**, *18*, 609. [CrossRef] [PubMed]
14. Qu, Z.; Jiang, P.; Zhang, W. Development and Application of Infrared Thermography Non-Destructive Testing Techniques. *Sensors* **2020**, *20*, 3851. [CrossRef] [PubMed]
15. Schroeder, J.A.; Ahmed, T.; Chaundry, B.; Shepard, S. Non-destructive testing of structural composites and adhesively bonded composite joints: Pulsed thermography. *Compos. Part A* **2002**, *33*, 1511–1517. [CrossRef]
16. Maldaque, X. *Theory and Practice of Infrared Technology for Nondestructive Testing*; John Wiley & Sons: New York, NY, USA, 2011; pp. 343–366.
17. Wen, B.; Zhou, Z.; Zeng, B.; Yang, C.; Fang, D.; Xu, Q.; Shao, Y.; Wan, C. Pulse-heating infrared thermography inspection of bonding defects on carbon fiber reinforced polymer composites. *Sci. Prog.* **2020**, *103*. [CrossRef] [PubMed]
18. Deane, S.; Avdelidis, N.P.; Castanedo, I.C.; Zhang, H.; Nezhad, H.Y.; Williamson, A.A.; Mackley, T.; Davis, M.J.; Maldague, X.; Tsourdos, A. Application of NDT thermographic imaging of aerospace structures. *Infrared Phys. Technol.* **2019**, *97*, 456–466. [CrossRef]
19. Vavilov, V.P.; Burleigh, D.D. Review of pulsed thermal NDT: Physical principles, theory and data processing. *NDT E Int.* **2015**, *73*, 28–52. [CrossRef]
20. Moustakidis, S.; Anagnostis, A.; Karlsson, P.; Hrissagis, K. Non-destructive inspection of aircraft composite materials using triple IR imaging. *IFAC-PapersOnLine* **2016**, *49*, 291–296. [CrossRef]
21. Chen, D.; Zhang, X.; Zhang, G.; Zhang, Y.; Li, X. Infrared Thermography and Its Applications in Aircraft Non-destructive Testing. In Proceedings of the 2016 International Conference on Identification, Information and Knowledge in the Internet of Things, Beijing, China, 20–21 October 2016; IIKI: Beijing, China, 2016; pp. 374–379. [CrossRef]
22. Shepard, S.M. Flash Thermography of Aerospace Composites. In *Proceedings of the IV Conferencia Panamericana de END*; AAENDE: Buenos Aires, Argentina, October 2007.
23. Rajic, N.; Rowlands, D.; Tsoi, K.A. An Australian Perspective on the Application of Infrared Thermography to the Inspection of Military Aircraft. In Proceedings of the 2nd International Symposium on NDT in Aerospace 2010, DGZfP, Hamburg, Germany, 22–24 November 2010.
24. Genest, M.; Castanedo, I.C.; Piau, J.M.; Guibert, S.; Susa, M.; Bendada, A.; Maldague, X.; Brothers, M.; Fahr, A. Comparison of Thermography Techniques for Inspection of F/A-18 Honeycomb Structures. In Proceedings of the Aircraft Aging 2007 Conference, Naples, Italy, 16–18 September 2007.
25. ASTM E1461-13. *Standard Test Method for Thermal Diffusivity by the Flash Method*; ASTM International: West Conshohocken, PA, USA, 2013; Available online: www.astm.org (accessed on 22 January 2021).
26. Clark, L.M.; Taylor, R.E. Radiation loss in the flash method for thermal diffusivity. *J. Appl. Phys.* **1975**, *46*, 714–719. [CrossRef]
27. Kim, S.; Kim, Y. Improvement of specific heat measurement by the flash method. *Thermochim. Acta* **2007**, *455*, 30–33. [CrossRef]
28. Benítez, H.D.; Castando, I.C.; Bendada, A.; Maldague, X.; Loaiza, H.; Caicedo, E. Definition of a new thermal contrast and pulse correction for defect quantification in pulsed thermography. *Infrared Phys. Technol.* **2008**, *51*, 160–167. [CrossRef]

29. Kostroun, T.; Pilař, J.; Dvořák, M. Design of active infrared thermography NDT system with optical excitation. In Proceedings of the 49th International Conference DEFEKTOSKOPIE 2019/NDE for Safety, České Budějovice, Czech Republic, 5–9 November 2019; VUT v Brně: Brno, Czech Republic, 2019; pp. 75–85.
30. Ospina, F.J.F.; Benitez, H.D. From local to global analysis of defect detectability in infrared non-destructive testing. *Infrared Phys. Technol.* **2014**, *63*, 211–221. [CrossRef]
31. Almond, P.D.; Peng, W. Thermal imaging of composites. *J. Microsc.* **2001**, *201*, 163–170. [CrossRef] [PubMed]
32. Air Accidents Investigation Institute. Available online: https://uzpln.cz/pdf/20171113113844.pdf (accessed on 20 December 2020).
33. German Federal Bureau of Aircraft Accidents Investigation. Available online: https://www.bfu-web.de/EN/Publications/Investigation%20Report/2003/Report_03_3X164-0-Heppenheim-DuoDiscus.pdf?__blob=publicationFile (accessed on 20 December 2020).

Article

High-Resolution Strain/Stress Measurements by Three-Axis Neutron Diffractometer

Pavol Mikula *, Vasyl Ryukhtin, Jan Šaroun and Pavel Strunz

Department of Neutron Physics, Nuclear Physics Institute ASCR, v.v.i., 250 68 Řež, Czech Republic; ryukhtin@ujf.cas.cz (V.R.); saroun@ujf.cas.cz (J.Š.); strunz@ujf.cas.cz (P.S.)
* Correspondence: mikula@ujf.cas.cz; Tel.: +420-2-6617-3159

Received: 2 November 2020; Accepted: 27 November 2020; Published: 30 November 2020

Abstract: Resolution properties of the unconventional high-resolution neutron diffraction three-axis setup for strain/stress measurements of large bulk polycrystalline samples are presented. Contrary to the conventional two-axis setups, in this case, the strain measurement on a sample situated on the second axis is carried out by rocking the bent perfect crystal (BPC) analyzer situated on the third axis of the diffractometer. Thus, the so-called rocking curve provides the sample diffraction profile. The neutron signal coming from the analyzer is registered by a point detector. This new setup provides a considerably higher resolution (at least by a factor of 5), which however, requires a much longer measurement time. The high-resolution neutron diffraction setting can be effectively used, namely, for bulk gauge volumes up to several cubic centimeters, and for plastic deformation studies on the basis of the analysis of diffraction line profiles, thus providing average values of microstructure characteristics over the irradiated gauge volume.

Keywords: residual stresses; neutron diffraction; three axis setting; high resolution; bent crystal monochromator; bent crystal analyzer

---

1. Introduction

Residual stresses are typical phenomena associated, e.g., with the welding of different kinds of structural material. Stresses are generally responsible for the deformation of structures during production and subsequently influencing the behavior of these structures during service [1–3]. However, they also occur when external load or some kind of shape forming is applied to the sample. X-ray diffraction and neutron diffraction are excellent nondestructive techniques for the determination of strain/stress fields in polycrystalline materials. X-ray diffraction, which due to a strong attenuation of X-rays in the material, is limited to sample surface measurements. On the other hand, neutron diffraction, thanks to a low attenuation of neutrons in most materials, is suitable for strain scanning of bulk samples. The conventional neutron diffraction method is based on the precise determination of the relative change in the d_{hkl}-spacing of particularly oriented crystalline grains in the gauge volume [1–3]. In neutron and X-ray diffraction, the angular positions of the diffraction lines are determined by the well-known Bragg condition $2dhkl \cdot \sin \theta hkl = \lambda$ (θhkl—Bragg angle, λ—the wavelength). The elastic strain ε is defined as $\varepsilon = \Delta d_{hkl}/d_{0,hkl}$. It is in fact the relative change in the lattice spacing with respect to the value $d_{0,hkl}$ corresponding to a strain-free sample element. The strain in the material is a tensor, and for its determination, at least three strain components should be obtained. However, only one component parallel to the scattering vector Q is determined from the measurement of one diffraction line (see the inset in Figure 1). Therefore, the strain measurements have to be carried out for three geometrical positions. It is clear that for strain evaluation, the determination of the $d_{0,hkl}$ value also represents a key task [1–3]. The differentiation of the Bragg condition provides a simple formula for the strain ε as $\varepsilon = -\cot \theta hkl \cdot \Delta \theta hkl$. According to this formula, the strain ε in the material brings about a change in the scattering angle $2\theta hkl$ by $\Delta(2\theta hkl)$ of the position of the diffraction line with respect to the position of the line of the strain-free element. Therefore, from the angular shift in the diffraction line $\Delta\theta hkl$, the average lattice macrostrain component over the irradiated gauge volume element can be determined. The neutron strain/stress instrument is a 2-axis diffractometer (see Figure 1) equipped with a position-sensitive detector (PSD), usually used for the mapping of residual strains inside polycrystalline materials. The spatial resolution mainly determined by the beam defining optics is usually of the order of millimeters. The diffractometer is usually equipped with a focusing monochromator (bent perfect crystal) that has the curvature optimized to achieve a maximum luminosity and resolution with respect to a chosen diffraction line [1–3]. Together with the focusing monochromator, the beam optics system includes slits before and after the sample that determine the dimensions of the gauge volume element and, of course, the position-sensitive detector for imaging the beam profile diffracted by the irradiated element [4–6]. The strain/stress scanner can also be equipped with an auxiliary machine, e.g., for thermo-mechanical manipulation with the sample. Then, the resolution of the instrument results from the combination of uncertainties coming from the thickness and radius of curvature of the bent crystal monochromator, the widths of the slits (in most cases of 0.5–2 mm), the divergence of the incoming and diffracted beam by the irradiated element, and the spatial resolution of the position-sensitive detector. Their combination results in a total uncertainty usually forming a Gaussian image in the detector with a *FWHM* of about (5–10) $\times 10^{-3}$ rad, which can be sufficient for the measurement of the angular shifts of diffraction lines generated by strains. However, such resolution is not sufficient for the investigation of the changes in the diffraction profile, which would, e.g., allow for microstructure characterization of plastically deformed polycrystalline samples.

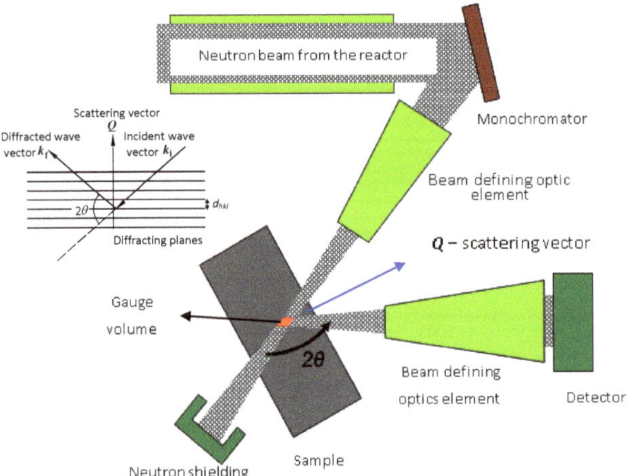

Figure 1. Scheme of the conventional strain scanner.

Encouraged by the first recent preliminary results [7], we have decided to study resolution properties of a new unconventional high-resolution neutron-diffraction three-axis setup for large bulk polycrystalline samples experimentally with the aim of a possible exploitation in the field of strain/stress measurements. The most important obtained results are presented.

2. Materials and Methods

Several years ago [8–10], the first attempts were made with a high-resolution three-axis setting, as schematically shown in Figure 2a. Following the drawing shown in Figure 2a (for small widths of the samples), the best resolution from this setting can be obtained when minimizing dispersion of the whole system. When solving the problem in momentum space, it means that the orientation of the Δk momentum elements related to the monochromator and analyzer should be matched to that of the sample, and this can be achieved by proper radii of curvatures of the bent perfect crystal monochromator and analyzer. Contrary to the conventional two-axis strain/stress scanner, the diffraction profiles with the three-axis setup are obtained by rocking the bent perfect crystal (BPC) analyzer situated on the third axis of the diffractometer, and the neutron signal is registered by the point detector (see Figure 2) [7,8]. By properly adjusting the curvature of the analyzer, the three-axis setting exploits the focusing in both real and momentum space. The resolution of this alternative setting was optimized by using a well-annealed α-Fe low-carbon steel rod (technically pure iron with C less than 1%) of the diameter of 8 mm (standard sample). For the experimental studies, the three-axis neutron optics diffractometer installed at the Řež research reactor LVR-15 (Research Centre eŘž, Husinec, Czech Republic) and equipped with Si(111)-monochromator and Ge(311)-analyzer single-crystal slabs of dimensions 200 × 40 × 4 and 20 × 40 × 1.3 mm^3 (length × width × thickness), respectively, was used. The Si(111) monochromator providing the neutron wavelength of 0.162 nm had a fixed curvature with the radius R_M of about 12 m, and the radius of curvature of the analyzer was changeable in the range from 3.6 to 36 m. Figure 3 shows some results of the optimization procedure related to the setting shown in Figure 2a, when searching for a minimum *FWHM* of the rocking curve, and the optimum curvature of the analyzer. The optimum radii of curvature of the Ge(311) analyzer for α-Fe(110) and α-Fe(211) reflections have been found as R_A = 9 m and R_A = 3.6 m, respectively. However, it has been found that high-resolution determination of the lattice changes can be achieved even on large irradiated gauge volumes (see Figure 2b), though slightly relaxed.

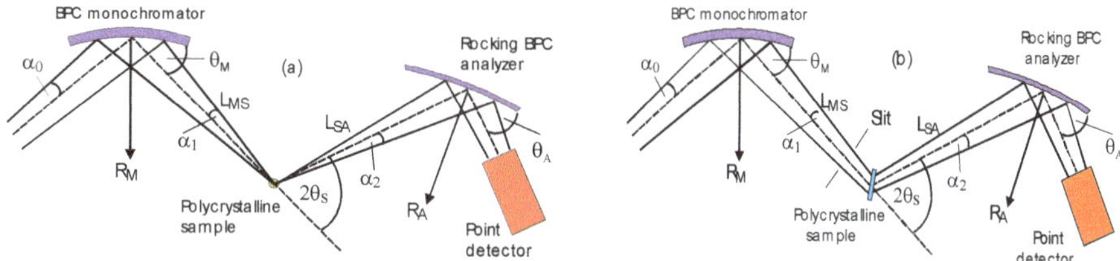

Figure 2. Three-axis diffractometer settings employing bent perfect crystal (BPC) monochromator and analyzer as used in the feasibility studies (R_M, R_A—radii of curvature, θ_M, θ_A—Bragg angles) for vertical (**a**) and horizontal (**b**) positions of a polycrystalline sample. This figure has been reprinted from Ref. [7].

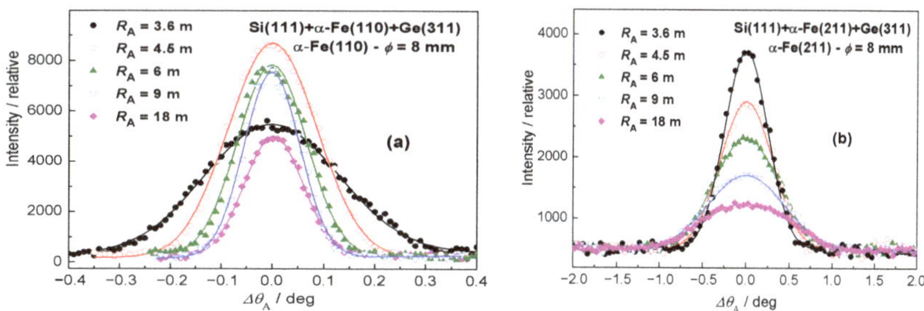

Figure 3. Analyzer rocking curves for different radii of curvature and for (**a**) α-Fe(110) and (**b**) α-Fe(211) reflections.

3. Results

After the adjustment of the diffractometer setting for optimum resolution, the first step was measuring the resolution of the experimental setting for different diameters of the well-annealed α-Fe(110) standard samples situated in the vertical position (see Figure 4). Concerning the resolution represented by the *FWHM* of the analyzer rocking curves, it can be seen from Figure 4 that the diameter of the sample (generally the thickness) plays an important role. In the next two steps, the influence of the slit width on the resolution properties for two diameters of the well-annealed sample of 4.9 and 8 mm in the horizontal position was studied (see Figures 5 and 6). In both cases, three slit widths of 5, 10, and 20 mm were used. This can be useful when comparing the effect of different widths of the irradiated sample, as the vertical dimension of the irradiated sample does not play a principal role with respect to the resolution of the experimental setting. As expected, it can be seen from Figures 5 and 6 that the irradiated width of the sample (represented in our case by the slit width) plays an important role. However, for the slit widths of 5 and 10 mm, the *FWHM* is also slightly influenced by the diameter of the sample. The inspection of Figures 4–6 reveals that with the exception of the slit width of 20 mm, in all other cases, the angular resolution was sufficiently high for observation of not only elastic strains, but also changes in the diffraction profiles for microstructural (microstrains, mean grain size) studies of plastically deformed samples on the basis of diffraction profile analysis [11,12]. Furthermore, the comparison of Figure 5c with Figure 6c shows that in the case when the slit width is much larger than the diameter of the sample, the diameter itself has a negligible influence on the *FWHM* of the rocking curve. Finally, a slightly deformed α-Fe(110) low-carbon steel rod of $\varphi = 4.9$ mm put in the horizontal position for two slit widths was measured. The corresponding experimental rocking curves are shown in Figure 7. It should be pointed out that for measurements related to Figures 4b and 5, the same sample was used, however, after annealing. If we compare Figure 7a,b with Figure 5a,b, the effect of plastic deformation on the *FWHM* of the rocking curve, as well as the diffraction profile itself, is evident and could be sufficient for a possible evaluation of the microstructure parameters on the basis of the diffraction profile analysis. Of course, if the samples are situated at the diffractometer with a high accuracy, the relative change in the lattice constant $\varepsilon = \Delta d_S/d_0$ (strain) can also be derived from the peak position shift $\Delta\theta_A$. The peak shift $\Delta\theta_A$ corresponds to the change in the scattering angle $\Delta(2\theta_S)$ as $\Delta\theta_A = -\Delta(2\theta_S)$. Then, with the help of the differentiated form of the Bragg condition $\Delta d_S/d_0 = -\Delta\theta_S \cot \theta_S$ (d_0 is the lattice spacing of the virgin sample), the average elastic strain values ε_R (radial component) and ε_L (longitudinal component) within the irradiated gauge volume could be evaluated.

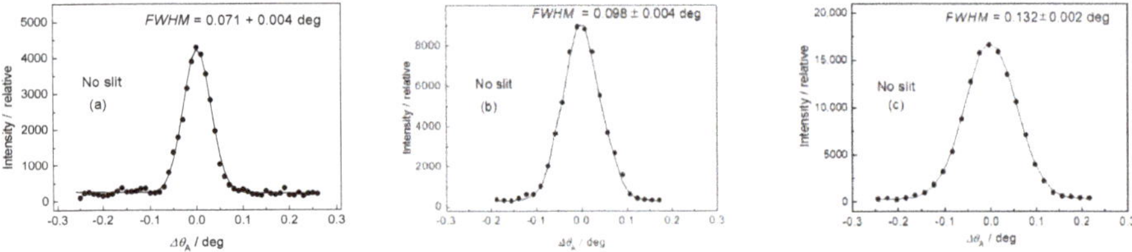

Figure 4. Rocking curves for the annealed α-Fe(110) low-carbon steel rods in vertical position: (**a**) $\varphi = 2$, (**b**) 4.9, and (**c**) 8 mm.

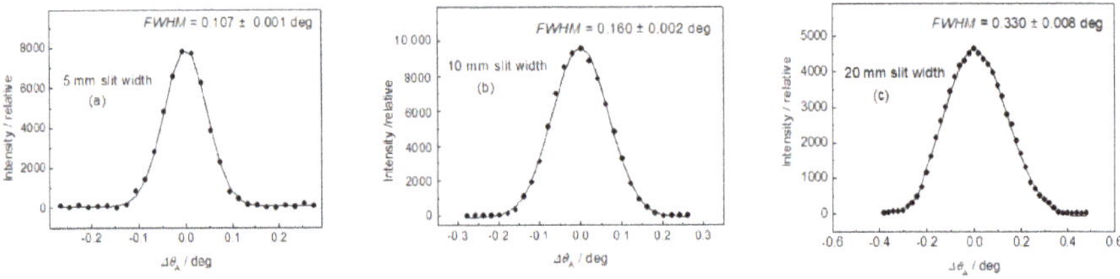

Figure 5. Rocking curves for the annealed α-Fe(110) low-carbon steel rods of $\varphi = 4.9$ mm in horizontal position: (**a**) 5, (**b**) 10, and (**c**) 20 mm Cd slits.

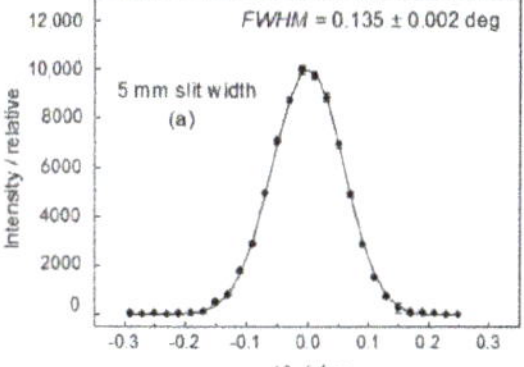

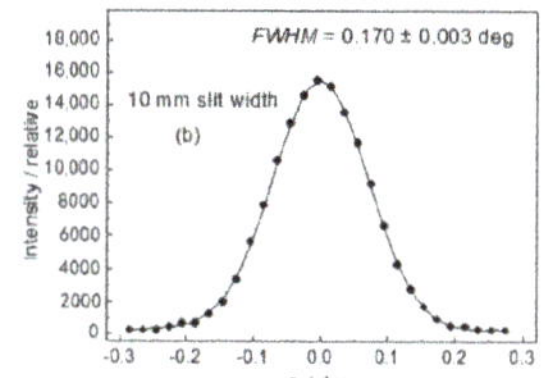

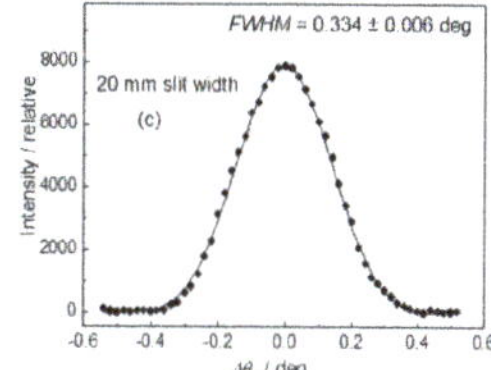

Figure 6. Rocking curves for the annealed α-Fe(110) low-carbon steel rod of φ = 8 mm in horizontal position: (**a**) 5, (**b**) 10, and (**c**) 20 mm Cd slits.

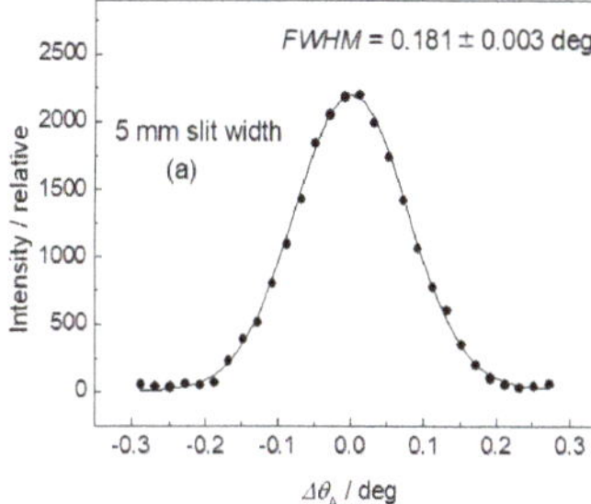

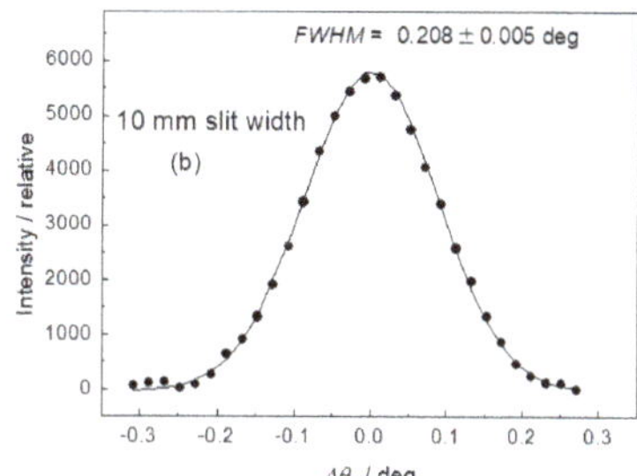

Figure 7. Rocking curves for the slightly deformed α-Fe(110) low-carbon steel rod of φ = 4.9 mm installed in horizontal position for (**a**) 5 mm slit width and (**b**) 10 mm slit width.

4. Discussion

The obtained results shown in Figures 5–7 prove that the presented diffraction method provides a sufficiently high angular resolution, represented by the *FWHM* of the diffraction profile, which allows for the possibility of studying some plastic deformation characteristics (root-mean-square microstrains, as well as the effective grain size), by applying shape analysis on the neutron diffraction peak profiles [11,12]. The advantage of this method permits the investigation of large volumes of polycrystalline samples (several cubic centimeters), thus providing average values of microstructure characteristics over the irradiated gauge volume, e.g., as a function of macroscopic strain loaded on the sample by an auxiliary instrument. Further application of this method can be found in strain/stress studies of textured samples with a large size of grains, as the conventional diffraction method requiring small irradiated gauge volume would be problematic. Naturally, it could also be used for strain scanning in the polycrystalline materials similarly to the conventional two-axis strain/stress scanners when evaluating strains in small gauge volumes (few cubic millimeters) from peak position shifts $\Delta\theta_S$. However, for such measurements, this method would not be practical, because the step-by-step rocking curve analysis would make it much more time-consuming. On the other hand, when using the sample with a width of about 10 mm (or more), the resolution of the conventional two-axis scanner would be so relaxed that it would not allow us to investigate the effect of elastic strains from the shifts in the diffraction profiles nor the effects of plastic deformation on their *FWHM* and the form. As can be seen in Figure 1, the width of the sample gauge volume introduces a decisive uncertainty to the resolution, i.e., to the *FWHM* of the diffraction profile imaged by the position-sensitive detector.

5. Conclusions

The presented high-resolution neutron diffraction method can be successfully used for the strain/stress measurements, namely, on bulk samples exposed to an external thermo-mechanical load, e.g., in a tension/compression rig and in an aging machine. The bulk samples in the form of a rod with a diameter of several millimeters can be investigated in the vertical, as well as horizontal, position. Similarly, the bars of a rectangular form could also be tested. In comparison to the conventional two-axis neutron diffraction strain/stress scanners, the three-axis alternative offers a considerably higher $\Delta d/d$ resolution. However, the presented method, if used for conventional elastic strain/stress scanning, would be much more time-consuming. On the other hand, when using a large-volume sample accompanied with a considerably higher detector signal, this drawback can be partly eliminated. Nevertheless, the presented method could be useful, namely, at high-flux neutron sources where the experimental results can be obtained within a reasonable measurement time. It can be stated that the presented method using the three-axis neutron diffraction setting can offer further complementary information to that achieved by the other methods commonly used.

Author Contributions: Conceptualization and project administration, P.S. and J.Š.; writing—original draft preparation, P.M.; neutron diffraction measurements, P.M. and V.R. All authors have read and agreed to the published version of the manuscript.

Funding: This research received no external funding.

Acknowledgments: Measurements were carried out in the CANAM NPI CAS Řež infrastructure (MŠMT project no. LM2015056). The presented results were also supported in the frame of LM2015074 infrastructure MŠMT project "Experimental nuclear reactors LVR-15 and LR-0". Furthermore, J. Šaroun and V. Ryukhtin acknowledge support from the Czech Academy of Sciences in the frame of the program "Strategie AV21, No.23" and P. Mikula acknowledges support from ESS participation of the Czech Republic—OP (CZ.02.1.01/0.0/0.0/16_013/0001794). The authors also thank B. Michalcová for significant assistance in measurements and basic data processing.

Conflicts of Interest: The authors declare no conflict of interest.

References

1. Noyan, I.C.; Cohen, J.B. *Residual Stress: Measurement by Diffraction and Interpretation*, 1st ed.; Springer: New York, NY, USA, 1987.
2. Hutchings, M.T.; Krawitz, A.D. Measurement of Residual and Applied Stress Using Neutron Diffraction. In *NATO ASI Series*; Hutchings, M.T., Krawitz, A.D., Eds.; Kluwer Academic Publisher: Amsterdam, The Netherlands, 1992; Volume 26.
3. Stelmukh, V.; Edwards, L.; Santisteban, J.R.; Ganguly, S.; Fitzpatrick, M.E. Weld stress mapping using neutron and synchrotron X-ray diffraction. *Mater. Sci. Forum* **2002**, *404–407*, 599–604. [CrossRef]
4. Mikula, P.; Vrána, M.; Lukáš, P.; Šaroun, J.; Wagner, V. High-resolution neutron powder diffractometry on samples of small dimensions. *Mater. Sci. Forum* **1996**, *228–231*, 269–274. [CrossRef]
5. Mikula, P.; Vrána, M.; Lukáš, P.; Šaroun, J.; Strunz, P.; Ullrich, H.J.; Wagner, V. Neutron diffractometer exploiting Bragg diffraction optics—A high resolution strain scanner. In Proceedings of the ICRS-5 Conference, Linköping, Sweden, 16–18 June 1997; Volume 2, pp. 721–725.
6. Seong, B.S.; Em, V.; Mikula, P.; Šaroun, J.; Kang, M.H. Unconventional performance of a highly luminous strain/stress scanner for high resolution studies. *Mater. Sci. Forum* **2011**, *681*, 426–430. [CrossRef]
7. Mikula, P.; Šaroun, J.; Ryukhtin, V.; Stammers, J. An alternative neutron diffractometer performance for strain/stress measurements. *Powder Diffr.* **2020**, *35*, 185–189. [CrossRef]
8. Vrána, M.; Lukáš, P.; Mikula, P.; Kulda, J. Bragg diffraction optics in high-resolution strain measurements. *Nucl. Instrum. Methods Phys. Res. A* **1994**, *338*, 125–131. [CrossRef]
9. Macek, K.; Lukáš, P.; Janovec, J.; Mikula, P.; Strunz, P.; Vrána, M.; Zaffagnini, M. Austenite content and dislocation density in electron beam welds of a stainless maraging steel. *Mater. Sci. Eng. A* **1996**, *208*, 131–138. [CrossRef]
10. Hirschi, K.; Ceretti, M.; Lukáš, P.; Ji, N.; Braham, C.; Lodini, A. Microstrain measurement in plastically deformed austenitic steel. *Textures Microstruct.* **1999**, *33*, 219–230. [CrossRef]
11. Delhez, R.; de Keijser, T.H.; Mittemeijer, E.J. Determination of crystallite size and lattice distortions through X-ray diffraction line profile analysis. *Fresenius' Z. Anal. Chem.* **1982**, *312*, 1–16. [CrossRef]

12. Davydov, V.; Lukáš, P.; Strunz, P.; Kužel, R. Single-line diffraction profile analysis method used for evaluation of microstructural parameters in the plain ferritic steel upon tensile straining. *Mater. Sci. Forum* **2008**, *571–572*, 181–188. [CrossRef]

Publisher's Note: MDPI stays neutral with regard to jurisdictional claims in published maps and institutional affiliations.

© 2020 by the authors. Licensee MDPI, Basel, Switzerland. This article is an open access article distributed under the terms and conditions of the Creative Commons Attribution (CC BY) license (http://creativecommons.org/licenses/by/4.0/).

Article

Parameters Identification of the Anand Material Model for 3D Printed Structures

Martin Fusek, Zbyněk Paška *, Jaroslav Rojíček and František Fojtík

Department of Applied Mechanics, Faculty of Mechanical Engineering, VŠB—Technical University of Ostrava, 708 00 Ostrava, Czech Republic; martin.fusek@vsb.cz (M.F.); jaroslav.rojicek@vsb.cz (J.R.); frantisek.fojtik@vsb.cz (F.F.)
* Correspondence: zbynek.paska@vsb.cz; Tel.: +420-597-325273

Abstract: Currently, there is an increasing use of machine parts manufactured using 3D printing technology. For the numerical prediction of the behavior of such printed parts, it is necessary to choose a suitable material model and the corresponding material parameters. This paper focuses on the determination of material parameters of the Anand material model for acrylonitrile butadiene styrene (ABS-M30) material. Material parameters were determined using the genetic algorithm (GA) method using finite element method (FEM) calculations. The FEM simulations were subsequently adjusted to experimental tests carried out to achieve the possible best agreement. Several experimental tensile and indentation tests were performed. The tests were set up in such a way that the relaxation and creep behaviors were at least partially captured. Experimental tests were performed at temperatures of 23 °C, 44 °C, 60 °C, and 80 °C. The results obtained suggest that the Anand material model can also be used for ABS-M30 plastic material, but only if the goal is not to detect anisotropic behavior. Future work will focus on the search for a suitable material model that would be able to capture the anisotropic behavior of printed plastic materials.

Keywords: Anand material model; material parameters; ABS-M30; indentation test; genetic algorithm

Citation: Fusek, M.; Paška, Z.; Rojíček, J.; Fojtík, F. Parameters Identification of the Anand Material Model for 3D Printed Structures. *Materials* **2021**, *14*, 587. https://doi.org/10.3390/ma14030587

Academic Editor: Juergen Stampfl
Received: 30 November 2020
Accepted: 22 January 2021
Published: 27 January 2021

Publisher's Note: MDPI stays neutral with regard to jurisdictional claims in published maps and institutional affiliations.

Copyright: © 2021 by the authors. Licensee MDPI, Basel, Switzerland. This article is an open access article distributed under the terms and conditions of the Creative Commons Attribution (CC BY) license (https://creativecommons.org/licenses/by/4.0/).

1. Introduction

The presented article is based on a conference paper [1]. The previous paper is extended with new experimental data and a new interpretation of the procedure used for the identification of material parameters.

The research project at VSB—Technical University of Ostrava is concerned, among other things, with the design of suitable robot arms and manipulators manufactured from plastic materials using 3D printing, one of the most advanced methods of component manufacture [2]. The individual arms (in general components made of printed plastic) are exposed to different types of mechanical loading.

The design of a structure is often determined by results of topological optimization analysis [3]. The structure obtained from the topological optimization process can also be produced using 3D printing [2]. A topological optimization methodology called the solid isotropic material with penalization (SIMP) method and the level set method [4] are often available in commercial software. The evolutionary structural optimization (ESO) method can be counted among the methods that are independent of the material model [5], which iteratively removes or adds a finite amount of material.

In order to predict the behavior of these printed components, knowledge of the appropriate material model and its parameters is necessary (in addition to the load) for FEM analysis. The parameters of printed 3D structures are highly dependent on the 3D printing technology, the laying of the filament, and the setting of the printing process (e.g., temperature). This article describes a procedure for the determination of the material parameters of a printed structure manufactured using 3D printing with a Fortus 450mc

(Stratasys Ltd., Eden Prairie, MN, USA) 3D printer [6]. The material used for printing is ABS-M30 [7].

Several material models for acrylonitrile butadiene styrene (ABS) are described in the literature. For the large strain deformation and fracture behavior, there are three material models with different levels of complexity: (1) the Drucker–Prager yield function; (2) the Raghava yield function; (3) the Gurson yield function. These three models are compared in [8]. In [9], the isotropic Drucker–Prager yield criterion was used for the behavioral simulation of three semicrystalline polymers: high-density polyethylene (HDPE), polypropylene (PP), and polyamide 6 (PA 6). The Johnson–Cook model was used in [10] for the impact test. A constitutive model for polymers can also be found in [11]. The Anand model [12] was developed for aluminium alloys with viscoplastic behavior and is available in most FEM programs, which makes it advantageous for practical engineering uses. Regarding the thermoplastics used in 3D printing technologies, several material models were tested, and the Anand model gave surprisingly good results. Although the Anand model was developed to describe the viscoplastic behavior of metallic materials, its formulation is not limited to one material group only. It does not use the classical description used in other models, but rather it works with deformation resistance as an internal variable.

Based on the previous study [1], the Anand material model [12,13] was selected for further use. The FEM solution was implemented in ANSYS© software (Ansys, Canonsburg, PA, USA) [14], which has an extensive library of material models. ANSYS© was used as the FEM solver and was executed via a module written in Python programming language [15] by the authors. This module is independent of the FEM solver and can also be used in other commercial software (e.g., MSC.Marc). MATLAB (The MathWorks, Inc., Natick, MA, USA) can also be used for the same purpose (see [16]). The so-called finite element model updating (FEMU) [17] method was used to determine the material parameters. Another possibility in material analysis is the use of a neural network (e.g., [18]). FEM results can then be used to train the neural network (see [16]).

In [19], results obtained using the inverse approach were compared for two methods: using the gradient-based method and the evolutionary algorithm for a thermoelastic–viscoplastic material model. The article used the gradient method, which combines the steepest descent gradient and the Levenberg–Marquardt algorithm. The evolutionary algorithm for the real search space was described in the article mentioned. Both methods were able to determine the material parameters. The evolutionary method involves smaller relative error levels compared to the gradient-based method. On the contrary, the gradient method involves lower calculation complexity. The group of evolutionary algorithms also includes the genetic algorithm (GA), which is used in modified form in this work.

The influence of a certain parameter on a result is measured by the sensitivity. The sensitivities are described in [20] as "the partial derivatives of the output functions with respect to the parameters". The modification of the sensitivity calculation was tested in the article mentioned and the resulting values were used to select parameters suitable for the identification. The time required for the process used for identification of the material parameters depends on the number of parameters identified and the number of experiments simulated. A low value for the sensitivity of a parameter may also indicate inappropriate or insufficient selection of experiments for the particular material model.

The proposed procedure is aimed at the identification of parameters in material models with larger numbers of material parameters and experiments. A procedure based on the selection of appropriate parameters for the identification process has not been published in the literature yet.

The main idea of the article is to demonstrate a modern approach for the determination of material parameters (not only 3D-printed materials) based on the use of a proper combination of standard laboratory tests at elevated temperatures and modern numerical optimization methods using custom and commercial software.

2. Experiments

Tensile tests, graded tensile tests, and indentation tests were performed. The experiments were performed on a Testometric M500-50CT (LABOR machine, s.r.o., Otice, Czech Republic) testing machine at four different temperatures (23 °C, 44 °C, 60 °C, and 80 °C). The testing machine used was equipped with a furnace (see Figure 1a). The machine was equipped with a strain gauge force sensor with a measuring range of 50 kN and a measuring accuracy of ±10 N. The maximum tensile force of the machine was 50 kN. The samples were clamped with pneumatic clamping jaws. The magnitude of the clamping force was set by the pressure in the system.

(a)

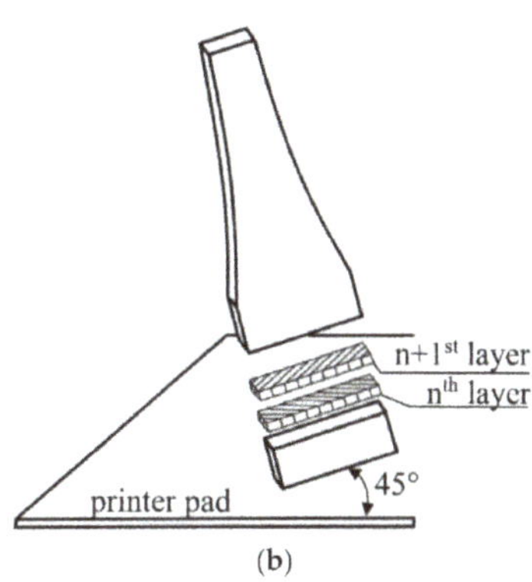

(b)

Figure 1. (**a**) The Testometric M500-50CT tensile testing machine equipped with a temperature chamber. (**b**) The position of specimens during 3D printing and the method used to layer the material.

The specimen shown in Figure 2 was used for the tensile tests, graded tensile tests, and indentation tests. Figure 1b schematically shows the position of all specimens in the printing chamber during 3D printing on a Fortus 450mc printer and the method used to lay individual layers of ABS-M30 material.

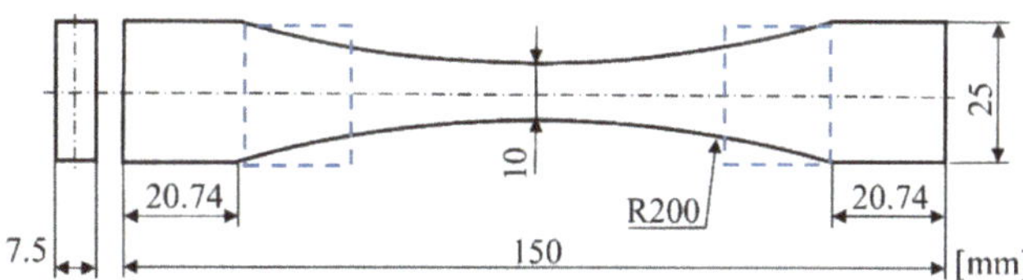

Figure 2. Specimen shape for simple and graded tensile tests.

After printing, the samples were stored in a dry and dark place for one month and then the necessary experiments were performed.

2.1. Tensile Tests

All tensile tests were deformation-controlled. Simple tensile tests were performed at four different temperatures (23 °C, 44 °C, 60 °C, and 80 °C) and at three different rates of deformation (0.017 mm s^{-1}, 0.167 mm s^{-1}, and 1.667 mm s^{-1}) until specimen failure. Figure 3a shows the force against the displacement at 44 °C, 60 °C, and 80 °C, at a constant deformation rate of 0.017 mm s^{-1} (the constant test parameters are shown in the upper right corner of the diagram). It can be seen from the figure that the maximum force value decreased with increasing temperature, however at the beginning of the test the courses of the force were quite similar. One experiment was performed at room temperature, but with the highest deformation rate of 1.667 mm s^{-1} (see Figure 3b). Figure 3c,d shows the force dependence at 44 °C and 60 °C and at the specified deformation rates. The magnitude of the rate of the deformation had no significant influence on the force course at the beginning of the tests, but led to higher maximum force values at the end of the tensile tests.

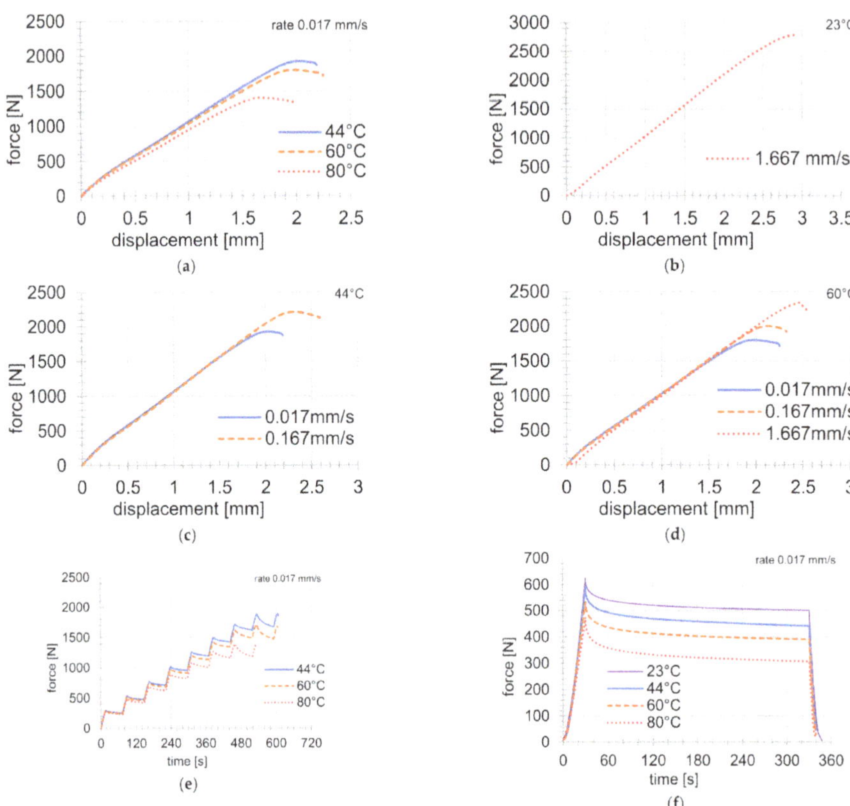

Figure 3. (**a**) Tensile tests at temperatures of 44 °C, 60 °C, and 80 °C, with constant rates of deformation at 0.017 mm s^{-1}. (**b**) Tensile tests at 23 °C and at a constant rate of deformation of 1.667 mm s^{-1}. (**c**) Tensile tests at a temperature of 44 °C, with two different rates of deformation. (**d**) Tensile tests at 60 °C, with three different rates of deformation. (**e**) Graduated tensile tests under three different temperatures of 44 °C, 60 °C, and 80 °C. (**f**) Indentation tests with time delay at temperatures of 23 °C, 44 °C, 60 °C, and 80 °C, with a constant indenter rate of 0.017 mm s^{-1}.

2.2. Graded Tensile Tests

Graded tensile tests were carried out at three different temperatures (44 °C, 60 °C, and 80 °C), with the same deformation step size of 0.25 mm (see Figure 3e). The specimen was elongated by 0.25 mm at each step, at a deformation rate of 0.017 mm s^{-1}. The time delay at the given strain value was always 60 s. This was done until the specimen failed. During the graded tensile test, the creep phenomenon was partially detected. Up to the value of the tensile force of approximately 800 N, there was no significant relaxation of the tension.

2.3. Indentation Tests

Indentation tests were carried out at 23 °C, 44 °C, 60 °C, and 80 °C (see Figure 3f) on the samples depicted in Figure 2. The area where the indentation test was carried out is shown by a dashed rectangle. The indenter was a steel sphere with a diameter of 5 mm. The indentation test consisted of three phases. First, the indenter was pressed into the material at a speed of 0.017 mm s^{-1} to a depth of 0.5 mm, then the time delay of 300 s followed, and finally the indenter returned to the starting position (relief) at the same speed.

3. Material Model

One of the material models used for viscoplastic materials is the Anand material model [12]. The Anand model was proposed for use in the analysis of the rate-dependent deformation of metals at high temperatures. The Anand viscoplastic model is pre-built in the commercial finite element software ANSYS©, and therefore it is much easier to use. The Anand model is typically used for solder alloys [21,22], however the previously mentioned benefits led the authors to test it for materials used for 3D printing (ABS-M30). The material model involves 11 material parameters: the Poisson ratio (u), Young's modulus (E), initial value of deformation resistance (s_0), activation energy/universal gas constant (Q/R), pre-exponential factor (A), stress multiplier (x_i), strain rate sensitivity of stress (m), hardening/softening constant (h_0), coefficient for deformation resistance saturation value (S), strain rate sensitivity of saturation (deformation resistance) value (n), and strain rate sensitivity of hardening or softening (a).

The Anand viscoplastic model was originally developed for material forming applications [12,13]. It is also applicable to general viscosity problems, which include the influence of the strain rate and temperature. Materials at elevated temperatures are highly dependent on the influence of the temperature magnitude and history, strain rate, and strain hardening. The Anand model is a complex material model that introduced an internal variable S (deformation resistance), a variable that represents the resistance against the plastic behavior of the material. This is different from other material models, and the decomposition of the individual components of the deformation is not straightforward.

The rate of plastic deformation is described using the following relationship:

$$\dot{\varepsilon}_{pl} = \dot{\varepsilon}^{a}{}_{pl}\left(\frac{3}{2}\frac{S}{q}\right), \tag{1}$$

where $\dot{\varepsilon}_{pl}$ is the tensor of the inelastic strain rate and $\dot{\varepsilon}^{a}{}_{pl}$ is the rate of accumulated equivalent plastic strain. Here, $\dot{\varepsilon}^{a}{}_{pl}$ is given by the equation:

$$\dot{\varepsilon}^{a}{}_{pl} = \left(\frac{2}{3}\dot{\varepsilon}_{pl} : \dot{\varepsilon}_{pl}\right)^{\frac{1}{2}}, \tag{2}$$

where the operator ":" stands for inner product of the tensors. S is the deviator of the Cauchy stress tensor, which can be expressed using the following relation:

$$S = \sigma - pI, \tag{3}$$

where σ is the Cauchy stress tensor and p is defined as one-third of the trace of the tensor matrix σ, as in the following relation:

$$p = \frac{1}{3}tr(\sigma). \tag{4}$$

Here, I represents a second-order unit tensor. The quantity q is the equivalent stress according to the following relation:

$$q = \left(\frac{3}{2}S : S\right)^{\frac{1}{2}}. \tag{5}$$

The rate of accumulated plastic deformation depends on q and on the internal state variable s. This dependence can be expressed by the following equation:

$$\dot{\varepsilon}^{a}{}_{pl} = Ae^{\left(-\frac{Q}{R\theta}\right)}\left\{\sinh \xi \frac{q}{s}\right\}^{\frac{1}{m}}, \tag{6}$$

where A, ξ, and m are the model constants; Q is the activation energy; R is the universal gas constant; θ is the absolute temperature; and s is the internal state variable.

Equation (6) indicates that plastic deformation occurs at any stress level. This contrasts with other theories of plasticity that use areas of plasticity (yield function). Classical theories assume that plasticity occurs above a certain stress value, otherwise the deformations are elastic.

The development of the internal state parameter s is described as follows:

$$\dot{s} = \oplus h_o \left|1 - \frac{s}{s^*}\right|^{a} \dot{\varepsilon}^{a}{}_{pl}, \tag{7}$$

where a and h_o are constants, while s^* represents the saturated value of the internal parameter. The $\oplus$ operator is defined to return $+1$ if $s \leq s^*$, otherwise it will return -1. The effect of softening or hardening is included in the model by this operator. The saturation values of s^* depend on the rate of equivalent plastic deformation $\dot{\varepsilon}^{a}{}_{pl}$ and can be expressed as follows:

$$s^* = \hat{s}\left\{\frac{\dot{\varepsilon}^{a}{}_{pl}}{A} e^{\left(\frac{Q}{R\theta}\right)}\right\}^{n}, \tag{8}$$

where $\hat{s}$ and n represent constants.

Expression (7) shows that the development of the parameter s depends on the rate of the equivalent plastic deformation, and at the same time on the current state of the internal state parameter s.

4. Theoretical Description of the Identification Procedure

The material parameter identification can be described mathematically as:

$$\begin{gathered} f(X) = \text{minimum}, \\ \text{subject to } g_j(X) > 0, \;\; j = \{1, 2, 3, \ldots, N_C\}, \end{gathered} \tag{9}$$

where f(X) is an objective function, X represents a vector of material parameters, $g_j(X)$ is j-th constrain function, and N_C is the number of constraint functions. For minimalization, the finite element model updating (FEMU) approach was used, which is described in [23]. Schematically, the parameters of the material model were determined by repeated optimization calculations with custom software written in Python programming language. The solution procedure was as follows: FE models of the experiments were created at first. The models were created in commercial software (ANSYS©) as boxes, where the inputs were values of the material parameters and the outputs were values corresponding to loaded or measured data (in this paper these were a force, a displacement, and a current time). The difference between the simulation outputs and the measured data defines the value of an objective function. The simulations were solved in a cycle, whereby the inputs were changed with respect to the minimized value of the objective function. All experiments were deformation-controlled, but forces were used for formulation of the objective function. The difference between experimental data and data from the simulation model for one experiment was solved as an individual objective function $f_i(X)$ for the i-th experiment:

$$f_i(X) = \frac{\sum_{j=1}^{N_i}\left|F_j^{EXP} - F_j^{FEM}(X)\right|}{\sum_{j=1}^{N_i}\left|F_j^{EXP}\right|}, \; i = \{1, 2, 3, \ldots, N\}, \tag{10}$$

where N_i is the number of measurement points for the i-th experiment, F_j^{EXP} is an experimental force, $F_j^{FEM}(X)$ is the force obtained from the simulation, and N is the number of experiments.

The value of objective function for all experiments (f(X)) was solved as:

$$f(X) = \sqrt{\frac{\sum_{i=1}^{N} f_i(X)^2}{N}}, \tag{11}$$

A genetic algorithm (GA) was used to identify the material parameters, as in [1]. A GA for the identification of material parameters was used in [24]. The GA was used because it is not dependent on the material model. A detailed description of the GA can be found in [25]. A chromosome is defined by a vector of genes:

$$X = \{p_i,\ i = \{1, 2, 3, \ldots, N_p\}\}, \tag{12}$$

where p_i is the i-th gene, which corresponds to the i-th material parameter, and N_p is the number of parameters.

An initial population was given by N_G solutions, which was generated randomly by using the hill-climbing algorithm [26] from an initial chromosome. For the i-th gene we used:

$$p_i^{New} = p_i^{Old} \cdot (1 + \mathrm{Rand}(-k_{HC}, k_{HC})),\ i = \{1, 2, 3, \ldots, N\}, \tag{13}$$

where p_i^{Old} is the original value of the i-th gene, p_i^{New} is the proposed value of the i-th gene, and the Rand function generates random values with uniform distribution from the interval given by $\pm k_{HC}$.

The constraint functions are defined as:

$$g_j(X) = p_j,\ j = \{1, 2, 3, \ldots, N_p\}, \tag{14}$$

where the number of constraint conditions N_c is the same as the number of parameters N_p.

In each cycle a new individual is created, its objective function value is calculated, and then the individual is added to the population. Regarding the quality of an individual, the solution is determined by the value of the objective function (Equation (11)). The new individual is created in two ways:

1. With a 40% probability using the hill-climbing algorithm for the best individual in the population (13);
2. With a 60% probability using crossover.

The child chromosome was created from three chromosomes of parents, whereby the parents were selected randomly (uniform distribution) from the population. The parents were sorted by the value of the objective function and are indicated by a superscript of 1, 2, or 3; parent 1 had the best (the lowest) objective function and parent 3 had the worst (the greatest) objective function. The child value of the gen (p_i^{Child}) is calculated as:

$$p_i^{Child} = p_i^{Best} + \Delta p_i.\ i = \{1, 2, 3, \ldots, N\}, \tag{15}$$

where Δp_i is the gene modification value. The modification value is calculated from parent genes as follow:

$$\begin{gathered} \Delta p_i = k\left(p_i^{\,1} - p_i^{\,2}\right), \text{ or} \\ \Delta p_i = k\left(p_i^{\,1} - p_i^{\,3}\right),\ i = \{1, 2, 3, \ldots, N\}, \end{gathered} \tag{16}$$

where equations are selected randomly (with 70% and 30% probability); $p_i^{\,1}, p_i^{\,2}, p_i^{\,3}$ are the gen values of the first parent, second parent, and third parent, respectively; k is a growth coefficient $(1-2)$.

The size of the resulting gene values is controlled. If the gene values change too little $\Delta p_i < p_i^{Best}(1 \pm 0.0001)$ or too much $\Delta p_i > p_i^{Best}(1 \pm 0.05)$, then the new gene values are generated randomly using the hill-climbing algorithm with $k_{HC} = 0.001$.

Parameters suitable for identification are selected using sensitivity analysis. This sensitivity analysis shows an effect of the parameter on the value of the objective function with respect to the other parameters. The sensitivity analysis is performed from $Np + 1$ calculations as follows:

$$f = f(X), \quad f_{+\Delta i} = f\left(\left\{p_1, p_2, \ldots, p_i(1 + k_{Sen}), \ldots, p_{Np}\right\}\right), \tag{17}$$

where k_{Sen} is a sensitivity coefficient $(0.05 - 0.001)$. The sensitivity value S_i is calculated according to the equation:

$$S_i = \frac{100}{k_{Sen}}\left|1 - \frac{f_{+\Delta i}}{f}\right| [\%],\ i = \{1, 2, 3, \ldots, Np\}. \tag{18}$$

Parameters whose sensitivity values are greater than the critical value S_{Krit} are selected for identification:

$$S_i > S_{Krit} \rightarrow \text{identify } p_i,\ i = \{1, 2, 3, \ldots, Np\}. \tag{19}$$

In later phases of identification, it is advisable to check if the given parameter p_i is not in the local minimum of the objective function. Here, the number of required calculations is $2Np + 1$:

$$f_{+\Delta i} > f < f_{-\Delta i},\ i = \{1, 2, 3, \ldots, Np\}. \tag{20}$$

5. Description of the Identification Process

Ten experiments were used for identification (see Figure 3a–f), however the experiments at 44 °C were excluded from the identification process and were used only for validation. The assignment of the experiments to the values of the individual objective functions is shown in Table 1.

Table 1. Assignment of the individual objective functions to individual experiments.

$f_1(X)$	$f_2(X)$	$f_3(X)$	$f_4(X)$	$f_5(X)$	$f_6(X)$	$f_7(X)$	$f_8(X)$	$f_9(X)$	$f_{10}(X)$
T_1, ID	$T_1, \dot{x}_3$	$T_3, \dot{x}_1$	$T_3, \dot{x}_2$	T_3, ID	$T_3, \dot{x}_3$	T_3, GT	$T_4, \dot{x}_1$	T_4, ID	T_4, GT

In Table 1, the temperatures are denoted as T_1 = 23 °C, T_2 = 44 °C, T_3 = 60 °C, and T_4 = 80 °C; the deformation rates are $\dot{x}_1 = 0.017$ mm s^{-1}, $\dot{x}_2 = 0.167$ mm s^{-1}, and $\dot{x}_3 = 1.667$ mm s^{-1}; ID denotes the indentation tests and GT denotes the graded tensile tests (see Chapter 2). Data from the tests at temperature T_2 were used to validate the parameters.

The initial values of parameters are shown in Table 2.

Table 2. Initial parameters of the Anand material model.

E [MPa]	μ [−]	s_0 [MPa]	Q/R [K]	A [1/s]	x_i	m [−]	h_0 [MPa]	$\hat{S}$ [MPa]	n [−]	a [−]
1780	0.33	19.54	8350	3134	5.18	0.2466	183,245	31.0	0.0098	1.524

The value of the modulus of elasticity E is temperature-dependent, which is shown in Figure 3a. For this reason, three values of the modulus of elasticity E_{20}, E_{50}, and E_{80} (for 20, 50, and 80 °C) were added to the material model. The temperature chamber made it impossible to use optical methods (e.g., the digital image correlation method (DIC)) [27] to measure the Poisson number, so this parameter was excluded from identification.

Chromosome (12) consists of these material parameters, whereby the individual genes correspond to the individual material parameters:

$$X = \left\{ E_{20}, E_{50}, E_{80}, s_0, \frac{Q}{R}, A, x_i, m, h_0, \hat{S}, n, a \right\}. \tag{21}$$

The hill-climbing algorithm was used with the parameter $k_{HC} = 0.001$, the genetic algorithm was used with the parameter $N_G = 10$, and the growth coefficient $k = 1.5$ was used in both cases. The sensitivity coefficient was $k_{Sen} = 0.01$. The sensitivity value shown in (18) and the criterion shown in (20) were recalculated. The way in which the model parameters were determined is described by the following three tables. Table 3 contains the parameter values during the solution.

Table 3. Parameter values with three moduli of elasticity during identification.

#	E_{20} [MPa]	E_{50} [MPa]	E_{80} [MPa]	μ [−]	s_0 [MPa]	Q/R [K]	A [1/s]	x_i	m [−]	h_0 [MPa]	$\hat{S}$ [MPa]	n [−]	a [−]
0	1780	1780	1780	0.33	19.5	8350	3134	5.18	0.247	183,245	31.0	0.0098	1.524
1	1175	1161	933	-	19.5	9486	3134	5.75	0.24	183,245	41.4	0.0098	1.897
2	1175	1161	933		19.5	9486	3134	5.82	0.241	169,305	43.6	0.012	2.936
3	1175	1203	960		19.5	9486	3134	5.82	0.224	169,305	43.6	0.0152	2.936
4	1196	1203	997		18.0	9486	3048	5.82	0.224	146,821	43.6	0.0186	3.527
5	1196	1224	997		18.0	9486	3263	8.82	0.213	138,175	43.6	0.0226	3.527

The values for the objective function during the solution are shown in Table 4 and the sensitivity values are shown in Table 5. The initial parameters for identification (values of the objective function) are given in line 0 of Table 3. The sensitivity was calculated from "line 0" values, and the resulting values are shown on line 1 of Table 5. The genetic algorithm was run 200 times in every identification cycle. This process was repeated in lines 2 to 5. Line 5 in Tables 4 and 5 shows the resulting parameter and error values. The fulfilment of criterion (20) is indicated in Table 5 as U. The resulting values (line 5, Table 4) were used to calculate the sensitivity (line 6* in Table 5). Values in line 6* were no longer used for further calculation. This shows that the minimum given by the coefficient k_{Sen} had not been reached yet. Table 5 shows in the last step (line 5) a decrease of the objective function value by 0.1% after 200 calculation cycles; the identification process was, therefore, completed.

Table 4. The objective function values during identification.

#	$f_1(X)$ [%]	$f_2(X)$ [%]	$f_3(X)$ [%]	$f_4(X)$ [%]	$f_5(X)$ [%]	$f_6(X)$ [%]	$f_7(X)$ [%]	$f_8(X)$ [%]	$f_9(X)$ [%]	$f_{10}(X)$ [%]	$f(X)$ [%]
0	27	28	20	23	28	27	23	21	23	25	25
1	7.1	7.9	6.3	7.2	4.9	6.6	8.5	9.9	3.7	7.0	7.1
2	5.5	7.6	5.7	7.3	3.7	6.9	5.6	9.5	4.7	5.4	6.4
3	5.7	7.7	4.6	6.5	3.7	6.2	6.3	8.4	4.5	5.1	6.0
4	5.2	7.9	4.4	6.4	4.0	6.3	4.9	7.0	4.4	4.5	5.6
5	4.2	7.9	4.2	6.3	3.4	6.3	4.9	7.0	4.5	4.0	5.5

Table 5. Sensitivity values during identification. Parameters that fulfil the criterion are shown in bold.

#	$S_{E_{20}}$	$S_{E_{50}}$	$S_{E_{80}}$	S_{s0}	S_{QR}	S_A	S_{xi}	S_m	S_{h0}	S_S	S_n	S_a	S_{Krit}
1	**13.4**	**34.3**	**42.0**	4.1	**115.0**	6.3	**42.1**	**30.8**	3.3	**47.2**	4.2	**12.7**	10
2	U	U	U	1.0	U	0.2	**33.9**	**15.2**	**7.3**	**16.6**	**8.3**	**23.4**	5
3	3.9	**97.2**	**102.4**	4.4	U	U	U	**8.4**	1.4	U	**9.2**	1.8	5
4	**3.8**	U	**5.2**	**8.5**	U	**4.5**	U	U	**7.0**	U	**7.7**	**30.5**	0.1
5	U	**0.3**	U	U	U	**0.9**	U	**8.6**	**1.6**	U	**16.7**	U	0.1
6*	30.0	U	0.1	7.5	U	3.2	U	U	3.0	U	4.2	6.8	-

6. Results

The obtained final parameter values of the Anand material model are summarized in Table 6 and the FEM analyses were performed with these material parameters. The value of the Poisson ratio was $\mu = 0.33$. Figures 4 and 5 show a comparison of experiments with results from the FEM analysis.

Table 6. Final parameter values.

E_{20} [MPa]	E_{50} [MPa]	E_{80} [MPa]	s_0 [MPa]	Q/R [K]	A [1/s]	x_i	m [−]	h_0 [MPa]	$\hat{S}$ [MPa]	n [−]	a [−]
1196	1224	997	18.0	9486	3263	8.82	0.213	138,175	43.6	0.0226	3.527

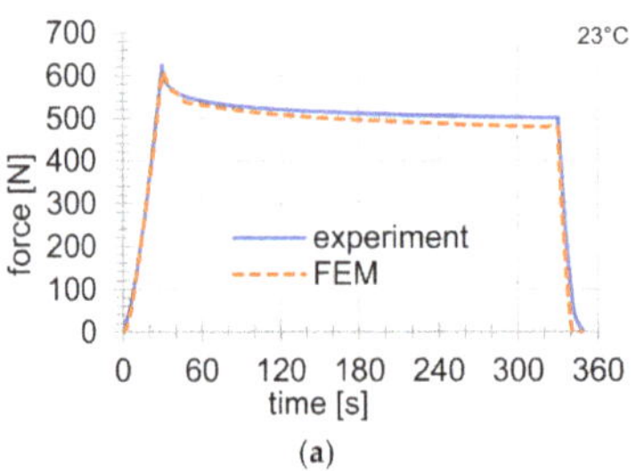

(a)

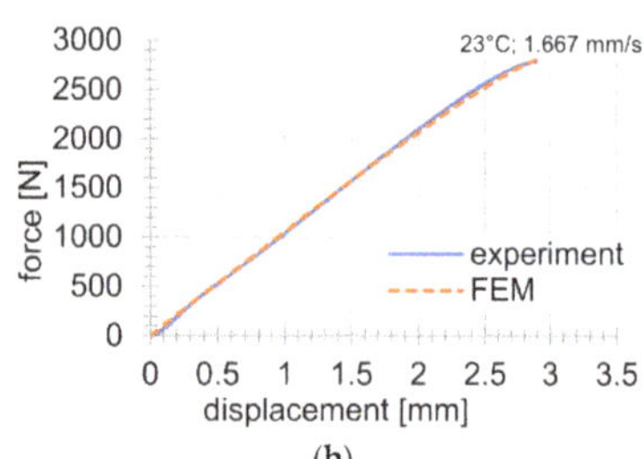

(b)

Figure 4. Comparison of experiments with FEM solutions: (**a**) indentation test at 23 °C; (**b**) tensile test at 23 °C and rate of deformation of 1.667 mm s^{-1}.

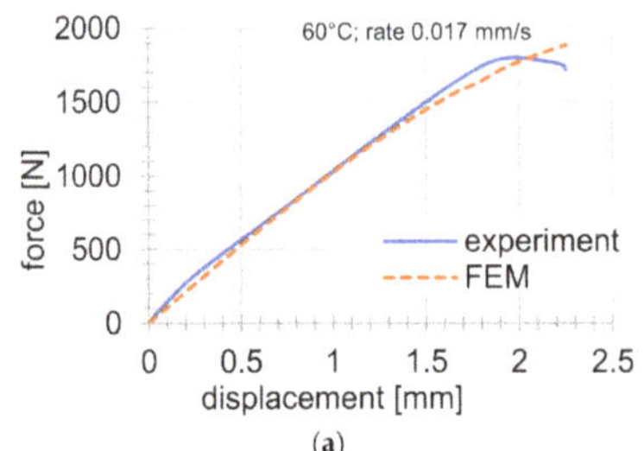

(a)

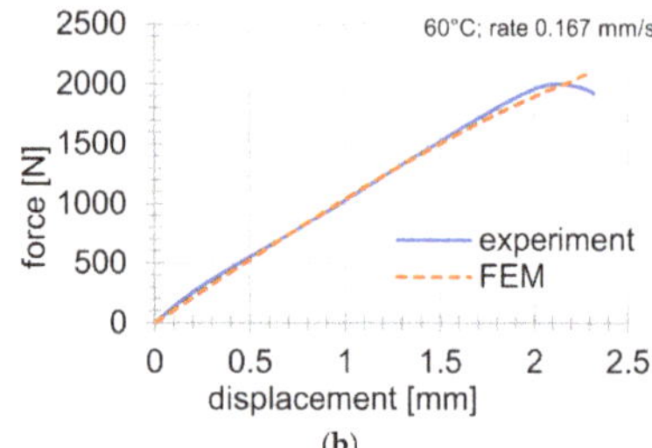

(b)

Figure 5. *Cont.*

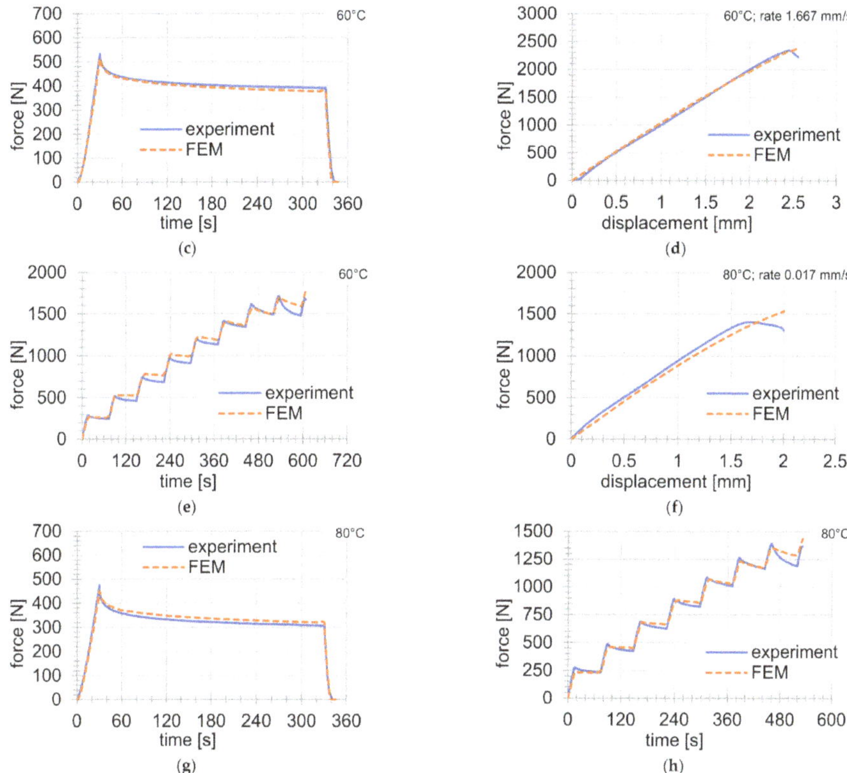

Figure 5. Comparison of experiments with FEM solutions (continuation): (**a**) tensile test at 60 °C and rate of deformation of 0.017 mm s^{-1}; (**b**) tensile test 60 °C and rate of deformation of 0.167 mm s^{-1}; (**c**) indentation test at 60 °C; (**d**) tensile test at 60 °C and rate of deformation of 1.667 mm s^{-1}; (**e**) graded tensile test at 60 °C; (**f**) tensile test at 80 °C and rate of deformation of 0.017 mm s^{-1}; (**g**) indentation test at 80 °C; (**h**) graded tensile test at 80 °C.

Very good agreement can be seen between the experiments and the FEM solution. Different behavior just before the failure of the specimen is shown in Figure 5a,b,d–f,h. This behavior is less significant under lower temperatures and higher deformation rates. The difference can be seen in Figure 5a,f especially (temperatures of 60 and 80 °C and deformation rate of 0.017 m s^{-1}). This effect was not captured in the simulations.

7. Validation

Validation testing was carried out on a series of experiments at 44 °C. Experiments at this temperature were not used for material parameter identification. The results are shown in Figure 6. In Figure 6, a very good agreement can be seen between FEM solutions and experiments. The greater error occurs just before the failure of the specimen, as also mentioned in the previous chapter. The conclusion is that the material models with those material parameters do not accurately describe the mechanical behavior shortly before the sample ruptures. In order to describe this area of difference better, more experiments will be necessary, especially focused on this small area. If no experiments of this kind are carried out, the influence of this small area on the value of the objective function is negligible.

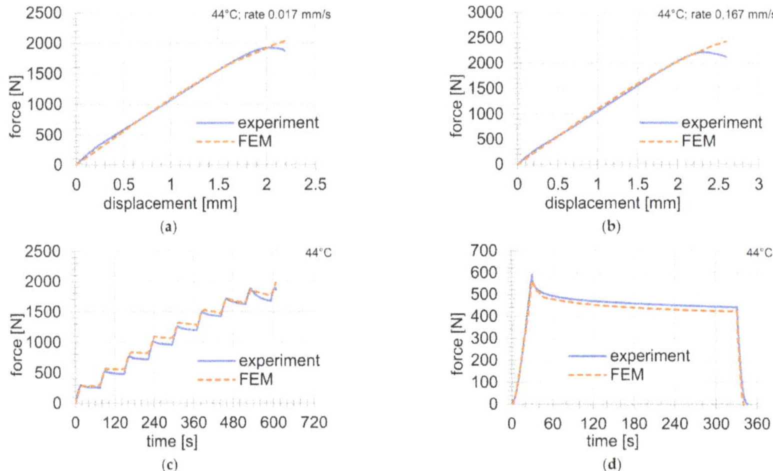

Figure 6. Comparison of FEM solutions with validation experiments at 44 °C: (**a**) simple tensile test at rate of deformation of 0.017 mm s^{-1}; (**b**) simple tensile test at rate of deformation of 0.167 mm s^{-1}; (**c**) graded tensile test; (**d**) indentation experiment.

8. Discussion

The GA modification was used to identify the parameter values. In order to reduce the number of calculation cycles and individuals in the GA population, the number of parameters was reduced by means of sensitivity control. The sensitivity calculation was performed at the beginning of the solution, and then again when the GA convergence rate decreases significantly.

Further sensitivity tests showed that if some parameters are changed there is still a decrease, while others are at their minima in the range tested (see table of resulting sensitivity U). In subsequent cycles, the value of the objective function gradually decreased in the range of 0.01% to 100 GA steps. This value decreased further in each cycle. The resulting effects of these calculations on the value of the objective function and the values of parameters were negligible, and therefore are not described in this contribution.

The Anand constitutive model used in this article contained 9 parameters and 3 moduli of elasticity, which were identified for temperatures of 20, 50, and 80 °C. Figure 3a shows the influence of temperatures on the value of E. This dependence appeared at all speeds tested, so it was included in the identification, although its effect was not significant. Input values of the material parameters were taken from [1] for the ABS-M30 material and determined at a temperature of 23 °C and displacement rate of 1 mm s^{-1} from tensile tests and indentation tests, while the Poisson's ratio was determined from tensile tests with DIC. These experiments were not used for identification in this contribution because they showed large differences (originally E = 1780 MPa, new value E = 1100 MPa) in the behavior compared to the results presented above. The authors believe that this was due to a difference in the supply of the material and different 3D printing settings.

In [1], indentation and DIC measurement data were also used to identify and validate the results. When using a temperature chamber, it was not possible to perform DIC measurements—more extensive modifications of the experimental equipment would be necessary. In the case of DIC, this mainly involves determining the value of the Poisson ratio, which was, therefore, taken from [1]. Indentation tests describe the behavior of multi-axial stresses in the area above the yield point in the compression area. Tensile tests carried out at lower temperatures and higher speeds (Figure 3b,c) show almost linear behavior.

The use of different types of experiments (e.g., tensile tests, indentation tests) for identification or at least for validation is the key for the further use of the identified parameters. Indentation, tensile, and graded tensile tests at 44 °C were used for validation (see [28]). Components produced using 3D printing are not only loaded by tension, so the authors assume that after modifying the experimental equipment, a different set of experiments will be performed. On the other hand, conventional components are loaded in the area corresponding to the first half of the loading curve, where tensile tests in particular show very good agreement with the FEM solution.

9. Conclusions

The obtained results show the possibility of finding the material model parameters using indentation or tensile tests. The results of the FEM calculations are in good agreement with the data obtained experimentally. Attention will be paid to the modification of the test equipment in order to enable DIC measurements. The Anand material model can be used for specimens loaded at different temperatures. It was found that the Anand material model describes the behavior of the ABS-M30 material very well with respect to different temperatures and loading speeds. On the other hand, the Anand model is not able to capture the anisotropic behavior of the investigated material sufficiently. For tests on specimens printed in the most unfavorable position (related to the load capacity of the specimen), the material model parameters identified as described above will give conservative results and the model can be used to design components. Studies on a more advanced material model that includes the effects of material anisotropy and a damage model are currently in progress. The proposed identification methodology is not dependent on the material model.

Author Contributions: Conceptualization, Z.P.; material model description, M.F.; planning and execution of experiments, F.F.; identification procedure and FEM, J.R.; resources, M.F., J.R., and F.F.; writing—original draft preparation, Z.P., M.F., and J.R.; writing—review and editing, Z.P., M.F., J.R., and F.F.; visualization, Z.P. and F.F. All authors have read and agreed to the published version of the manuscript.

Funding: This research received no external funding.

Institutional Review Board Statement: Not applicable.

Informed Consent Statement: Not applicable.

Data Availability Statement: Data can be provided upon request from the correspondent author.

Acknowledgments: This article was elaborated under support of the Research Center of Advanced Mechatronic Systems, reg. no. CZ.02.1.01/0.0/0.0/16_019/0000867, within the framework of the Operational Program for Research, Development, and Education.

Conflicts of Interest: The authors declare no conflict of interest.

References

1. Fusek, M.; Paška, Z.; Rojíček, J. Identification of material parameters and material model for 3D printing structure. In Proceedings of the 58th International Scientific conference on Experimental Stress Analysis 2020, VSB-Technical University of Ostrava, Ostrava, Czech Republic, 19–22 October 2020; Volume 98, p. 106, ISBN 978-80-248-4451-0.2020.
2. Paska, Z.; Rojicek, J. Methodology of arm design for mobile robot manipulator using topological optimization. *MM Sci. J.* **2020**, *2020*, 3918–3925. [CrossRef]
3. Yao, P.; Zhou, K.; Lin, Y.; Tang, Y. Light-Weight Topological Optimization for Upper Arm of an Industrial Welding Robot. *Metals* **2019**, *9*, 1020, ISSN 2075-4701. [CrossRef]
4. Andreasen, C.S.; Elingaard, M.O.; Aage, N. Level set topology and shape optimization by density methods using cut elements with length scale control. *Struct. Multidiscip. Optim.* **2020**, *62*, 685–707, ISSN 1615-147X. [CrossRef]
5. Xie, Y.M.; Steven, G.P. A simple evolutionary procedure for structural optimization. *Comput. Struct.* **1993**, *49*, 885–896. [CrossRef]
6. Fortus 380mc and Fortus 450mc. Available online: https://www.stratasys.com/3d-printers/fortus-380mc-450mc (accessed on 28 February 2020).
7. Dizon, J.R.C.; Espera, A.H.; Chen, Q.; Advincula, R.C. Mechanical characterization of 3D-printed polymers. *Addit. Manuf.* **2018**, *20*, 44–67. [CrossRef]
8. Hund, J.; Naumann, J.; Seelig, T. An experimental and constitutive modeling study on the large strain deformation and fracture behavior of PC/ABS blends. *Mech. Mater.* **2018**, *124*, 132–142, ISSN 01676636. [CrossRef]

9. Manaia, J.P.; Pires, F.A.; de Jesus, A.M.P.; Wu, S. Mechanical response of three semi crystalline polymers under different stress states. *Polym. Test.* **2020**, *81*, 132–142, ISSN 01429418. [CrossRef]
10. Louche, H.; Piette-Coudol, F.; Arrieux, R.; Issartel, J. An experimental and modeling study of the thermomechanical behavior of an ABS polymer structural component during an impact test. *Int. J. Impact Eng.* **2009**, *36*, 847–861, ISSN 0734743X. [CrossRef]
11. Duan, Y.; Saigal, A.; Greif, R.; Zimmerman, M.A. A uniform phenomenological constitutive model for glassy and semicrystalline polymers. *Int. J. Impact Eng.* **2001**, *41*, 1322–1328, ISSN 0032-3888. [CrossRef]
12. Anand, L. Constitutive Equations for Hot-Working of Metals. *Int. J. Plast.* **1985**, *1*, 213–231. [CrossRef]
13. Brown, S.; Kim, K.; Anand, L. An internal variable constitutive model for hot working of metals. *Int. J. Plast.* **1989**, *5*, 95–130, ISSN 07496419. [CrossRef]
14. Ansys. Southpointe (USA). 2020. Available online: https://www.ansys.com/ (accessed on 30 November 2020).
15. Python. Beaverton (USA). 2020. Available online: https://www.python.org/ (accessed on 30 November 2020).
16. Papazafeiropoulos, G.; Muñiz-Calvente, M.; Martínez-Pañeda, E.; Zimmerman, M.A. Abaqus2Matlab: A suitable tool for finite element post-processing. *Adv. Eng. Softw.* **2017**, *105*, 9–16, ISSN 09659978. [CrossRef]
17. Rahmani, B.; Mortazavi, F.; Villemure, I.; Levesque, M. A new approach to inverse identification of mechanical properties of composite materials: Regularized model updating. *Compos. Struct.* **2013**, *105*, 116–125, ISSN 02638223. [CrossRef]
18. Waszczyszyn, Z.; Ziemiański, L. Neural networks in mechanics of structures and materials–new results and prospects of applications. *Comput. Struct.* **2001**, *79*, 2261–2276, ISSN 00457949. [CrossRef]
19. Andrade-Campos, A.; Thuillier, S.; Pilvin, P.; Teixeira-Dias, F. On the determination of material parameters for internal variable thermoelastic–viscoplastic constitutive models. *Int. J. Plast.* **2007**, *23*, 1349–1379, ISSN 07496419. [CrossRef]
20. Van Griensven, A.; Meixner, T.; Grunwald, S.; Bishop, T.; Diluzio, M.; Srinivasan, R. A global sensitivity analysis tool for the parameters of multi-variable catchment models. *J. Hydrol.* **2006**, *324*, 10–23, ISSN 00221694. [CrossRef]
21. Cheng, Z.; Wang, G.; Chen, L.; Wilde, J.; Becker, K. Viscoplastic Anand model for solder alloys and its application. *Solder. Surf. Mt. Technol.* **2000**, *12*, 31–36. [CrossRef]
22. Motalab, M.; Cai, Z.; Suhling, J.C.; Lall, P. Determination of Anand constants for SAC solders using stress-strain or creep data. In Proceedings of the 13th InterSociety Conference on Thermal and Thermomechanical Phenomena in Electronic Systems, San Diego, CA, USA, 30 May–1 June 2012; IEEE: New York, NY, USA, 2012.
23. Rojíček, J. Identification of material parameters by FEM. *Mod. Mach. Sci. J.* **2010**, *2*, 185–188.
24. Dusunceli, N.; Colak, O.U. Determination of material parameters of a viscoplastic model by genetic algorithm. *Mater. Des.* **2010**, *31*, 1250–1255. [CrossRef]
25. Vose, M.D. *The Simple Genetic Algorithm: Foundations and Theory*; Mit Press: Cambridge, MA, USA, 1999.
26. Simon, D. *Evolutionary Optimization Algorithms: Biologically-Inspired and Population-Based Approaches to Computer Intelligence*, 1st ed.; John Wiley & Sons: Hoboken: Hoboken, NJ, USA, 2013; ISBN 978-0-470-93741-9.
27. Schreier, H.; Orteu, J.J.; Sutton, M.A. *Image Correlation for Shape, Motion and Deformation Measurements*; Springer: Boston, MA, USA, 2009; ISBN 978-0-387-78746-6. [CrossRef]
28. Hartmann, S.; Gibmeier, J.; Scholtes, B. Experiments and Material Parameter Identification Using Finite Elements. Uniaxial Tests and Validation Using Instrumented Indentation Tests. *Exp. Mech.* **2006**, *46*, 5–18, ISSN 0014-4851. [CrossRef]

 materials

MDPI

Article

Effect of the Test Procedure and Thermoplastic Composite Resin Type on the Curved Beam Strength

Robin Hron *, Martin Kadlec and Roman Růžek

Materials and Technologies Department, Aviation Division, VZLU-Czech Aerospace Research Centre, 199 05 Prague, Czech Republic; kadlec@vzlu.cz (M.K.); ruzek@vzlu.cz (R.R.)
* Correspondence: hron@vzlu.cz; Tel.: +420-604-829-041

Abstract: The application of thermoplastic composites (TPCs) in aircraft construction is growing. This paper presents a study of the effect of an applied methodology (standards) on out-of-plane interlaminar strength characterization. Additionally, the mechanical behaviour of three carbon fibre-reinforced thermoplastic composites was compared using the curved beam strength test. Data evaluated using different standards gave statistically significantly different results. The study also showed that the relatively new polyaryletherketone (PAEK) composite had significantly better performance than the older and commonly used polyphenylensulfid (PPS) and polyetheretherketone (PEEK). Furthermore, considering the lower processing temperature of PAEK than PEEK, the former material has good potential to be used in serial aerospace production.

Keywords: composite; thermoplastic; interlaminar strength; polyphenylensulfid; polyetheretherketone; polyaryletherketone; curved beam

Citation: Hron, R.; Kadlec, M.; Růžek, R. Effect of the Test Procedure and Thermoplastic Composite Resin Type on the Curved Beam Strength. *Materials* **2021**, *14*, 352. https://doi.org/10.3390/ma14020352

Received: 20 November 2020
Accepted: 7 January 2021
Published: 12 January 2021

Publisher's Note: MDPI stays neutral with regard to jurisdictional claims in published maps and institutional affiliations.

Copyright: © 2021 by the authors. Licensee MDPI, Basel, Switzerland. This article is an open access article distributed under the terms and conditions of the Creative Commons Attribution (CC BY) license (https://creativecommons.org/licenses/by/4.0/).

1. Introduction

More than 95 percent of composites used in the aerospace industry are thermosets [1]. However, the share of high-performance thermoplastic composites (TPCs) in the aeronautical industry is rising year after year even at the expense of these thermosets. It is given by attractive properties such as fracture resistance [2–4], formability [5,6], welding [7,8], self-healing possibilities [9,10] and finally recyclability [11]. With regard to modern trends and requirements, we can say that the recyclability of composites belongs (compared to metals) among their weakest aspects. Thanks to thermoplastics, today, we can talk about real recycling of composites. Thermoplastics can be heated and moulded repeatedly without negatively affecting the material's physical properties. The curing process is completely reversible. These polymers are already polymerized and do not "cure". They are melt-processable and, due to a lack of cross links, recyclable (reformable) at temperatures above their glass transition temperature. Softening by heating further enables welding of subcomponents. This leads to the elimination of fasteners and adhesives, as is showed in [12,13]. The requested performance of structural TPC parts can be easily achieved by stacking tailored blanks with a combination of the thermoforming process—this is demonstrated, for example, on the thermoplastic rib in [14]. The biggest advantages are that TPCs have a short curing time (compared to thermosets); neither absorb water nor degrade when exposed to moisture; and have excellent fire, smoke and toxicity (FST) properties. The disadvantages of thermoplastics include their higher purchase price. However, the total cost of a component may be less than the thermosetting component due to lower production and storage costs. The most used TPC in aircraft construction is polyfenylensulfid (PPS), and other commonly used thermoplastic matrices are polyetheretherketone (PEEK) and polyetherketoneketone (PEKK). For excellent thermal stability, thermoplastics can be used even when there are higher operating temperatures. The growing production of thermoplastics goes hand in hand with their development. New types of thermoplastic matrices are being introduced into the market. TPCs have been used in advanced airframes, for example, on the horizontal tailplane of AW 169, on the weapon bay doors of F-22, on the rudder of Boeing Phantom Eye, and on the rudder and elevators of G650.

The mechanical behaviour of composite materials is commonly characterized by tensile or compressive strength, by the impact damage and by the environment. There are several material characteristics and methods which can be defined as a barrier for the failure mechanism of composite materials. Two of these characteristics are interlaminar shear strength (ILSS) based on shear loading and interlaminar tensile strength (ILT) based on testing of curved beams. The authors discussed these characteristics previously in References [15,16]. The comparison of ILSS properties on the different types of thermoplastics matrices was formerly studied considering creep and stress relaxation [17], interlaminar shear strength [15] and impact resistance [18]. It could be stated that curved beams better conform to real stress–strain conditions of real curved structures used in composite structures. Additionally, curved beams are sensitive to delamination at locations with high interlaminar stresses. Unfortunately, both the ILSS and ILT values are not readily available (are not included in the material sheets as a standard).

In general, one of the major barriers to accurate failure prediction for polymer-matrix composites is the lack of matrix-dominated material properties, which could be used as a basis for the development of failure criteria [19,20].

The ILT strength generally represents the weakest point of a laminated composite system. At the same time, ILT strength is one of the most difficult material strength properties to characterize [21].

An accurate evaluation of ILT strength is needed to define delamination failure. Delamination is one of the primary failure modes that occur in aerospace composite structures. Currently, the ASTM D6415 [22] and AITM1-0069 [23] curved-beam (CB) methods are standard practices for measurements of ILT strength. Figure 1 shows a typical CB test setup for a 4.95-mm thick fabric PPS material.

Figure 1. Curved-beam test setup and delamination failure of a polyphenylensulfid (PPS) specimen.

The typical failure mode is tensile delamination. Failure starts in the beam radius area at about two-thirds of the thickness. It corresponds to the maximum ILT stress location. Subsequently, delamination quickly propagates through the beam flanges. ASTM D6415 provides equations for ILT strength calculation. Makeev et al. [21] measured the ASTM D6415 CB strength for multiple unidirectional carbon fibre and glass fibre-reinforced epoxy-matrix prepreg tape composites. Based on their experience, the manufacturing process to produce CB coupons with uniform radius and thickness should be preferred. However, it is not possible generally for several practical reasons associated with specific structure design. Additionally, the CB strength data typically exhibit large scatter. For example, Makeev et al. [21] shows that, for 0.26-inch thick CB coupons manufactured from Hexcel IM7/8552 unidirectional tape and cured per manufacturer's specifications under nominal cure pressure, the average ASTM D6415 ILT strength varies between 68.9 MPa and 82.7 MPa and the coefficient of variation (COV), defined as the ratio of standard deviation to the average value, is usually higher than 20% [21]. The question is whether the ASTM D6415 CB interlaminar strength data, including the large scatter, are coupon-specific. The CB strength is not a coupon-independent material property, suggesting that ASTM D6415 is not an adequate approach to measure the ILT strength of materials. The AITM 1-0069 standard is a very similarly procedure to the ASTM D6415 test and evaluation methods. A comparison of these methods and results evaluated based on defined procedures is discussed by the authors of this paper hereafter.

Another method used for the evaluation of ILT strength is ASTM D7291 [24]. This method applies a tensile force normal to the plane of the composite laminate using adhesively bonded thick metal end-tabs. It was noted in ASTM standard D7291 that thickness strength results using this method will in general not be comparable to ASTM D6415 or AITM 1-0069 since ASTM D7291 subjects a relatively large volume of material to an almost uniform stress field while ASTM D6415 and AITM 1-0069 subject a small volume of material to a nonuniform stress field. It seems that characterization of ILT strength using ASTM D7291 is more representative than ASTM D6415. The reason is the possibility of different failure modes occurring—the failure could occur not only in the composite material but also at the bond lines between the composite and the metal end-tabs. End-tabs are used with the aim of ILT load transfer to the composite.

Formerly, Jackson and Martin [25] studied carbon/epoxy CB specimen configurations to establish a method and specimen for assessing ILT strength. They concluded that specimens with curved geometries include manufacturing problems that cannot be described by flat panels. Failure modes and strengths defined based on curved beam specimens with manufacturing flaws correspond to those in the actual structure with similar flaws. In the case of specimens not containing any significant flaws, a true material property can be defined. Jackson and Martin [25] observed CB strength reduction (up to a factor of four) in low-quality CB specimens containing macroscopic voids detected using fractography analysis. However, it could not explain the large scatter in the strength data observed in high-quality CB specimens. They have not made available the detailed non-destructive inspection (NDI) techniques.

Makeev et al. [21] focused their work on ILT failure and did not address the in-ply transverse tensile failure (matrix ply cracking), which is different from the ILT failure (delamination) discussed by See O'Brien et al. [26,27] for measurement of in-ply transverse tensile material properties. ASTM D6415 significantly underestimates ILT strength in the case where the CB coupon contains porosity even at a low-porosity content. Better values of strength properties can be evaluated after refinement of the ASTM D6415 procedure. This includes measurement of the critical voids in the CB radius area and transition of the defective information into a finite element stress analysis model. The ILT material strength results of the modified CB tests presented by Makeev [21] for the unidirectional IM7/8552 carbon/epoxy tape composite were in excellent agreement with the short-beam tests. The CB tests can be used for assessment of the effects of porosity defects on ILT performance by the refined CB method proposed by Makeev [21]. Hao et al. [28] investigated deformation and strength of CB specimens with various thicknesses and radius–thickness ratios. Strength increased with increases in the thickness and radius–thickness ratio.

The objectives of the presented work are to compare ILT strength of three present-day thermoplastic composite (two commonly used—PPS and PEEK—and one relatively new—polyaryletherketone (PAEK)) materials and to analyse the procedures defined in the ASTM D6415 and AITM 1-0064 standards with the aim of evaluating potentially strength result dissimilarities. Except the matrix type, the effect of the test temperature on ILT strength was evaluated. A comparison using out-of-plane interlaminar strength has been proposed with respect to the fact that interlaminar strength can be two orders of magnitude less than the tensile strength in fibre direction and that even a small load applied in the through thickness direction can lead to the delamination. The basic mechanical and physical properties of the materials used are given in Tables 1 and 2. The materials were selected on the basis of the manufacturer's experience and their actual/planned use in the construction of aircrafts.

2. Experiment

2.1. Material

PPS, PEEK and PAEK thermoplastic polymer melts with T300 3K, 5HS, 280 gm-2 FAW (Fabric Area Weight), 43% RC (Resin Content) (50% by volume) carbon fabric (280 gsm) were compared. The coupons were manufactured by Latecoere Czech Republic, Prague using thermoforming technology. Thermoforming is used to convert a flat consolidated continuous fibre-reinforced laminate into a complex shape with no change in original laminate thickness. The laminates were heated to the required temperature and then quickly formed under ambient pressure with a few minutes dwell time (Figure 2).

Figure 2. Curved-beam strength samples.

Indicative properties of the compared matrices and used carbon laminates with these matrices are shown in the Tables 1 and 2. Figure 3 graphically illustrates the thermal properties of the compared thermoplastics. By comparing the properties of individual matrices in Table 1, we can see that the PAEK matrix has the highest flexural strength and elongation. On the other hand, it has the lowest compressive strength. The highest compressive strength has a PPS matrix. In tensile strength, the differences are not so large (less than 10%). The greatest differences show thermal properties. The PPS thermoplastics has glass-transition temperature (T_g) and melt temperature (T_m) that are significantly lower than the two remaining matrices. These have almost identical T_g, but PEEK has a T_m 38 °C higher than the PAEK matrix. When comparing the fibre-reinforced laminate with PPS, PEEK and PAEK, we find that most properties differ at the minimum. The greatest differences are in the strength properties in the 90° direction and in-plane shear strength; see Table 2.

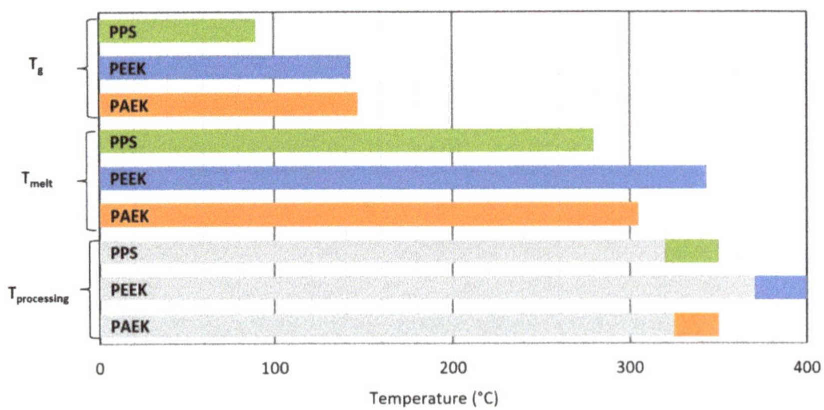

Figure 3. Compared thermal properties of the PPS, PEEK and PAEK matrices [29].

Table 1. Physical, mechanical and thermal properties of the compared matrices [30–32].

Property	PPS	PEEK	PAEK
Specific gravity (g/cm^3)	1.35	1.3	1.4
T_g (°C)	90	143	147
Melt temperature T_m (°C)	280	343	305
Moisture absorption (%)	0.02	0.2	0.2
Tensile strength (MPa)	90.3	97.2	95
Tensile modulus (GPa)	3.8	3.59	3.7
Elongation at yield (%)	3	3	4.5
Compression strength (MPa)	148	120	117
Compression modulus (GPa)	3.0	-	-
Flexural strength (MPa)	125	138	141
Flexural modulus (GPa)	3.7	4.1	4.2
Processing temperature (°C)	320–350	370–400	325–350

Table 2. Mechanical properties of the carbon laminates with the PPS, polyetheretherketone (PEEK) and polyaryletherketone (PAEK) polymer melts [29–31].

Property	PPS	PEEK	PAEK
Tensile strength 0° (MPa)	757	776	805
Tensile modulus 0° (GPa)	55.8	56.1	58
Tensile strength 90° (MPa)	754	827	739
Tensile modulus 90° (GPa)	53.8	55.6	59
Compressive strength 0° (MPa)	643	585	628
Compressive modulus 0° (GPa)	51.7	51.6	52
Compressive strength 90° (MPa)	637	595	676
Compressive modulus 90° (GPa)	51.7	49.7	53
In Plane Shear Strength (MPa)	119	155	147
In Plane Shear Modulus (GPa)	4.4	4.5	4.1
Flexural strength 0° (MPa)	1027	-	1040
Flexural modulus 0° (GPa)	60	-	60
Flexural strength 90° (MPa)	831	859	879
Flexural modulus 90° (GPa)	44.8	46.3	48

The samples were divided according to Table 3 as the samples for testing at room temperature (RT) and the samples for testing at a cold temperature of −55 °C (CT). A temperature of −55 °C represents the typical operating temperature in aerospace. Show material properties at this temperature are important for airworthiness.

Table 3. Overview of the tested sets.

Set (Resin)	Fabric	Lay-Up	Ø Width (mm)	Ø Thickness (mm)	Ø α (Deg)	Number of Samples	
						RT	CT
PEEK	T300JB 3K, 5HS, 280 gsm FAW, 42% RC (50% BV)	$[(0,90)/(\pm45)]_4/(0,90)$	25.31	2.78	91.0	6	5
PPS	T300 3K, 5HS, 280 gsm FAW, 43% RC (50% BV)	$[[(0,90)/(\pm45)]_4]$	25.14	4.95	89.5	5	5
PAEK	T300JB 3K, 5HS, 277 gsm FAW, 42% RC (50% BV)	$[[(0,90)/(\pm45)]_4]$	25.24	4.65	90.7	5	5

2.2. Material Structure Analysis

Metallographic analysis was performed on the non-tested spare samples for all thermoplastic types. The analysed samples contain only minor porosities (<20 μm in the radius and <200 μm in the flat parts). The void size and quantity were not sufficient to visualize a negative effect on interlaminar strength in cases where the specimens were exposed to a quasi-static loading. No significant deviations between the sets were observed. A typical cross section before testing of PAEK thermoplastic is shown in Figure 4.

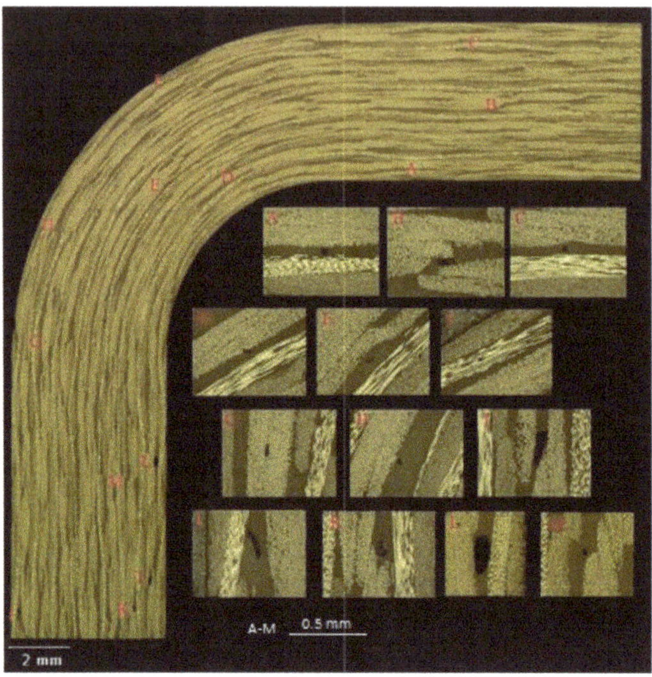

Figure 4. Example of the cross section before testing: PAEK.

2.3. *Test Method*

The objective of the curved beam strength test is to determine the strength characteristics of the composite material in the out-of-plane (z) direction. Radial tensile stress in this direction of the composite (through the thickness of the material) is induced in the curved region of the test specimen when bending is applied. The bending load is applied using a four-point bending fixture with two pairs of cylindrical supports with different span lengths (l_t and l_b).

Before the tests started, a comparison of the two most used test methods was performed (ASTM D6415 [2,3]—the standard test method for measuring the curved-beam strength of fibre-reinforced polymer-matrix composites—and AITM 1-0069 [24]— determination of curved-beam failure loads). This method of loading induces a constant bending moment in the curved region of the specimen. The main motivation was to compare the effect of the spans. Three configurations were prepared: the first was per the ASTM standard, where fix values of span were used (l_t = 75 mm and l_b = 100 mm); the second configuration used a modified ASTM (ASTM mod) span (l_t = 45 mm and l_b = 75 mm); and the third was per the AITM standard. In this method, the spans are calculated based on the sample geometry in Equations (1)–(4). Based on these calculations, l_t = 26.4 mm and l_b = 40.6 mm were set.

$$l_t > 2 \cdot \left(\left(R_i + t + \frac{D}{2} \right) \cdot \sin(\varphi) + \left(\frac{t}{4} + 1 \right) \cdot \cos(\varphi) \right) \pm 0.5 \tag{1}$$

$$l_t > 2 \cdot \left(\left(R_i + t + \frac{D}{2} \right) \cdot \sin(\varphi) + \left(\frac{t}{4} + 1 \right) \cdot \cos(\varphi) \right) \pm 0.5 \tag{2}$$

$$l_b > l_t + t + 10 \pm 0.5 \tag{3}$$

$$l_b > l_t + t + 20 \pm 0.5 \tag{4}$$

In these equations, l_t denotes the span of the top fixture, l_b is the span of the bottom fixture, R_i is the inner radius, t is the thickness of the sample, D is the roller diameter and φ is the angle from the horizontal of the sample legs.

Tests were performed on a static load machine Instron 55R1185 (Norwood, MA 02062-2643, 825 University Ave, USA) with an installed load cell with a capacity of ±10 kN and with control system Instron K5178. Recording of the force, displacement and extensometer data was ensured by the software Bluehill 3. The test setup is shown in Figure 5. A test specimen was placed on the bottom cylindrical bars. Then, the extensometer Instron 2620-604 with a base of 50 mm was installed. The extensometer recorded the axial displacement between the upper and lower parts of the fixture. The specimen was loaded by a constant crosshead speed of 2 mm/min and the test ended when the loading rapidly decreased (approximately a 30-percent drop).

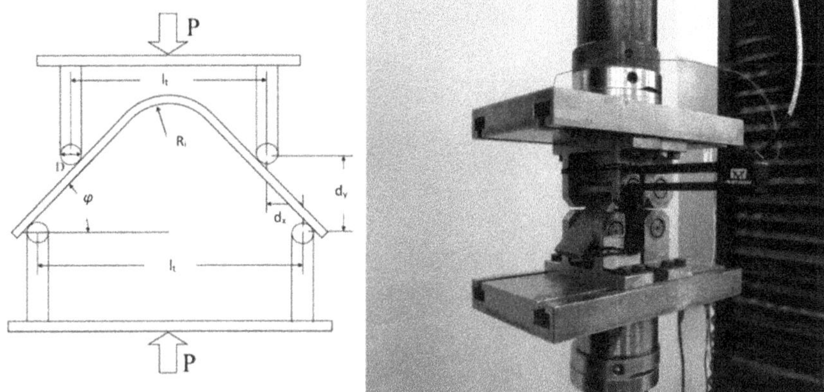

Figure 5. Curved-beam strength test.

2.4. Statistical Analysis

Numerous techniques exist for statistical assessment of experiments. For this paper, two types of evaluation were used: (1) T-test for the test method effect where a single factor for two sets is assessed by evaluation of the *p*-value, which is compared to significance level $\alpha = 0.05$; where for a *p*-value lower than α, the effect is statistically significant; and where the data sets have different mean values and (2) the Taguchi technique of design of experiments (DOE). DOE is the experimental strategy that facilitates the study of multiple factors at different levels. Questions concerning the influence of these factors on the variation of results can only be obtained by performing an analysis of variance (ANOVA).

In this ANOVA design, 2 factors representing both the materials and test temperature were chosen. Three qualitative levels were set for thermoplastic type (A), and two levels were set for the temperature (B). The full factorial experiment made it possible to also investigate the interactions AB of these two factors.

The full model was

$$Y_{ijk} = \mu_g + \alpha_i + \beta_j + (\alpha\beta)_{ij} + \varepsilon_{ijk} \tag{5}$$

$$\varepsilon_{ijk} \sim N\left(0, \sigma^2\right)\ i = 1,\ 2,\ 3.;\ j = 1, 2.;\ k = 1, 2, \ldots, 6. \tag{6}$$

where μ_g represents a grand mean term common to all observations, α_i is the effect of the *i*th level of A, β_j is the effect of the *j*th level of B, and $(\alpha\beta)_{ij}$ is the interaction effect of level *i* of A and level *j* of B combined. Also, a test for normality of residuals ε needs to be done.

3. Results and Discussion

3.1. Test Method Evaluation

A comparison of the methods was performed on the PPS material; see Table 4 and Figure 6. The statistical analysis of data using multiple t-tests showed that the AITM test method has a statistically significant effect (p-value = 0.011) when compared to pooled values of the two ASTM methods. It means that the AITM data were different from the other two sets. The results of the ASTM and ASTM mod procedures were not statistically different due to scatter of the data. The results confirmed the general rule that, with shorter distances (span lengths), greater forces are required for samples to fail. In general, it is not always possible to have one geometry available. Dimensions may be based on the actual design of the construction and simulation of a real loading.

Table 4. Comparison test method influence on interlaminar strength, σ_r (MPa).

σ_r (MPa)	PPS		
	ASTM	ASTM Mod	AITM
Mean	71.3	75.1	81.4
S.D.	4.09	6.36	3.33
C.V.	5.73	8.47	4.09
Min.	65.4	68.9	84.7
Max.	78.8	85.6	77.4

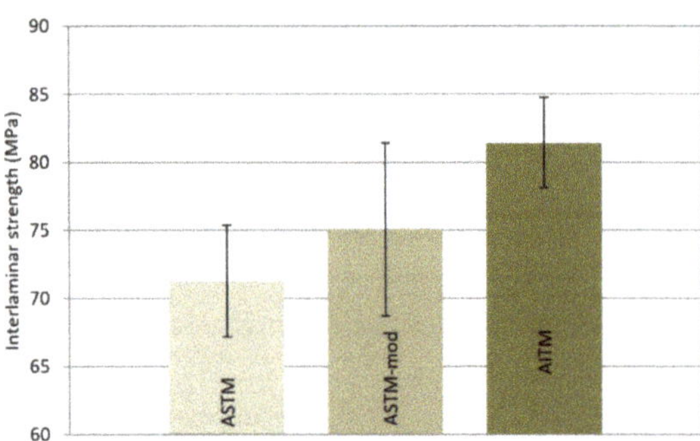

Figure 6. Measured interlaminar strength, σ_r (MPa): the test method comparison using statistic evaluation showed a significant difference of the AITM method.

For the following experiments, the AITM test method was chosen. The main reason was to take sample geometry into account when setting the test fixture (span length).

3.2. Thermoplastic Type and Temperature Evaluation

The highest interlaminar strength was achieved on the set with the PAEK thermoplastic; see Table 5 and Figure 7.

For room temperature (RT), the average interlaminar strength of the PAEK set was to 18% higher than that for PPS and 16% higher than that for the PEEK set. The PAEK set also showed the smallest variance in measured values. Coefficient of variation (CV) for both test temperatures was less than 3%. The greatest variance of the measured values was evaluated on the PPS set (11%).

For cold temperatures (CT), compared to RT, the strength increased by approx. 10% for PPS, by 8% for PEEK and by 6% for PAEK. The values measured on the PAEK sets showed a very small coefficient of variation (less than 3%). For the PPS and PEEK sets, CV was about 10 %.

Table 5. Measured interlaminar strength, σ_r (MPa).

	Interlaminar Strength (MPa)					
	RT			CT		
	PPS	PEEK	PAEK	PPS	PEEK	PAEK
	83.6	79.8	91.8	89.8	87.3	99.5
-	84.7	86.4	88.3	89.5	78.2	95.0
	80.0	73.3	92.0	95.3	78.1	99.7
	63.5	78.6	96.0	73.1	87.0	94.6
	77.4	82.3	92.7	77.8	98.2	100.1
	-	77.0	-	-	-	-
Mean	**77.9**	**79.6**	**92.2**	**85.1**	**85.8**	**97.8**
S.D.	8.52	4.49	2.73	9.26	8.29	2.75
C.V.	10.95	5.65	2.97	10.89	9.66	2.81
Min.	63.5	73.3	88.3	73.1	78.1	94.6
Max.	84.7	86.4	96.0	95.3	98.2	100.1

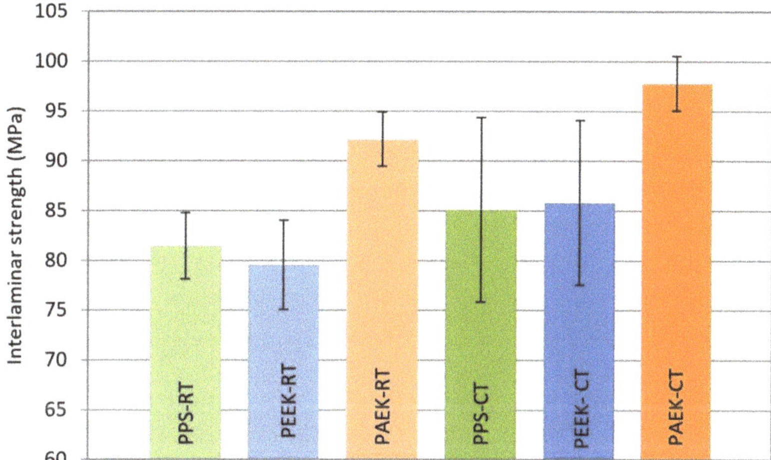

Figure 7. Comparison of the interlaminar strength for 6 data sets, *p*-value 0.05.

A series of statistical analyses to compare the sets was performed. First, all measured data were analysed by ANOVA using DOE++ ReliaSoft (Version 1.0.7; ReliaSoft Corporation, MI, USA). The purpose was to evaluate the main factors and their effects on the results using a general full factorial design with multiple level factors—temperature and thermoplastic type. The Anderson–Darling test of residuals for model (5) proved the normality. The results of analysis for each factor in model (5) are shown in Table 6 in following order: A: thermoplastic type, B: temperature and AB: interaction interactions between both factors. Statistically significant effects of both testing temperature (B) and thermoplastic type (A) on measured strength values were found (p-values < 0.05). The interaction of effects (AB) was not proven. In a graphical form, the results are illustrated in Figure 8.

A complete evaluation of the differences for each parameter is shown in Table 7. *t*-tests applied for individual sets revealed that the PAEK interlaminar strength was higher than that for both the PEEK and PPS sets for both temperatures. No difference was proven between the PPS and PEEK sets. Cold conditions at −55 °C increased interlaminar strength in all investigated cases compared with RT conditions.

Table 6. ANOVA analysis of thermoplastic type and temperature influence on data results.

Source of Variation	Degrees of Freedom	Sum of Squares (Partial)	Mean Squares (Partial)	F Ratio	*p*-Value
Model	5	1468.419	293.6838	6.8664	0.0004
A: thermoplastic type	2	1128.575	564.2875	13.1932	0.0001
B: Temperature	1	312.3451	312.3451	7.3027	0.0122
AB	2	3.4559	1.7279	0.0404	0.9605
Residual	25	1069.2783	42.7711		
Pure Error	25	1069.2783	42.7711		
Total	30	2537.6973			

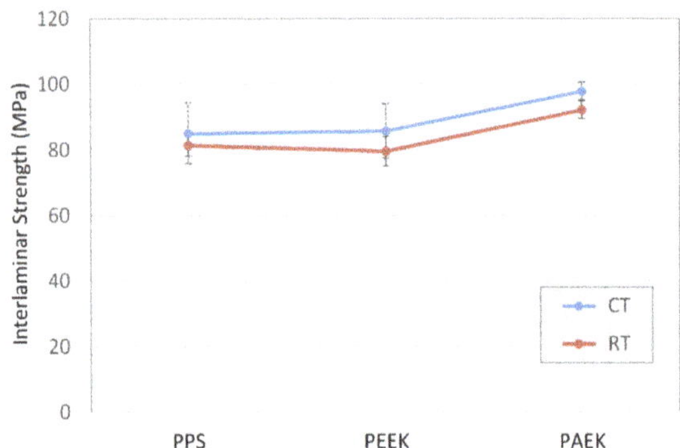

Figure 8. Graphical comparison of the interlaminar strength results: temperature and thermoplastic-type influence.

Table 7. Statistical comparison of individual files using *t*-tests. D, files are different; ND, files are not different.

	-	RT PPS	RT PEEK	RT PAEK	CT PPS	CT PEEK	CT PAEK
RT	PPS	-	ND	D	ND	ND	D
	PEEK	ND	-	D	ND	ND	D
	PAEK	D	D	-	ND	ND	D
CT	PPS	ND	ND	ND	-	ND	D
	PEEK	ND	ND	ND	ND	-	D
	PAEK	D	D	D	D	D	-

For better failure identification, the edges of the sample were painted in white colour. Valid failure occurred for all samples: a delamination in curvature occurred. Figure 9 shows the PPS-CT set failure modes. This set has a relatively high coefficient of variation (<10%). It can be seen from the figure that, at the lowest values, the failure occurred locally, and, in this set, specifically in the middle (sample 9) or at the outer radius (sample 10). For the other samples (6–8), the failure was over the entire thickness of the test sample. Figure 10 shows the PAEK-CT set samples. This set had a small coefficient of variation (2.8%), and similar failures were noted. Localization of the failures could be caused by clustering of the porosity in one place, by a manufacturing defect, or as a natural property of the material.

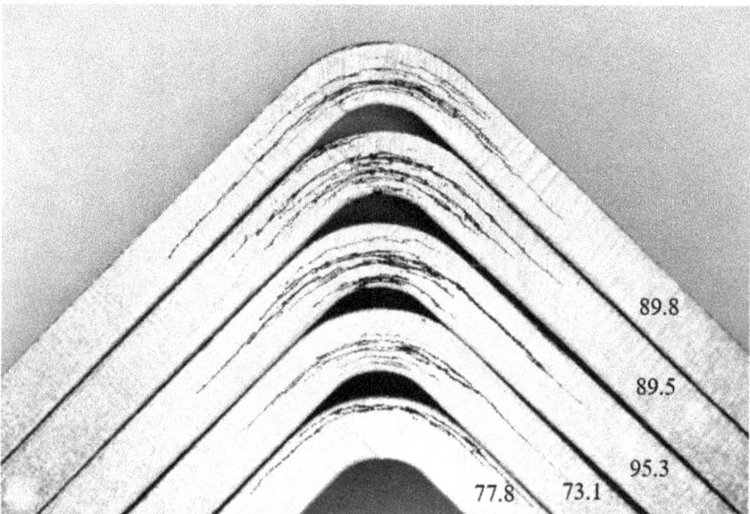

Figure 9. Examples of a typical failure mode for set PPS-CT: the set with the highest coefficient of variation, with the values given in the figure being strengths in MPa.

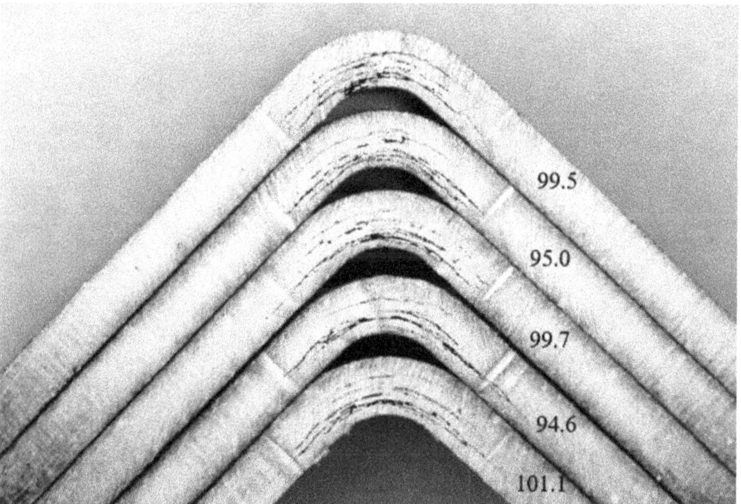

Figure 10. Examples of a typical failure mode for set PAEK-CT: the set with the lowest coefficient of variation, with the values given in the figure being strengths in MPa.

4. Conclusions

Three methods for interlaminar strength evaluation were analysed. Data evaluated using the AITM standard give significantly different results from the data based on the ASTM procedures. This implies that results from different methods cannot be directly compared and used for numerical analyses.

On the basis of statistical evaluation of three different composite materials, it can be stated that the choice of thermoplastic type can have a significant effect on the values of interlaminar strength. The values measured on the PAEK samples were statistically significantly higher than those measured on the PPS and PEEK samples (valid for both test temperatures).

The metallographic analysis performed on the samples showed similar homogeneity of all investigated materials, and the void size and quantity were not sufficient for a negative effect on interlaminar strength.

Cold conditions at −55 °C increased interlaminar strength in all investigated cases compared with RT conditions. The PAEK material gives significantly higher interlaminar shear strength in comparison with the PPS and PEEK materials. Additionally, the PAEK material has the lowest scatter in measured data. The PAEK thermoplastic material is a new composite material planned for application in future aircrafts.

Further tests on the coupon with TPC matrix are planned for the future in order to expand the material database. Based on these tests, a material will be selected that could replace the thermosetting materials used in aircraft structures for interior panels.

Author Contributions: Conceptualization, R.H., M.K. and R.R.; Methodology, R.H., M.K. and R.R.; Validation, R.H., M.K. and R.R.; Formal analysis, R.H., M.K. and R.R., Investigation, R.H., M.K. and R.R.; Resources R.H., M.K. and R.R.; Data curation, R.H., M.K. and R.R.; Writing—original draft preparation, R.H., M.K. and R.R.; Writing—reviewing and editing R.H., M.K. and R.R.; Visualization, R.H., M.K. and R.R.; Supervision, R.H., M.K. and R.R.; Funding acquisition, R.H., M.K. and R.R. All authors have read and agreed to the published version of the manuscript.

Funding: This work was funded by the Ministry of Industry and Trade of the Czech Republic in the frame of "TRIO" call. Project No. FV 30033—Design and technological development of primary aircraft parts of advanced shapes made of composites with a thermoplastic matrix.

Data Availability Statement: Data sharing is not applicable to this article.

Conflicts of Interest: The authors declare no conflict of interest.

References

1. Dale, B. Thermosets vs. Thermoplastics: Is the Battle Over? 2015. Available online: www.compositesword.com (accessed on 10 February 2020).
2. Ye, L.; Beehag, A.; Friedrich, K. Mesostructural Aspects of Interlaminar Fracture in Thermoplastic Composites: Is Crystallinity a Key? *Compos. Sci. Technol.* **1995**, *53*, 167–173. [CrossRef]
3. Brown, K.A.; Brooks, R.; Warrior, N. Characterizing the Strain Rate Sensitivity of the Tensile Mechanical Properties of a Thermoplastic Composite. *JOM* **2009**, *61*, 43–46. [CrossRef]
4. Čoban, O.; Bora, M.Ö.; Sinmazçelik, T.; Cürgül, İ.; Günay, V. Fracture Morphology and Deformation Characteristics of Repeatedly Impacted Thermoplastic Matrix Composites. *Mater. Des.* **2009**, *30*, 628–634. [CrossRef]
5. Lee, J.H.; Vogel, J.H. An Investigation of the Formability of Long Fiber Thermoplastic Composite Sheets. *J. Eng. Mater. Technol.* **1995**, *117*, 127–132. [CrossRef]
6. Martin, T.A.; Bhattacharyya, D.; Pipes, R.B. Deformation Characteristics and Formability of Fibre-reinforced Thermoplastic Sheets. *Compos. Manuf.* **1992**, *3*, 165–172. [CrossRef]
7. Ageorges, C.; Ye, L. Resistance Welding of Thermosetting Composite/Thermoplastic Composite Joints. *Compos. Part A* **2001**, *32*, 1603–1612. [CrossRef]
8. Yousefpour, A.; Hojjati, M.; Immarigeon, J.P. Fusion Bonding/Welding of Thermoplastic Composites. *J. Thermoplast. Compos. Mater.* **2004**, *17*, 303–341. [CrossRef]
9. Yuan, Y.C.; Yin, T.; Rong, M.Z.; Zhang, M.Q. Self Healing in Polymers and Polymer Composites. Concepts, Realization and Outlook: A Review. *eXPRESS Polym. Lett.* **2008**, *2*, 238–250. [CrossRef]
10. Yao, L.; Rong, M.Z.; Zhang, M.Q.; Yuan, Y.C. Self-healing of Thermoplastics via Reversible Addition—Fragmentation Chain Transfer Polymerization. *J. Mater. Chem.* **2011**, *21*, 9060–9065. [CrossRef]
11. Schinner, G.; Brandt, J.; Richter, H. Recycling Carbon-fiber-reinforced Thermoplastic Composites. *J. Thermoplast. Compos. Mater.* **1996**, *9*, 239–245. [CrossRef]
12. Šedek, J.; Hron, R.; Kadlec, M. Bond Joint Analysis of Thermoplastic Composite Made from Stacked Tailored Blanks. *Appl. Mech. Mater.* **2016**, *827*, 161–168. [CrossRef]
13. Kruse, T.; Körwien, T.; Růžek, R. Fatigue behaviour and damage tolerant design of bonded joints for aerospace application. In Proceeding of the ECCM 2016 17th European Conference on Composite Materials, ECCM 2016, Munich, Germany, 26–30 June 2016; ISBN 978-300053387-7.
14. Růžek, R.; Šedek, J.; Kadlec, M.; Kucharský, P. Mechanical behavior of thermoplastic rib under loading representing real structure conditions. In Proceedings of the EAN 2016—54th International Conference on Experimental Stress Analysis, Srní, Czech Republic, 30 May–2 June 2016.

15. Kadlec, M.; Nováková, L.; Růžek, R. An Experimental Investigation of Factors Considered for the Short Beam Shear Strength Evaluation of Carbon Fiber–reinforced Thermoplastic Laminates. *J. Test. Eval.* **2014**, *42*, 580–592. [CrossRef]
16. Hron, R.; Kadlec, M.; Růžek, R. Effect of Temperature on the Interlaminar Strength of Carbon Fibre Reinforced Thermoplastic. In Proceedings of the EAN 2020—58th International Conference on Experimental Stress Analysis, Online, Czech Republic, 19–22 October 2020.
17. Guo, Y.; Bradshaw, R.D. Isothermal physical aging characterization of Polyether-ether-ketone (PEEK) and Polyphenylene sulfide (PPS) films by creep and stress relaxation. *Mech. Time Depend. Mater.* **2017**, *11*, 61–89. [CrossRef]
18. Escale, L.; de Almeida, O.; Bernhart, G.; Ferrero, G.J. Comparison of the Impast Resistance of Carbon/ Epoxy and Carbon /PEEK Composite Laminates. In Proceedings of the ECCM15—15th European Conference on Composite Materials, Venice, Italy, 24–28 June 2012.
19. Nikishkov, Y.; Makeev, A.; Seon, G. Simulation of damage in composites based on solid finite elements. *J. Am. Helicopter Soc.* **2010**, *55*, 042009. [CrossRef]
20. Nikishkov, Y.; Makeev, A.; Seon, G. Progressive fatigue damage simulation method for composites. *Int. J. Fatigue* **2013**, *48*, 266–279. [CrossRef]
21. Makeev, A.; Seon, G.; Nikishkov, Y.; Lee, E. Methods for assessment of interlaminar tensile strength of composite materials. *J. Compos. Mater.* **2015**, *49*, 783–794. [CrossRef]
22. *ASTM D6415-06a Standard Test Method for Measuring the Curved Beam Strength of Fiber-Reinforced Polymer-Matrix Composite;* ASTM International: West Conshohocken, PA, USA, 2013.
23. Airbus, S.A.S. *AITM1-00069, Issue2, Determination of Curved-Beam Failure;* Engineering Directorate: Blagnac, France, 2011.
24. *ASTM D7291-15 Standard Test Method for Through-Thickness "Flatwise" Tensile Strength and Elastic Modulus of a Fiber-Reinforced Polymer Matrix Composite Material;* ASTM International: West Conshohocken, PA, USA, 2015.
25. Jackson, W.C.; Martin, R.H. An interlaminar tensile strength specimen. *Compos. Mater. Test. Des.* **1993**, *11*, 333–354.
26. Armanios, E.; Ronald, B.; Dale, W.; O'Brien, T.; Chawan, A.; Demarco, K.; Paris, I. Influence of specimen configuration and size on composite transverse tensile strength and scatter measured through flexure testing. *J. Compos Technol. Res.* **2003**, *25*, 3–21. [CrossRef]
27. O'Brien, T.K.; Chawan, A.D.; Krueger, R.; Paris, I. Transverse tension fatigue life characterization through flexure testing of composite materials. *Int. J. Fatigue* **2002**, *24*, 127–145. [CrossRef]
28. Hao, W.; Ge, W.; Ma, Y.; Yao, X.; Shi, Y. Experimental investigation on deformation and strength of carbon/epoxy laminated curved beams. *Polym. Test.* **2012**, *31*, 520–526. [CrossRef]
29. *Tencate Advanced Composites, Introducing Tencate Cetex®TC1225;* Toray Advanced Composites: Morgan Hill, CA, USA, 2016.
30. *Tencate Advanced Composites, TenCate Cetex®TC1200 PEEK Resin System, Product Data Sheet;* Toray Advanced Composites: Morgan Hill, CA, USA, 2017.
31. *Tencate Advanced Composites, TenCate Cetex®TC1100 PPS Resin System, Product Data Sheet;* Toray Advanced Composites: Morgan Hill, CA, USA, 2016.
32. *Tencate Advanced Composites, TenCate Cetex®TC1225 PAEK Resin System, Product Data Sheet;* Toray Advanced Composites: Morgan Hill, CA, USA, 2017.

Article

Design of a Unique Device for Residual Stresses Quantification by the Drilling Method Combining the PhotoStress and Digital Image Correlation

Miroslav Pástor [1], Martin Hagara [1], Ivan Virgala [2,*], Adam Kal'avský [1], Alžbeta Sapietová [3] and Lenka Hagarová [4]

1 Department of Applied Mechanics and Mechanical Engineering, Faculty of Mechanical Engineering, Technical University of Košice, Letná 9, 04200 Košice, Slovakia; miroslav.pastor@tuke.sk (M.P.); martin.hagara@tuke.sk (M.H.); adam.kalavsky@tuke.sk (A.K.)

2 Department of Mechatronics, Faculty of Mechanical Engineering, Technical University of Košice, Park Komenského 8, 04200 Košice, Slovakia

3 Department of Applied Mechanics, Faculty of Mechanical Engineering, University of Žilina, Univerzitná 8215/1, 01026 Žilina, Slovakia; alzbeta.sapietova@fstroj.uniza.sk

4 Institute of Geotechnics of Slovak Academy of Sciences, Watsonova 45, 04001 Košice, Slovakia; hagarova@saske.sk

* Correspondence: ivan.virgala@tuke.sk; Tel.: +421-55-602-2588

Abstract: This paper presents a uniquely designed device combining the hole-drilling technique with two optical systems based on the PhotoStress and digital image correlation (DIC) method, where the digital image correlation system moves with the cutting tool. The authors aimed to verify whether the accuracy of the drilled hole according to ASTM E837-13a standard and the positioning accuracy of the device were sufficient to achieve accurate results. The experimental testing was performed on a thin specimen made from strain sensitive coating PS-1D, which allowed comparison of the results obtained by both methods. Although application of the PhotoStress method allows analysis of the strains at the edge of the cut hole, it requires a lot of experimenter's practical skills to assess the results correctly. On the other hand, the DIC method allows digital processing of the measured data. However, the problem is not only to determine the data at the edge of the hole, the results also significantly depend on the smoothing levels used. The quantitative comparison of the results obtained was performed using finite element analysis.

Keywords: residual stresses; hole-drilling; PhotoStress; digital image correlation; experimental analysis; finite element analysis

Citation: Pástor, M.; Hagara, M.; Virgala, I.; Kal'avský, A.; Sapietová, A.; Hagarová, L. Design of a Unique Device for Residual Stresses Quantification by the Drilling Method Combining the PhotoStress and Digital Image Correlation. *Materials* **2021**, *14*, 314. https://doi.org/10.3390/ma14020314

Received: 30 November 2020
Accepted: 7 January 2021
Published: 9 January 2021

Publisher's Note: MDPI stays neutral with regard to jurisdictional claims in published maps and institutional affiliations.

Copyright: © 2021 by the authors. Licensee MDPI, Basel, Switzerland. This article is an open access article distributed under the terms and conditions of the Creative Commons Attribution (CC BY) license (https://creativecommons.org/licenses/by/4.0/).

1. Introduction

Residual stresses are present in almost all materials. They may arise during the manufacturing process or over the life of a material. Their quantification process is essential because their superposition with stresses occurring in the material due to the external loading can lead to material failure. The measurement techniques used for residual stress quantification can be divided into three main groups: non-destructive, semi-destructive, and destructive.

In a production environment, determination of residual stresses in the material is usually based on the standardized hole-drilling strain gauge method allowing identification of in-plane residual stresses near the measured specimen surface made from an isotropic linear-elastic material.

The methodology of residual stresses quantification using the hole-drilling strain gauge method involves several steps:

1. smoothing of the specimen surface using, e.g., chemical etching (manufacturing technologies such as abrading or grinding have to be avoided);
2. attaching a strain gauge rosette with special measuring grids to the examined location;

3. drilling a through-hole at the geometric center of the strain gauge rosette in one step to the specimen whose thickness is much less than the diameter of the hole (denoted as thin) or a blind hole in a series of steps to the specimen whose thickness is much greater than the diameter of the hole (denoted as thick);
4. measuring the resulting relieved strains whose values depend on the residual stresses existing in the material of the hole;
5. determining the residual stresses in the removed material using mathematical relations based on linear elasticity theory.

Stresses that remain approximately constant along the depth are defined as uniform stresses. If the stresses vary significantly with depth, they are known as non-uniform stresses. If the residual stress in a thin specimen material is investigated, the uniform stress measurement is specified. Both uniform and non-uniform stress measurements are specified for thick specimens. The accuracy of the results is satisfying if the residual stresses quantified in a thick specimen material do not exceed about 80% of the material yield stress. On the other hand, the residual stresses determined in a thin specimen material are less than approximately 50% of the material yield stress. Experimental measurements described in the following parts of the paper will be applied to the investigation of strains relieved in the area surrounding the hole cut into the thin specimen loaded by uniaxial tension. Due to similarity, the methodology for residual stress quantification performed in a thin specimen, drilling a through-hole as described in detail in ASTM E837-13a standard [1], can be found in Appendix A.

Many commercial devices allow control of the hole or the core milling process and evaluate the results using their own software. Over recent years, some modifications in the milling process have occurred, too, e.g., translational motion of the milling cutter has been replaced by circular motion.

Measurement techniques such as moiré interferometry, electronic speckle pattern interferometry, shearography, photoelasticity as well as digital image correlation belong to the group of non-contact experimental methods, which gradually replace conventional strain gauges used for the strain analysis at a point (or its near surroundings) and provide full-field information about the strain distribution. Some of them have already been adapted for residual stress quantification.

The moiré interferometry (MI) method works on the principle of diffraction grating interferometry, where interference fringes occur after each deformation of the object surface. The grating can be physically attached or just virtually generated on the object surface by two symmetric beams of light transmitted from the same source of coherent laser emission. Key characteristics relating to the use of MI for residual stress analysis are high displacement and spatial resolution, the possibility for micro-scaled measurements and incremental hole-drilling investigation. Ya et al. [2] designed a unique measuring system based on phase-shifted MI combined with the hole-drilling method, and analyzed the residual stresses on an aluminum plate. The use of MI in the determination of residual stresses, e.g., in the AS10OU3NG material with one surface shot-peened [2], composite materials [3], welded titanium alloy plate [4] as well as a uniaxially tensioned plate made from black polymethylmethacrylate [5] suggest the possibility of using this method in a wide range of materials.

Electronic speckle-pattern interferometry (ESPI) is based on the interference of two monochromatic laser beams, i.e., a subject beam reflected from the object surface and a reference beam creating reference speckle-effect on the image plane of the CCD (charge-coupled device) camera. The method does not require to create grid or speckles, has a high displacement resolution [6,7], allows performing measurements outside the laboratory [8], and removes the effect of rigid-body motion [9,10]. Analysis can be done on curved and rough surfaces using the incremental hole-drilling method. A portable measuring system with one symmetrical dual-beam illumination constructed by Viotti et al. [11] is another specially designed device that provides automatic calculation of residual stresses from measured displacements using the least square method. In 2017, Lothhammer et al. [12]

constructed a measuring device designed for the residual stress quantification on the resistance-welded pipes. According to the authors, their device provides a more effective way to quantify the non-uniform stress distribution obtained at the same quality level compared to the conventional techniques.

Digital image correlation (DIC) is based on the comparison of digital images captured during the loading of an analyzed object. It is not as sensitive to ambient vibrations as the MI or ESPI methods, and therefore it is suitable for use in a production environment. In all the following references, specially adapted devices combining DIC and the core- or hole-drilling technique were used to quantify residual stresses. In 2005, one of the first residual stresses measurements using DIC was carried out by McGinnis et al. [13], when the authors investigated steel plates and reached less than 7% error in normal stress, which established the robustness of 3D DIC to capture the expected displacements. In 2006, Nelson et al. [14] calculated the residual stresses from displacements using dimensionless relations derived from numerical analysis, and good correspondence with the results acquired by the holographic method was achieved. In 2008, a hole-drilling device with an integrated camera allowing measurement of displacements on the object surface of 7 × 5.6 mm was developed [15]. The accuracy of the device was tested by the analysis of compression residual stresses simulated on the pipe loaded by the testing machine, whereby the acquired results similarly correspond to the results obtained by strain gauge rosette. In 2017, Baldi and Bertolino [16] developed a low-cost residual stress measuring instrument with integrated digital image correlation (iDIC). Using the proposed approach, the rigid-body motion is easily compensated for and no decorrelation problem results from large translations. The use of 2D iDIC was described in [17], where Baldi carried out measurements on specimens made from orthotropic material. The obtained results confirm that his proposed device is as accurate as previously developed optical techniques, however, its measuring range is significantly larger. In 2017, Rief et al. [18] designed a 3D DIC system connected to the hole-drilling device by a rigid frame ensuring that there was no sideward movement of the drill during the drilling process. The advantage of their device is that it is pre-calibrated since the camera's distance to the object surface is constant. In 2020, Brynk et al. [19] also used a fixed stereo-camera system placed symmetrically to the milling cutter attached to the mechanism allowing its movement in the direction perpendicular to the specimen surface and analyzed residual stresses in the LVM316 steel. The milling process was performed by a stepper motor and steered with an Arduino microcontroller allowing precise drilling of holes to desired depth as well as removing of the drilling head from the cameras' observation field. The current research done by Babaeeian and Mohammadimehr [20] aims to analyze the influence of the time elapsed effect on the levels of quantified residual stresses in composites.

Quantitative comparison of the displacement resolution and disadvantages of the aforementioned methods are shown in Table 1.

Table 1. Displacement resolution and disadvantages of the optical methods described.

Method	Displacement Resolution	Disadvantages
MI	10 nm with a pitch of 1 μm and a phase resolution of $2\pi/100$	Requirement for a creation/installation of a grating and interferometric stability
ESPI	Typically <50 nm depending on the laser wavelength and the geometrical arrangement	Requirement for an interferometric stability
DIC	±0.01–0.02 of a pixel for the in-plane components, $Z/50{,}000$ * for the out-of-plane component	Requirement for creation of very fine speckles

* Note: Z is the distance from the object to the camera for a typical stereo-camera system.

In this paper, which is an extension of the conference contribution [21], the unique device combines hole-drilling with two optical methods and differs from the devices described previously in that the cameras of the correlation system (used as a tool for the full-field displacement/strain analysis) move with the cutting tool. As the digital image correlation method is based on the correlation of digital images, it was necessary to verify whether the achieved positioning accuracy of the proposed drilling device was sufficient for reliable strain analysis. To eliminate the human failure factor during the measurement process, control software for adjustment of several parameters was created. The most important parameters adjusted were the shift and velocity of the block to which the milling tool was attached.

The initial testing of the drilling device found undesired mechanical clearances, which could significantly influence the measured data and, therefore, the device's design was improved several times. The results obtained from testing of the optimized prototype showed that the aforementioned technical problem was eliminated.

2. Development and Testing of the Unique Hole-Drilling Device

In the development of the unique drilling device, the authors used their own experiences obtained from using commercially produced devices SINT MTS 3000 (SINT Technology, Calenzano, Italy), SINT MTS 3000 Ring-Core (SINT Technology), and RS 200 (Vishay Precision Group, Malvern, PA, USA) not only in laboratory conditions but also by solving problems in technical practice [22–24]. The common feature of the aforementioned devices is adjusting the position of the milling cutter in the center of the strain gauge rosette. Positioning of milling cutter in RS200 and SINT MTS 3000 is performed manually using X-Y adjusting screws. In SINT MTS 3000 Ring-Core, semi-automatic positioning is undertaken because the center of the milled core is adjusted through a camera, which is not further used in the residual stresses quantification procedure. After drilling the hole or the core, its real position towards the center of the strain gauge rosette has to be determined. If eccentricity occurs, its value is entered into the evaluation software for the correction of the residual stresses calculation. Analysis of the eccentricity effect in the measurement of the hole-drilling residual stresses was carried out, e.g., by Ajovalasit [25], Beghini et al. [26] or Barsanti et al. [27]. As mentioned in the introduction, the topic of residual stresses determination using optical full-field methods in combination with the hole-drilling method is very up-to-date. Many authors devote themselves to develop their own drilling devices, where the optical system is mostly stationary, and the movement is undertaken by the milling cutter. There are two types of milling tool movements used—translational performed in the milling direction or a combination of translational and rotational movement.

In this paper, the design of such a drilling device is presented in which not only the milling tool moves to cut the hole, but also the optical system can be attached, used to determine displacement/strain fields, to the moving position (Figure 1). Achieving highly accurate positioning of the drilling device's mechanical parts (like 10^{-3} mm) and desired repeatability of the measurements become crucial parameters required for the application of the above-mentioned methodology. Commercially produced drilling devices are also designed in such a way as to ensure the motion of their mechanical parts as accurately as possible.

In the design of the unique drilling device, the authors considered technical parameters of the optical systems used at the authors' workplace, namely LF/Z-2 reflective polariscope (Vishay Precision Group, Malvern, PA, USA), as well as single-camera (2D) and stereo-camera (3D) digital image correlation system Q-400 (Dantec Dynamics A/S, Skovlunde, Denmark) (Figure 1).

(a)

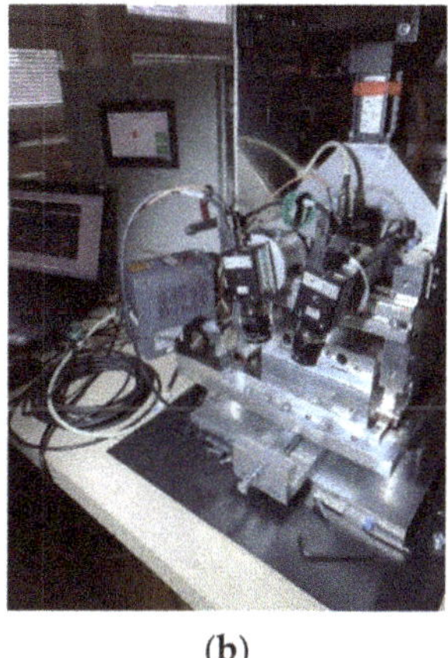

(b)

Figure 1. The prototype of the unique drilling device with: (**a**) a single-camera (2D) Q-400 digital image correlation (DIC) system and polariscope LF/Z-2; (**b**) a stereo-camera (3D) Q-400 DIC system.

2.1. Mechanical Design

When developing the drilling device, the authors were inspired by the construction of the SINT MTS 3000 Ring-Core system. However, the DIC system used in the designed device was, among other things, also used for the deformation analysis. The advantage of the strain analysis performed by the full-field optical methods is that the correction of the hole position carried out at the end of the measurement is unnecessary. If there occurs an insignificant change in the position of the drilled hole, ultimately it means that the deformation analysis will be made at a distance very close to the planned one. In the case of residual stresses determination, deviation from the analyzed location up to 1 mm does not cause a significant measurement error in the investigated stress levels. In respect of the experimental procedure, it is required to keep the essential principles valid for the commonly used hole-drilling method, e.g., regular changing of milling tools whose failure may lead to negative influenced results [28].

The designed mechanism allows motion in vertical and horizontal directions through the ball screws with pitch accuracy T5 ($\pm$0.023/300 mm). The ball screws are driven by B&R servomotors 8LVA23.B1015D100-0 controlled by servo driver 8EI4 $\times$ 5MWD10.0600-1. The servomotors have a nominal speed of 1500 rpm, a maximum speed of 6600 rpm, and a nominal torque of 1.33 Nm. The servomotors are connected to the ball screw by a flexible coupling. They use EnDat encoders with resolution 18/16 bits single turn/bits multi-turn, respectively. The pitch screw is 5 mm per round. The designed device uses a spindle with power of 850 W for drilling controlled by a frequency converter. It is possible to attach the single- or stereo-camera digital image correlation system to the lateral static part of the device, i.e., the camera will only move in the horizontal direction or to the block the milling tool is attached to. In such a case, the camera moves in two mutually perpendicular directions, i.e., horizontal and vertical.

2.2. Control System Design

The system is controlled by PLC 4PPC70.0702-20B with an implemented touch screen (power panel) for a HMI (human–machine interface). The flow of the information is shown in Figure 2. There are two working modes: manual and automatic. The manual mode allows the user to control the motion in both axes either by constant velocity motion or positioning the axis by setting the position value. This mode is used for the adjustment of the milling cutter with respect to the surface of the analyzed specimen. On the other hand, the automatic mode performs strictly predefined cycles of the drilling process defined by the ASTM E837-13a standard. In automatic mode, there are several parameters such as number of cycles, steps of drilling, required positions in both axes, delay after each cycle, velocity of drilling and the velocity of motion in both axes, which need to be set.

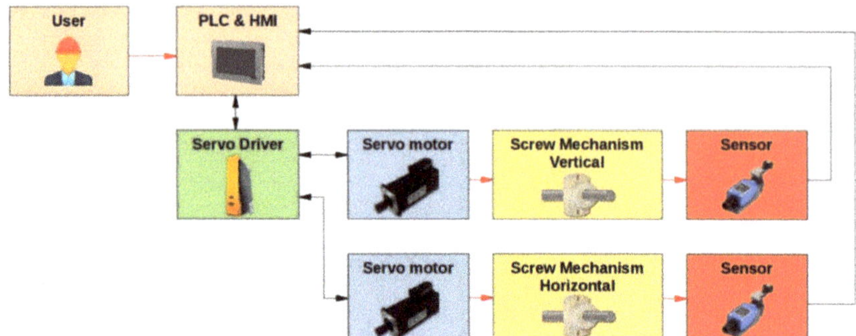

Figure 2. Information flow of control system.

The algorithm of automatic mode is as follows (Algorithm 1).

Algorithm 1 Automatic Mode

```
Setting of the required drilling parameters
Motion to the reference position in vertical and horizontal axes
WHILE (actualCycle ≠ requiredCycles)
  Motion in vertical and horizontal axes upon the pattern
  IF (milling machine = turn off) THEN
    milling machine -> turn on
  END_IF
  WHILE (drillingTime ≠ requiredDrillingTime)
    Motion in y-axis: cycle*drillingStep
    Counting of drillingTime
  END_WHILE
  Motion in y-axis: -cycle*drillingStep
  Motion in y-axis and x-axis to reference position
  Delay after the end of the cycle
  cycle = cycle + 1
END_WHILE
milling machine -> turn off
```

2.3. *Testing of the Positioning Accuracy*

The measurement process of the designed device with the DIC system is based on a mutual change in the position of the DIC system and the milling cutter. At first, the position of the milling cutter (MC) is adjusted to the analyzed area A1 of the specimen (see Figure 3). Subsequently, the milling tool is shifted to a new position A2, where distance L corresponds to the distance between the camera(s) and the axis of the milling cutter. In this position, the DIC system captures a reference image of the analyzed specimen area, after which the milling cutter is returned to A1 position, and the drilling process starts. After drilling the first step, the milling cutter moves to A2 position again, and the digital image corresponding to the first drilling cycle is captured. The process is repeated until the desired depth of the hole is achieved.

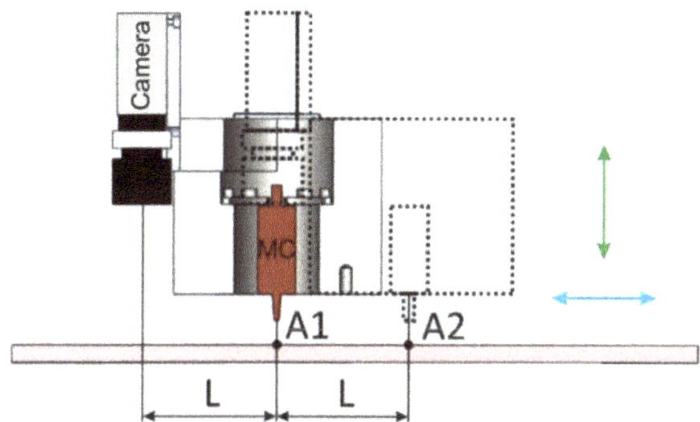

Figure 3. Principal scheme of the device positioning used for drilling of the hole and capturing digital images of the analyzed area by DIC in a series of steps.

Several methods based on different physical principles were used for the validation of the device positioning accuracy. The testing aimed to verify the real change of the milling tool position adjusted by the control software (by servo driver). The experimental procedures were performed in two stages:

- testing of the horizontal positioning (Figure 4a),
- testing of the vertical positioning (Figure 4b).

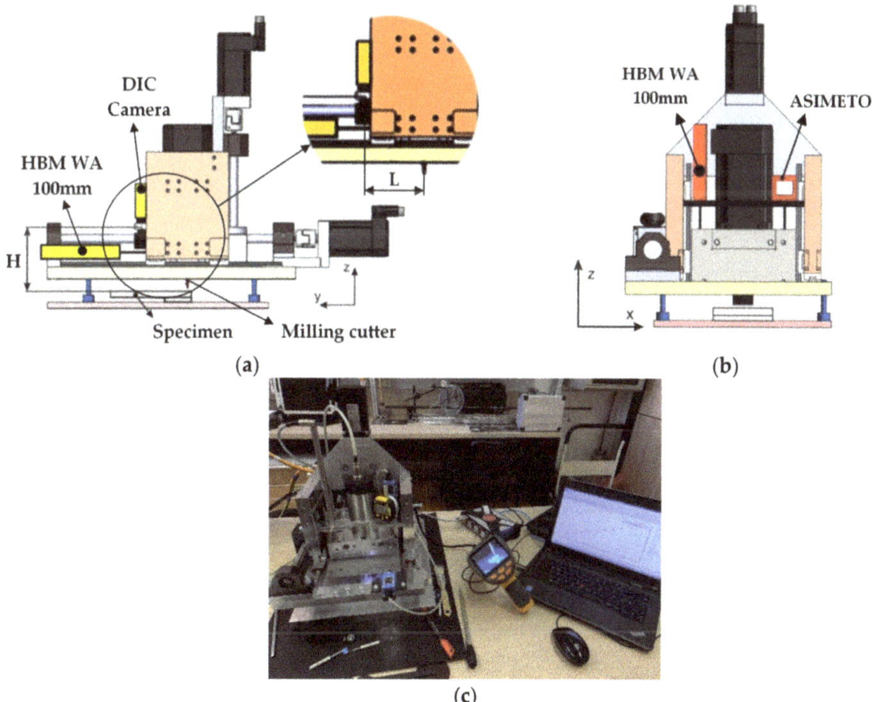

Figure 4. Testing of the positioning accuracy in: (**a**) horizontal direction; (**b**) vertical direction; (**c**) measuring string.

The measurement in which the mutual position between the camera of the 2D DIC system and the milling cutter changed 30× was performed to obtain information about the accuracy and repeatability of the device positioning. The horizontal direction shift was adjusted to the value L = 81 mm corresponding to the distance between the lens of the 2D DIC system and the milling cutter axis (Figure 3). The results obtained, which showed high accuracy of the desired value and achieved repeatability, are given in Figure 5. The detailed analysis suggested that deviations from the reference device position (after its move to a new position at a distance of 81 mm adjusted by the control software and return) were 10^{-3} mm. The same graph shows that the deviations registered by the inductive displacement transducer WA-100 mm (HBM, Darmstadt, Germany) and 2D DIC system Q-400 (Dantec Dynamics A/S, Skovlunde, Denmark) are similar.

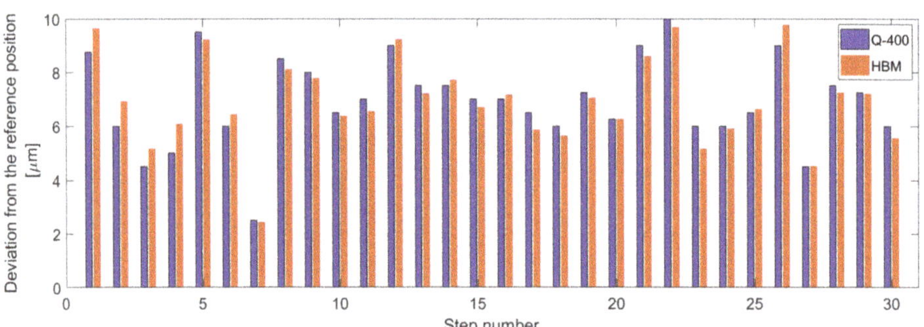

Figure 5. Deviations from the reference position of the device obtained for 30 repetitions of its movement to a new position at the distance 81 mm and subsequent return captured by the inductive displacement transducer WA-100 mm and 2D DIC.

Testing of the developed hole-drilling device positioning accuracy in a vertical direction, and analyzing the shape of the drilled hole were carried out on the specimen made from EN AW-5083 material, in which the blind holes were drilled using milling cutters provided with the RS 200 device, concretely the miller with a 3.2 mm tool diameter (Figure 6a). Measurements were performed according to the methodology set forth in the ASTM E837-13a standard, i.e., blind hole drilling up to 2 mm deep performed in 20 steps [1]. In this phase, all the steps were set up with the same increment of 0.1 mm (used to examine stresses uniformly distributed over the depth of the specimen).

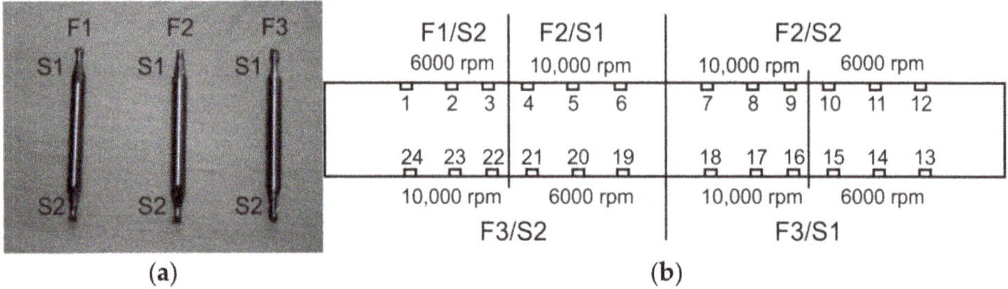

(a) **(b)**

Figure 6. Testing of the positioning accuracy: (**a**) milling cutters F1, F2, F3 with a diameter of 3.2 mm and denotation of the sides; (**b**) location of the drilled blind holes.

For the measurement of the drilling device vertical shift, the inductive displacement transducer WA-100 mm and digital indicator ASIMETO series 405 (ASIMETO, Weissbach, Germany) were used (Figure 4c). In several cases, the above-mentioned transducers captured any higher deviations towards the value adjusted in the control software than in the horizontal direction (see Table 2).

Table 2. Experimentally determined values of depths and diameters of the blind holes drilled by the designed prototype of the drilling device.

Location	Milling Cutter	Cutting Speed (rpm)	Average Value of the Blind Hole Diameter		Average Value of the Blind Hole Depth *		
			Direction X (mm)	Direction Y (mm)	HBM (mm)	ASIMETO (mm)	Microscope (mm)
1	F1/S1	6000	3.271	3.243	1.987	1.978	1.995
2	F1/S1	6000	3.276	3.250	1.995	1.990	1.994
3	F1/S1	6000	3.235	3.255	2.009	2.015	2.006
4	F2/S1	10,000	3.274	3.281	2.000	1.995	2.004
5	F2/S1	10,000	3.257	3.257	1.989	1.991	2.003
6	F2/S1	10,000	3.252	3.259	1.998	2.003	2.000
7	F2/S2	10,000	3.200	3.203	2.005	1.993	2.007
8	F2/S2	10,000	3.202	3.202	2.001	2.004	2.004
9	F2/S2	10,000	3.200	3.202	2.004	2.001	2.003
10	F2/S2	6000	3.201	3.205	2.006	2.006	2.005
11	F2/S2	6000	3.202	3.199	2.021	2.014	2.003
12	F2/S2	6000	3.199	3.201	2.010	2.008	2.001
13	F3/S1	6000	3.245	3.246	1.999	1.998	2.000
14	F3/S1	6000	3.249	3.225	2.001	1.996	2.001
15	F3/S1	6000	3.251	3.246	2.003	2.010	1.993
16	F3/S1	10,000	3.242	3.255	2.011	2.015	2.003
17	F3/S1	10,000	3.243	3.239	2.008	2.007	2.005
18	F3/S1	10,000	3.214	3.238	2.013	2.004	2.001
19	F3/S2	6000	3.257	3.245	2.006	2.008	2.006
20	F3/S2	6000	3.242	3.240	2.008	2.001	2.000
21	F3/S2	6000	3.237	3.252	2.012	2.020	1.999
22	F3/S2	10,000	3.252	3.255	2.005	2.011	1.999
23	F3/S2	10,000	3.251	3.255	2.007	2.009	2.003
24	F3/S2	10,000	3.243	3.264	2.011	2.010	2.002
AV **	-	-	3.237	3.238	2.004	2.004	2.002
SD ***	-	-	0.0145	0.0130	0.0013	0.0021	0.0003

* Note: The final depth of the drilled blind hole adjusted by the control software was 2 mm. ** AV—Average value (mm), *** SD—Standard deviation (mm).

The analysis of dimensions was carried out on a series of drilled blind holes (Figure 6b), whereby the influence of the cutting speed and the milling cutter wear was reviewed. It was experimentally proved that the developed drilling device can cut blind holes in accordance with the requirements specified in ASTM E837-13a standard. Examination of the shape of blind holes, i.e., analysis of their circularity achieved after the last (twentieth) drilling step and their cylindrical shape along the depth was undertaken using a specially adapted microscope TM-505B (Mitutoyo, Kanagawa, Japan), (Figure 7). The averaged distances of two antipodal points measured three-times in two mutually perpendicular directions are listed in Table 2.

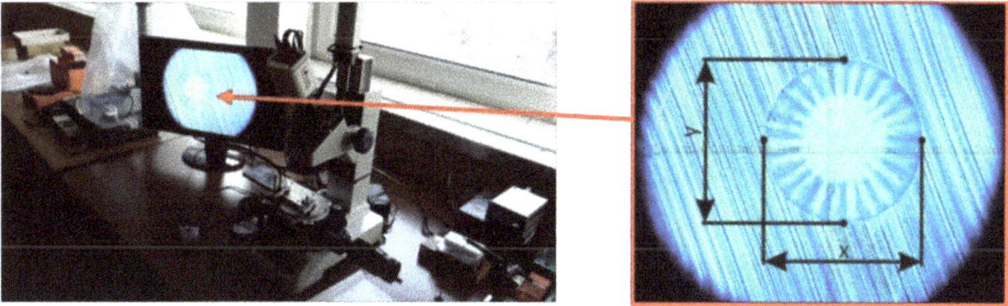

Figure 7. Measurement of the blind holes circularity in two mutually perpendicular directions.

The circularity of the cut holes was validated by checking the cylindrical part shape of the blind holes carried out on a split specimen using a specially adapted microscope TM-505B (Figure 8). These measurements also provide information about the real depth of the drilled blind hole, which was set to 2 mm by the control software. Table 2 shows that the real depth (registered by microscope) of the drilled blind hole is 2.002 ± 0.0003 mm and, thus, the achieved accuracy is equal to the top commercially produced drilling devices.

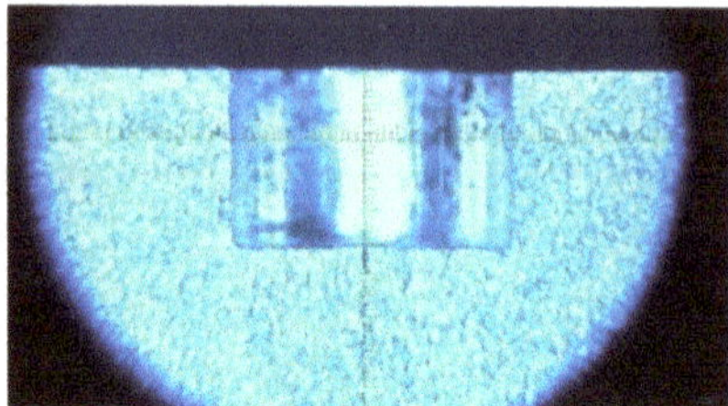

Figure 8. Measurement of the blind holes depth performed on the split specimen.

It can be stated that if the control of the vertical shift will be realized by the inductive displacement transducer WA-100 mm, the real depth of the drilled blind hole could be assessed based on the results in Table 2. It has to be noted that the control measurements of the depth of the drilled hole using an electron microscope cannot be used for real structures.

3. Experimental Measurements

After performing testing measurements with satisfying results, the authors carried out analysis of the displacement/strain fields evaluated in the vicinity of the hole cut by the designed prototype of the drilling device into the testing specimen loaded by a known loading.

The full-field strain analysis was undertaken on the specimen made from the PS-1D material (Vishay Precision Group, Malvern, PA, USA) loaded by uniaxial tension loading (Figure 9a) registered by the RSCC-50 kg sensor (HBM, Darmstadt, Germany) (Figure 9b). The given mechanical/optical properties of the analyzed strain-sensitive plastic coating are: Young's modulus of elasticity $E = 2500$ MPa, Poisson's ratio $\mu = 0.38$, thickness $t = 0.5$ mm and the fringe value of coating $f = 3790$ µm/m/fringe. The dimensions of the specimen are given in Figure 9a. The through-hole lying on the specimen's longitudinal axis was cut when the specimen was loaded by the uniaxial tension force of 250 N, and the relieved strains were observed using a polariscope based on the PhotoStress method as well as 3D DIC system. Measurement performed on any specimen in the near vicinity of the cut through-holes (due to the specimen's thickness) was performed under the same cutting conditions (cutting speed and velocity of horizontal and vertical positioning). The advantage of such a measurement method is that the results in the form of the relieved strain fields obtained by two different optical methods are observed on the specimen made from the same material and can be further analyzed (compared).

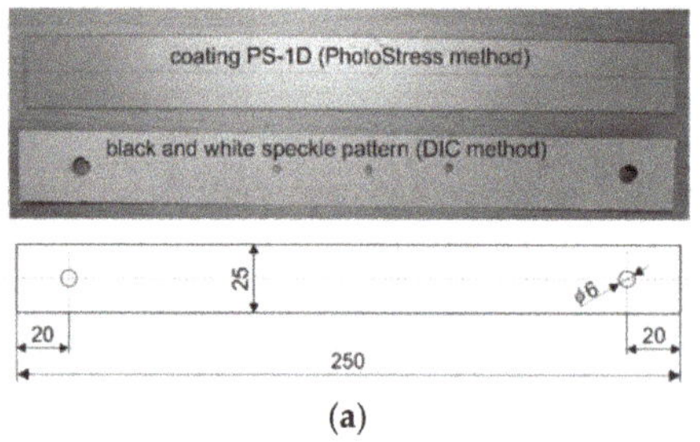

(**a**)

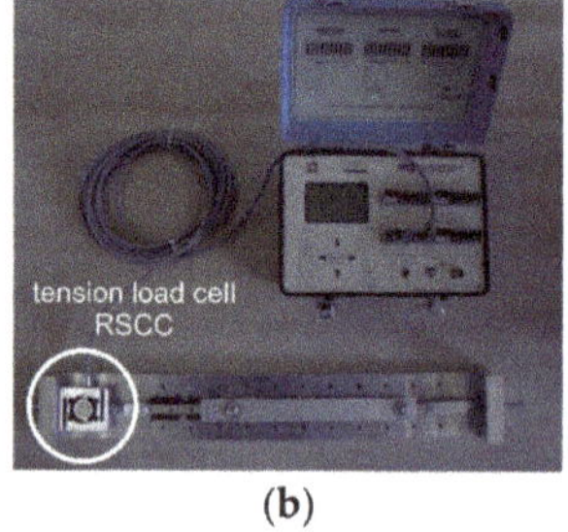

(**b**)

Figure 9. Testing specimen made from the PS-1D material: (**a**) the overall view with dimensions; (**b**) loading mechanism.

3.1. PhotoStress Method

The PhotoStress method is a non-contact full-field measurement technique used to determine surface strains and transform them into stresses occurring in a structure. To use this method, a special strain-sensitive plastic coating needs to be bonded to the analyzed structure. Subsequently, after applying the loading to the structure, its strains are transmitted to the coating assuming the same strain condition as the part of the structure, which it is bonded to. After illuminating the coating by polarized light emitted from a reflection polariscope, a colorful pattern that is viewed through the polariscope occurs due to strain. There are two types of color pattern investigated known as isoclinic and isochromatic fringes. At every point on an isoclinic, the directions of principal strains are parallel to the direction of the analyzer and polarizer polarization. The photoelastic strain pattern appears as a series of successive and contiguous, variously colored bands known as isochromatic fringes representing a different level of the principal stresses difference [29,30]. When the loading is applied to the structure, the coated part color changes from black characterizing the no-loading state of the structure and first colors appear in the areas of the highest stress. After increasing the loading, the color fringes spread throughout the coated part (Figure 10), additional fringes are generated in the highly stressed areas of the investigated structure and move towards the regions of zero or low stress until the maximum load is achieved.

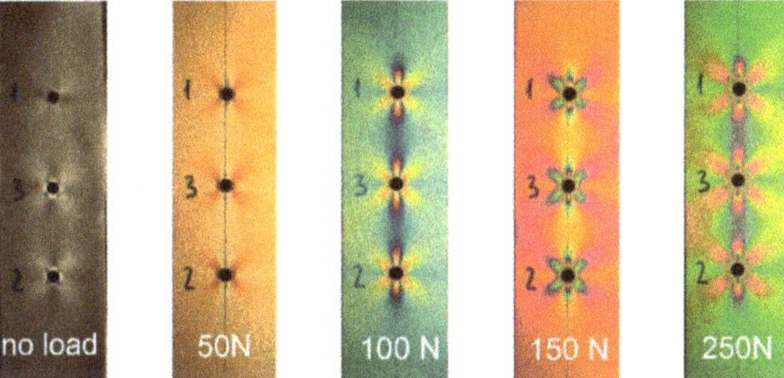

Figure 10. Characteristic change of the isochromatic fringes observed in PhotoStress method during increasing uniaxial loading.

The quantitative analysis of principal strain difference (strain intensity) occurring at any point of the coating can be quickly performed using a digital compensator attached to the polariscope. Strain intensity analysis depends only on the recognition of the fringe order defined by a color and understanding of the relationship between the fringe order and strain intensity as follows:

$$\varepsilon_1 - \varepsilon_2 = N \cdot \frac{\lambda}{2 \cdot t_c \cdot K} = N \cdot f \quad (1)$$

where $\varepsilon_1, \varepsilon_2$ are the principal strains, N is the fringe order, λ is the wavelength of the light emitted by polariscope, t_c is the thickness of the coating, K is the strain optical coefficient of the coating and f is the fringe value of coating.

The advantage of the PhotoStress method is its high sensitivity for small strain/stress levels. On the other hand, in the areas with high strain/stress gradient, a problem can occur by quantifying the values (for the higher-order fringe recognition, the microscope is required to be used).

3.2. Digital Image Correlation Method

The deformation analysis was also carried out by a low-speed digital image correlation system Q-400 (Dantec Dynamics A/S, Skovlunde, Denmark) working on the principle of the DIC method (some detailed information are presented in Appendix B). The digital image correlation principle is based on the correlation of digital images captured during loading of the analyzed object, where the images are not compared as whole units, but as small image elements called facets. The shape of the facet used by correlation systems Dantec Dynamics is squared, and each facet commonly comprises a group of pixels ranging in size from 15 × 15 px to 30 × 30 px. Depending on the type of analysis, the facet size can be adjusted (reduced or enlarged). The facets may touch or overlap, but there must not be an empty area between them. Since the information about the displacements of the analyzed object is obtained at the nodes of the virtual grid, which correspond in position to the centers of the facets, the overlap of the facets (Figure 11) is one way to increase the data resolution, i.e., to obtain a more considerable amount of data and, thus, to better reconstruct the surface of the analyzed object (especially around the edges). The manufacturer of correlation devices, Dantec Dynamics, recommends overlapping the facets up to 1/3 of the facet size because with such an overlap the data points are still independent.

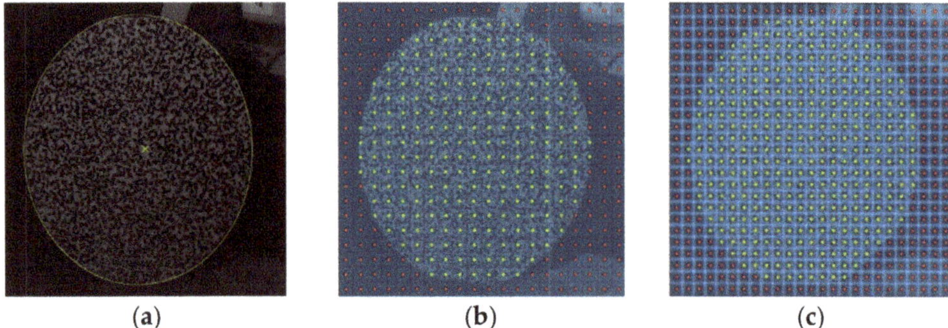

Figure 11. Illustrative example of the increase of the data resolution caused by the facet overlap: (**a**) defined area of interest; (**b**) measuring points (green dots) when the facets are in touch, (**c**) measuring points (green dots) when the facets are overlapped.

Dantec Dynamics correlation devices use an algorithm based on a pseudo-affine transformation to obtain information on the transformation coordinates of the analyzed object surface points. If transformation parameters of possible displacement, elongation, shear, and distortion of the facet $a_0 - a_7$ as shown in Figure 12 are considered, using the aforementioned algorithm the transformation coordinates $(u,\ v)$ can be calculated as follows:

$$\begin{aligned} u(a_0, a_1, a_2, a_3, \tilde{x}, \tilde{y}) &= a_0 + a_1 \cdot \tilde{x} + a_2 \cdot \tilde{y} + a_3 \cdot \tilde{x} \cdot \tilde{y}, \\ v(a_4, a_5, a_6, a_7, \tilde{x}, \tilde{y}) &= a_4 + a_5 \cdot \tilde{x} + a_6 \cdot \tilde{y} + a_7 \cdot \tilde{x} \cdot \tilde{y}, \end{aligned} \tag{2}$$

where $(\tilde{x}, \tilde{y})^T$ are the lens-distorted 2D coordinates of the point in the normalized image plane [31].

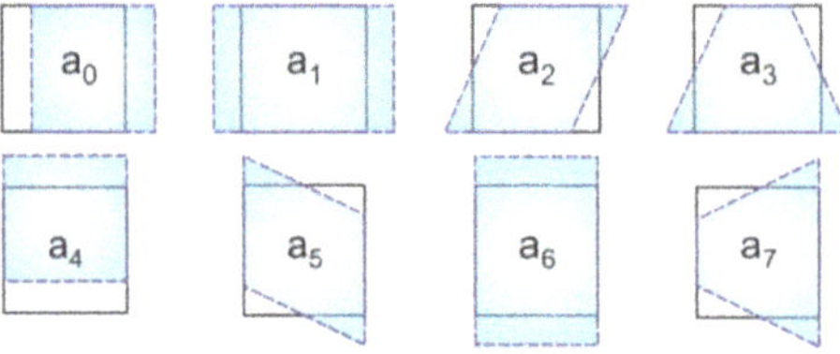

Figure 12. Transformation parameters used in the algorithm based on pseudo-affine transformation.

There are some relevant differences between the measurements of relieved deformations performed by 2D DIC and 3D DIC, respectively. Firstly, if the hole-drilling causes normal displacement of the analyzed specimen surface (this displacement component cannot be analyzed by a single-camera system), it leads to a correlation error and affects the analysis results. Another advantage of the 3D correlation system is the possibility to measure all the three displacement components not only of a flat but also curved specimen. Higher distortion of the lenses, which are not directed perpendicularly to the analyzed object surface, can be considered a small drawback of the 3D systems. Also, the requirement for a proper arrangement of the cameras towards the analyzed object means that the analysis performed by the 3D correlation system is done from a greater distance to the specimen. However, this limitation can be reduced using quality CCD cameras with sufficient image resolution of the sensors.

The accuracy of the results of the deformation analysis for both types of the analysis (2D or 3D) can be affected, for example, by:

- The quality of the speckle-pattern created on the analyzed object surface and illumination

For correct image correlation, it is essential to make sure that each facet is unique, i.e., it contains a unique pixel distribution with different levels of intensity of grey color. As the aim of the analysis was to use image resolution of the sensors as well as possible and to perform evaluation of the results near the edge of the drilled hole, it was necessary to create the black-and-white speckle-pattern with very fine speckles (Figure 13). The illumination of the specimen surface should be homogenous and reach high sharpness and sufficient contrast of speckles. Compliance with this requirement was ensured using the Dedolight DLH400DT (Dedo Weigert Film GmbH, München, Germany) halogen reflector with white light.

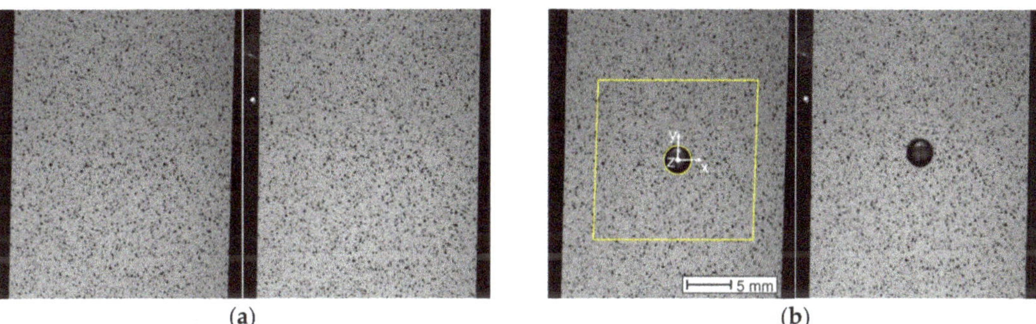

Figure 13. Pairs of digital images captured by 3D Q-400 Dantec Dynamics: (**a**) reference image; (**b**) evaluated image with a defined mask of evaluation.

- Calibration parameters

The correctness of the coordinates transformation of the object points from 3D world coordinate system into 2D sensor system depends on the accuracy of the obtained external (i.e., mutual position and rotation of the cameras) as well as internal (focal length, principal point coordinates, tangential and radial distortion of the lenses) calibration parameters of the cameras. The calibration of Dantec Dynamics correlation systems is automatized, and based on the Zhang algorithm [32]. Information about the calibration accuracy is provided by the so-called calibration residuum, whose value should not exceed 0.5 px. The size of the calibration target should approximately correspond to the size of the analyzed specimen. Therefore, calibration of the stereo-camera correlation system was done using a calibration target comprising 9 × 9 checkboard fields with a precisely defined distance of 3 mm. The residuum obtained by the cameras calibration reached the value of 0.279 px, which is, according to the aforementioned criterion, considered as accurate calibration.

- Correlation parameters and the related levels of smoothing

Both Q-400 Dantec Dynamics cameras captured the digital images with image resolution of 1800 × 2056 px. The pixel density was approximately 64 px/mm. During the analysis a couple of reference images (Figure 13a) captured by the maximum loading force of 250 N were taken in the Istra4D ver. 4.3.0 control software. The pair of images capturing the deformation of the analyzed specimen surface area after drilling of the hole (Figure 13b), was correlated with the reference one according to the following aspects. The facet size as one of the correlation parameters set up by the evaluation of the measurement was adapted to the fact that the drilled hole's edge would be reconstructed as accurately as possible if the facet was as small as possible. However, facets are required to be unique, i.e., all the facets have to comprise randomly distributed pixels with a high range of gray values. Therefore, the measurement was evaluated with the facet size set to 23 × 23 px, and their overlapping of 6 px ensured the increase of the data points (points in which the displacements and strains were evaluated) resolution.

The results of the deformation analysis performed by the Dantec Dynamics correlation systems are due largely to the properly set smoothing level. Istra4D ver. 4.3.0 contains two types of smoothing. The first known as *local regression* should be used mainly in cases if a high deformation gradient is expected in the deformation analysis results. The second is called *smoothing spline* and is used to smooth the deformation field with approximately homogeneously distributed deformation levels. As with the hole-drilling method the stress concentrator occurs in the specimen, the authors used a filter of local regression. Several studies described, e.g., in [33,34] were conducted on the proper setting of the local regression level. For the illustration, the authors point to the effect of local regression in Figure 14, which shows the relieved strain intensity fields obtained by different levels of smoothing. Figure 14a shows the relieved strain intensity field obtained by default settings of smoothing, i.e., without smoothing. The other three relieved strain intensity fields correspond to the results obtained by the settings of kernel size to 7 × 7 (Figure 14b), 15 × 15 (Figure 14c), and 31 × 31 (Figure 14d).

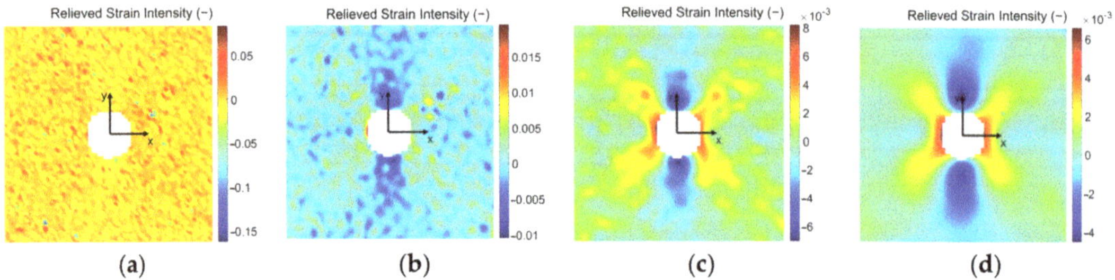

Figure 14. Influence of smoothing on the strain intensity field obtained by Q-400: (**a**) without loading; (**b**) kernel size of 7 × 7; (**c**) kernel size of 15 × 15; (**d**) kernel size of 31 × 31.

According to the manufacturers of Dantec Dynamics correlation systems, the strains are computed only from the local curvatures of facets when the kernel size is set to 3 × 3. The higher the kernel size, the higher the influence of the deformation gradient on the strains. For the kernel size set to 9 × 9 up to 31 × 31 (i.e., the highest level of smoothing based on local regression in Istra4D ver. 4.3.0), the strains are computed only from the deformation gradients. According to the authors' analysis, the kernel size of 15 × 15 for which the results obtained are presented in Figure 14c corresponds to the optimal setting of smoothing for the described type of evaluation. The effect of oversize smoothing can be observed in Figure 14d.

4. Results

Although the aforementioned measurement was conducted by two different non-contact optical methods providing the full-field information about the strains, the quantitative comparison of the results obtained from both techniques is not simple. While the results of the measurement by the DIC method provide quantitative information about the displacements as well as strains in the center of each facet, for the quantification of the results obtained using PhotoStress method, the use of a digital compensator is necessarily required in each evaluated location of the specimen.

In many cases, experimental testing results serve to verify the results obtained numerically, e.g., using software based on the finite element method (FEM) [35]. The above-described strain analysis in the vicinity of the hole drilled in the flat specimen loaded by uniaxial tension, is a typical analysis performed by the finite element analysis (FEA). For that reason, the authors reversed the standard approach and, thus, verified the experimentally obtained results by numerical analysis. Such an approach has already been used by the authors, and the results obtained were published in several scientific publications, e.g., [36–39]. Moreover, in conventional quantification of residual stresses by the hole-drilling technique the FEA is usually used, e.g., the correlation parameters $\overline{a}$, $\overline{b}$ used in the formulas for the computation of stress components from the relieved strains are determined mainly in a numerical way.

4.1. Finite Element Analysis

The numerical model of the analyzed specimen was created in Abaqus/CAE 2020 (SIMULIA, Johnston, RI, USA) software. The analysis was focused on the determination of the relieved displacements/strains occurring in the specimen (with dimensions, mechanical properties, constraints and loading described in head 3) after milling a through-hole lying on its longitudinal axis. Detailed information on the procedure of numerical analysis is presented in Appendix C.

4.2. PhotoStress Method

The relieved strain analysis performed in the near vicinity of the cut through-hole was performed by studying isochromatic fringes distribution observed on the specimen loaded by uniaxial tension. The isochromatic fringes captured after the milling of the through-hole lying on the longitudinal axis of the specimen loaded by the tension force of 250 N are shown in Figure 15a.

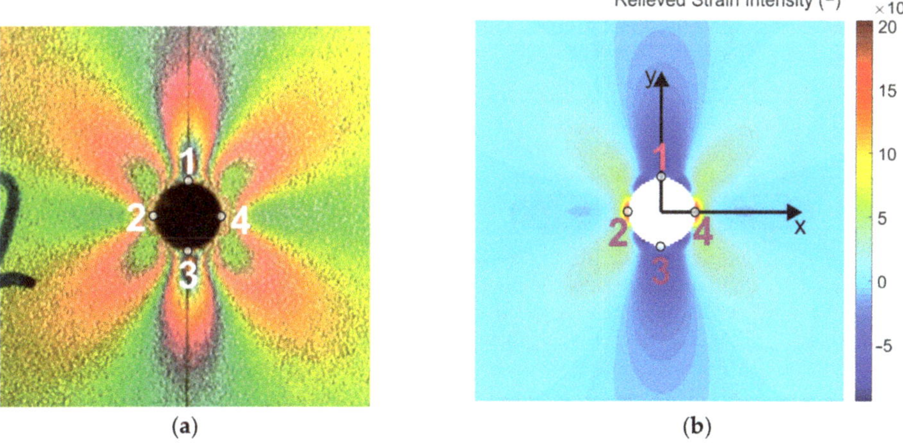

Figure 15. Relieved strain intensity analysis: (**a**) field of isochromatic fringes (PhotoStress); (**b**) finite element analysis (FEA).

Validation of the experimentally obtained results was done by comparison of the isochromatic fringes (interpreting the principal strain difference/strain intensity) with the strain intensity field obtained in Matlab (Figure 15b) as a consequence of stress-relieving (caused by the milling of the through-hole). The results obtained by the numerical analysis are symmetrical along the longitudinal axis of the specimen. However, the numerical analysis is carried out on an ideal model and with ideal boundary conditions. By comparing the results, the influence of the accuracy of the drilled through-hole on the measured values can be confirmed.

Four points (marked as 1, 2, 3 and 4) were chosen at the edge of the drilled through-hole (Figure 15a), in which the fringe order was assessed using digital compensator model 832 (Figure 16). Values of the relieved strain intensity in the aforementioned four points given in Table 3 were calculated according to Equation. (1).

(**a**)

(**b**)

Figure 16. The use of the digital compensator: (**a**) isochromatic fringes observed after cutting the through-hole using the compensator; (**b**) fringe order assessment.

Table 3. Comparison of the values of relieved strain intensity obtained in four selected points located at the circumference of the drilled through-hole with a diameter of 3.2 mm.

Point	PhotoStress		FEA
	Fringe Order	$\varepsilon_1 - \varepsilon_3$	$\varepsilon_1 - \varepsilon_3$
1	1.21	−0.00459	−0.00112
2	3.26	0.01235	0.02039
3	1.20	−0.00455	−0.00112
4	3.01	0.01141	0.0204

As the process of fringe order assessment is not automatized, correct determination of the fringe order value is significantly dependent on the practical skills and experience of the experimenter. According to Table 3, it is evident that the differences in the obtained results are significant. As the highest gradient of relieved strain intensity occurs at 0.1D distance to the edge of the drilled through-hole (D = 3.2 mm is the diameter of the drilled through-hole), the difference in results can also be caused by the assessment of the fringe order near the edge of the hole. To obtain more precise results, it would be necessary to arrange the elements allowing micro-scaled measurement into the measuring device or to use the digital photoelasticity method, which results of application are described in detail, e.g., in the paper of Ramesh and Sasikumar [40]. As the authors' workplace does not dispose of any of the aforementioned possibilities, the quantitative results obtained by the PhotoStress method need to be considered only as additional.

The second (comparative) analysis in the surrounding of two other through-holes drilled to the same specimen loaded by the identical uniaxial loading was realized to qualify the sensitivity of the PhotoStress method. According to Figure 17, it is evident that the through-hole denoted as 1 was drilled on the longitudinal axis of the specimen loaded by the uniaxial tension. The center of the through-hole denoted as 2 was moderately biased from the axis of the specimen (see Figure 17b). At first glance, the isochromatic fringe patterns observed in the vicinity of both through-holes seem to be identical. The moderate differences in color patterns (mainly in areas near the edges) can be observed through a thorough visual analysis. In the vicinity of the through-holes, the differences are not significant.

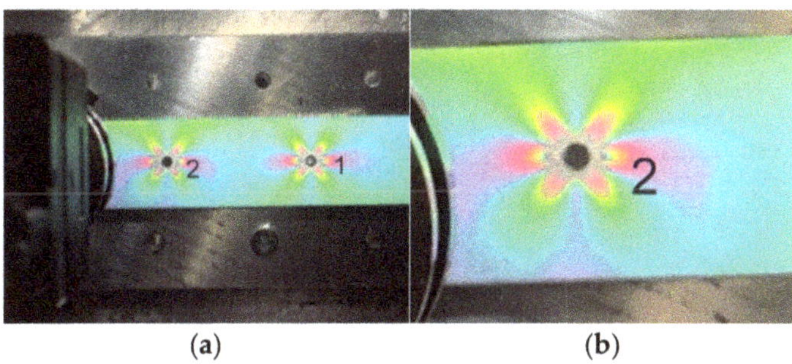

Figure 17. Demonstration of the PhotoStress method sensitivity—the through-hole denoted as 2 is not drilled at the axis of the specimen: (**a**) the overall view; (**b**) the detailed view.

Regular evaluation of the strain/stress analysis performed in the vicinity of the stress concentrator (close to of the drilled hole) is necessary to investigate residual stresses using the PhotoStress method. It can be stated that the results obtained from the optical method in combination with the hole-drilling method are directly dependent on the technical factors of both experimental techniques.

4.3. Digital Image Correlation (DIC) Method

The relieved strain fields obtained from the analysis performed by Q-400 Dantec Dynamics and described in part 3.2 are shown in Figure 18a. Concerning the principle of Dantec Dynamics correlation systems, which evaluate the data at the centers of facets, it is not possible to reconstruct the contour of the drilled hole having a diameter 3.2 mm. The hole obtained by the correlation of the images using the correlation parameters described above is of a diameter of approximately 4 mm. To compare the results obtained experimentally and numerically both in qualitative and quantitative way, it was necessary to remove the corresponding elements lying on the circles concentric to the edge of the hole from the resulting strain fields and to visualize them again (Figure 18b). As can be seen in Figure 18, the strain fields' distribution obtained by DIC corresponds with the results obtained numerically.

As the sensitivity of the DIC method is not as high as the PhotoStress method, the smaller the level of strain relieved is the higher differences in the results can be expected. In the case of small strains, as the drilled through-hole causes the stress concentration, it is convenient to perform the analysis as close as possible to the edge of the through-hole, where the highest levels of strains are relieved. On the other hand, the full-field comparison of the results obtained by FEA and DIC (Figure 19c), respectively, shows that the highest absolute difference between the results occurs just at the vicinity of the reconstructed through-hole edge.

For the quantitative analysis of the results obtained by DIC and FEA, 4 points (marked as 1, 2, 3, and 4) lying on the circumference of the circle of radius 4.5 mm were chosen (Figure 20).

The strain intensity values obtained in selected points for 4 different levels of smoothing (see Figure 14) are listed in Table 4. As can be seen, the difference in the resulting values is significant. However, using the local regression smoothing with the kernel size set up to 15 × 15 (according to the previous investigations of the authors, such an adjusted level of smoothing should be the optimal one for the facet size of 23 × 23 px and their overlapping of 6 px) the relative differences between the results obtained in four selected points by FEA and DIC achieved values of approximately 14%.

Table 4. Comparison of the relieved strain intensity obtained in four chosen points using DIC by 4 different levels of smoothing and FEA.

Point	Relieved Strain Intensity				
	DIC				FEA
	Without Smoothing	Kernel Size 7 × 7	Kernel Size 15 × 15	Kernel Size 31 × 31	
1	−0.01095	−0,00734	−0.00685	−0.00296	−0.00793
2	0.00327	−0.00058	0.00347	0.00374	0.00307
3	−0.00394	−0.00831	−0.00692	−0.00253	0.00793
4	0.00639	0.00089	0.00326	0.00327	0.00307

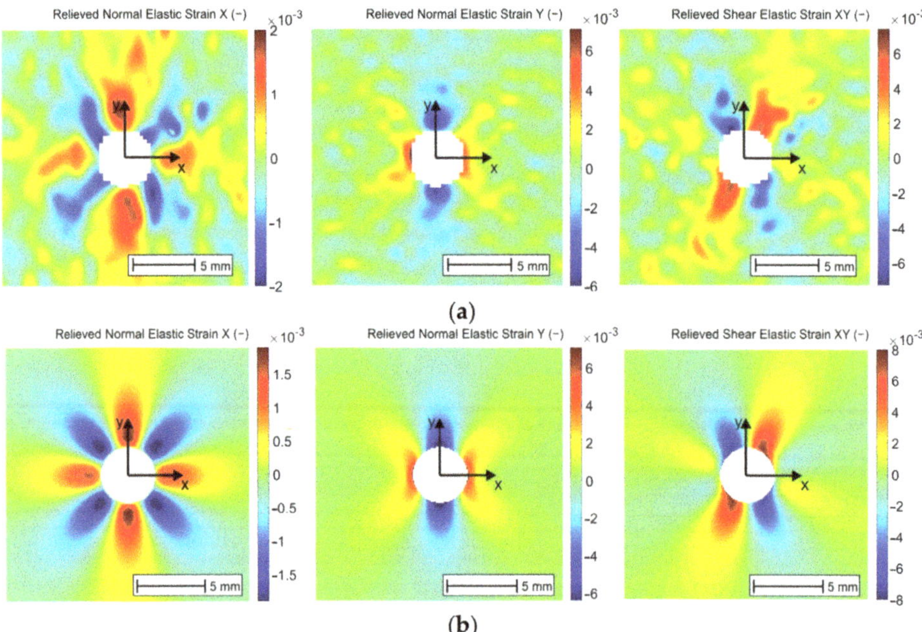

Figure 18. Qualitative comparison of the results obtained by (**a**) DIC; (**b**) FEA.

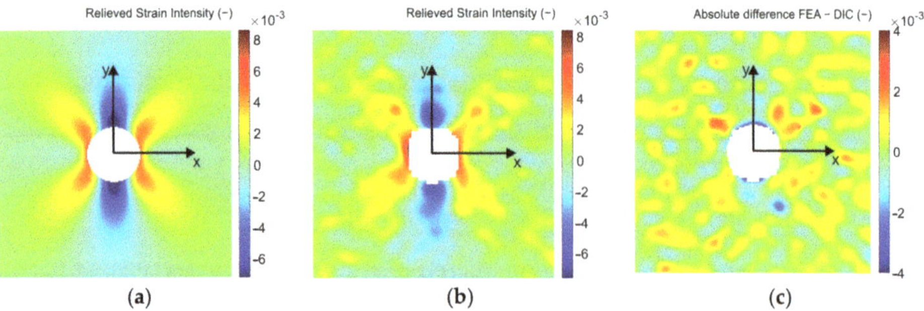

Figure 19. Relieved strain intensity fields obtained by: (**a**) FEA; (**b**) DIC; (**c**) absolute difference between the results obtained by FEA and DIC.

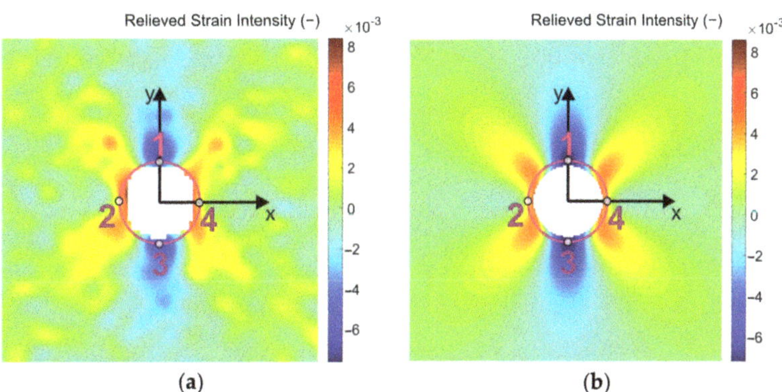

Figure 20. Location of 4 chosen (comparator) points and the comparison of the relieved strain intensity fields obtained by: (**a**) DIC, (**b**) FEA.

Another possibility to minimize correlation errors is to take into account the displacement fields (Figure 21), which are not so significantly affected by noise occurring by small deformation levels as the strains. The possibility of filtering out the effect of rigid-body motion is also one of the advantages of work with the displacement fields of the DIC method. While 2D DIC allows obtaining two displacement components in the analyzed object surface plane, 3D DIC provides displacements in three mutually perpendicular directions. One of the methodologies for calculation of the residual stresses from the displacement fields was developed by Makino and Nelson in 1994 [41].

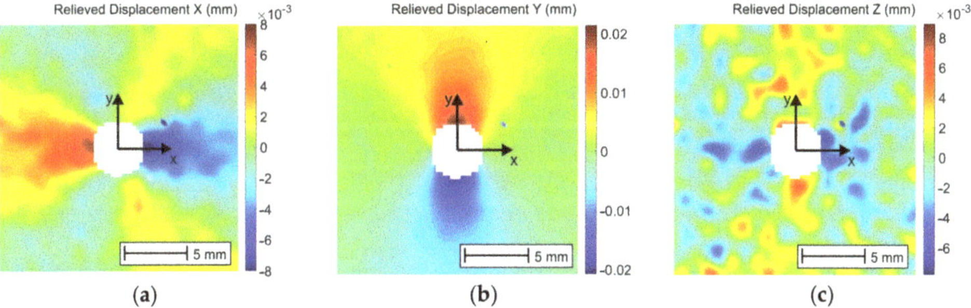

Figure 21. Relieved displacement fields obtained after filtering out the effect of the rigid-body motion: (**a**) displacement X, (**b**) displacement Y, (**c**) displacement Z.

The full-field comparison of the results obtained by FEA and DIC was carried out on the relieved displacement total fields (Figure 22). Figure 23a shows their absolute difference field. As most of the developed methodologies are based on quantifying residual stresses from displacements, it was necessary to determine the relative deviation of the measured data from the reference data, i.e., FEA data. For that reason, the mean relative differences between the relieved total displacements obtained in points located at the area of concentric annuli of width 0.5 mm (see Figure 23a) were calculated as follows:

$$diff_{relative}^{mean} = mean\left(abs\left(\frac{dispT_{FEA}(i,j) - dispT_{DIC}(i,j)}{dispT_{FEA}(i,j)}\right)\cdot 100\%\right) \quad (3)$$

where i, j are the coordinates of the corresponding points, in which the data were compared.

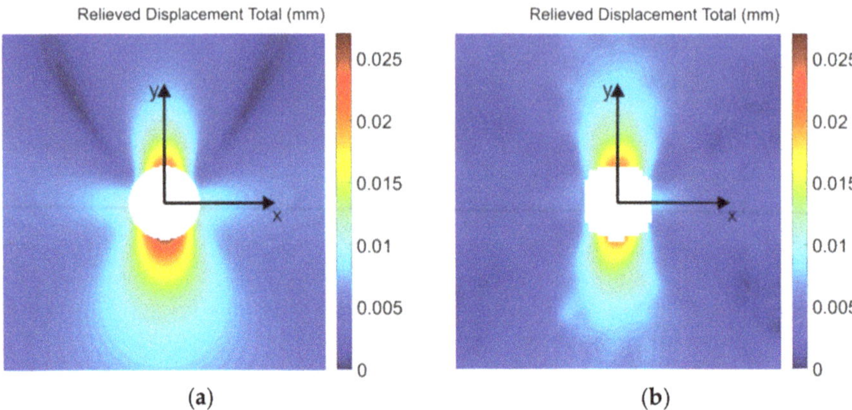

Figure 22. Relieved displacement total fields obtained by: (**a**) FEA, (**b**) DIC.

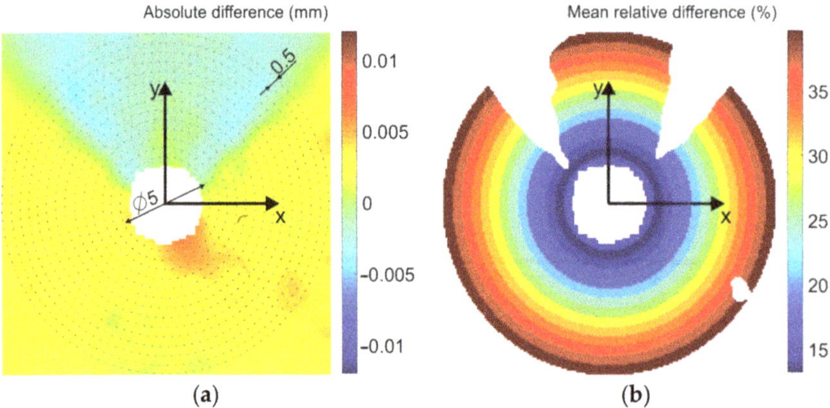

Figure 23. Difference between the relieved displacement total fields obtained by FEA and DIC: (**a**) absolute difference, (**b**) mean relative difference.

To avoid the results being influenced, the significant relative difference occurred (mainly in two areas, where the results obtained by FEA approached zero) were removed from the calculation (see the empty areas in Figure 23b). It can be stated that the relative difference increases with the distance from the center of the cut through-hole. However, for the area of diameter approximately 9 mm, the maximum relative difference was 21% corresponding to the accuracy of DIC achieved by residual stress quantification [42].

In some measurements (mainly by the repeated use of the milling cutter) performed by the authors, a small part of the speckle-pattern located near the edge of the drilled through-hole was corrupted (see Figure 24).

As this phenomenon was not visible with the naked eye (the authors observed it only in the digital images obtained by the DIC system), the cause was analyzed using scanning electron microscope FE SEM MIRA 3 (TESCAN, Brno, Czech) by 60× magnification (Figure 24c). It was found that the worn cutting miller caused small indents at the edge of the through-hole cut into the coating made from PS-1D material and damaged the background (white) color layers. Correlating such problem areas can lead to correlation problems/errors. For this case, the authors recommend that the evaluation mask is defined outside the damaged area.

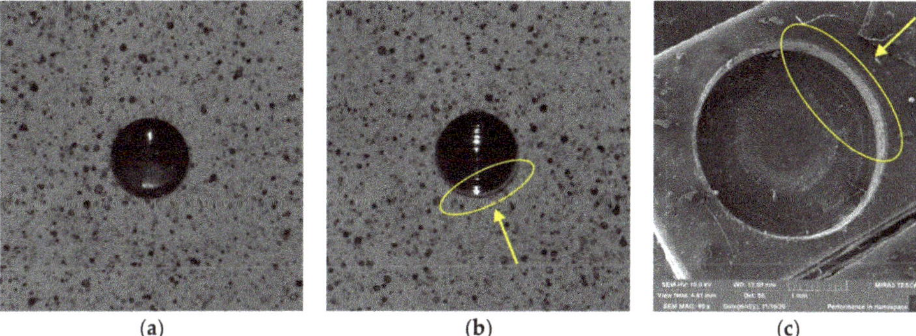

Figure 24. Milling the through-hole to the coating: (a) correctly milled through-hole without damage of the speckle-pattern; (b) damage of the speckle-pattern near the edge of the milled through-hole caused by the worn cutting miller; (c) analysis of the through-hole surrounding using an electron microscope.

5. Discussion

In this paper, the development of the device designed for the strain and stress analysis performed on the specimen made from various kinds of material (e.g., plastics, metals, composites, etc.) using the hole-drilling technique and combining the full-field optical methods (PhotoStress, DIC) is described. As the technique of digital image correlation is based on the correlation of digital images captured during the measurement process, high positioning accuracy is required. Through the experimental testing, the measurement series were performed and showed the following results:

- positioning accuracy reached like 10^{-3} mm in both horizontal and vertical direction,
- high repeatability of the measurement comparable to the repeatability obtained with top commercially produced drilling devices (see Figure 5).

As most of the authors do not provide any detailed information about their devices, to compare our device with the devices developed and described in the introduction to this paper is a relatively complex issue. Despite it, the pros and cons of the designed device can be mentioned. The main pros are:

- the measurement realized using two optical systems (based on PhotoStress and DIC method) simultaneously, which can lead not only to the verification of the results but also to further improvements of methodology for the residual stresses quantification;
- quick and accurate change of the DIC system position towards the analyzed specimen allowing capturing of the images from the desired distance;
- the possibility to analyze and compare the results obtained by using single- and stereo-camera DIC system, which can improve the methodology for the realization of measurements by the DIC method.

The dimensions of the designed device can be considered as its disadvantage. Compared with the commercially produced devices, which have been developed and optimized for many years, our prototype is more massive and, thus, portable and manipulable with more difficulty. The reason for this is that our drilling device's design has considered the minimum required dimensions of the working space. The measuring systems (single- or stereo DIC system, PhotoStress—reflective polariscope) work on optical principles and, thus, the analyzed specimen needs to be illuminated properly. The proposed prototype of the hole-drilling device was optimized so that the negative influence of the mechanical clearances was eliminated. In the future, the optimization of its dimensions in terms of weight minimization (by maintaining the sufficient stiffness requirement) is planned. To minimize the other components (servomotors, linear guides, etc.), it is necessary to consider their utility by operating load.

The methodology of evaluating the results in the experimental part by comparing the relieved strain fields obtained by the optical methods and numerical modeling is described in the second part of the paper. According to the results obtained, it can be stated that the PhotoStress method provides immediate information about the strain distribution with the sensitivity similar the other interference-based full-field methods, e.g., ESPI and moiré interferometry that allows its using also for the quantification of the residual stresses of smaller magnitudes. Comparison of the relieved strain intensity fields obtained by FEA and the PhotoStress method showed that using this optical method for full-field qualitative analysis is possible. However, the quantitative analysis carried out in four selected points located at the edge of the drilled through-hole suggests the need for farther improvement of the measurement methodology. To eliminate the measurement error, the authors plan to perform the sensitivity analysis of the results obtained in the area located close to the edge of a drilled hole.

The results obtained from the digital image correlation method are in the form of displacement and strain fields. The software Istra4D ver. 4.3.0 provided with the Q-400 correlation system Dantec Dynamics allows correlation parameters to be set up, i.e., facet size and overlapping, which need to be adapted to the size of speckles created on the analyzed specimen surface. According to the manufacturer, the strains are calculated from the local curvatures of the facets or deformation gradient depending on the smoothing adjustment level. In the paper, the effect of various levels of local regression smoothing is presented (Figure 13). With default settings, i.e., without smoothing, the results obtained in four selected points compared with numerical ones differ by approximately 400%. However, with the optimal level of smoothing determined by the authors the results differ by approximately 14% (Table 4). It has to be noted that the aforementioned relative difference cannot be achieved in the entire strain field, but only in the locations of such strain levels, which can be evaluated by the correlation system with sufficient accuracy. For that reason, it is convenient to work with the displacement fields, which are not as influenced by the smoothing used. A full-field comparison of relative differences between the total displacement fields obtained by FEA and DIC shows that the highest accuracy of the results is achieved in the vicinity of the cut through-hole (approximately to the 5 mm distance from the center of the hole). In such a case, DIC results differ from the results obtained numerically maximally by 20%, which is the expected accuracy of DIC in residual stresses analysis.

There is no known universal procedure applicable to set up all the evaluating software parameters used to determine stress components from strains/displacements. For that reason, before the analysis is made on the real structures, it is convenient to take measurements in the laboratory. To realize a series of laboratory measurements leading to the optimization of the methodology for quantifying residual stresses determined from the strains obtained by the PhotoStress method or by displacements/strains evaluated by the DIC method is one of the authors' future aims. For their validation, the commercially produced devices RS200 and SINT MTS 3000 will be used, with which the authors have a lot of experience gained by solving technical problems in practice. Finally, the authors plan to use the developed prototype for the residual stresses analysis outside the laboratory.

Author Contributions: Conceptualization, M.P., M.H. and A.S.; methodology, M.P., M.H. and L.H.; software, I.V. and A.K.; validation, M.P., M.H. and A.K.; formal analysis, M.P., M.H. and A.S.; investigation, M.P., M.H. and L.H.; resources, M.H., A.K.; writing—original draft preparation, M.P., A.S. and M.H.; visualization, M.H., M.P., L.H. and I.V.; project administration, M.P. and M.H. All authors have read and agreed to the published version of the manuscript.

Funding: This research was funded by Slovak Grant Agency APVV 15-0435, KEGA 030TUKE-4/2020, VEGA 1/0141/20, VEGA 1/0355/18 and ITMS 26220220141.

Institutional Review Board Statement: Not applicable.

Informed Consent Statement: Not applicable.

Data Availability Statement: Data sharing is not applicable to this article.

Acknowledgments: The authors would like to thank the Slovak Grant Agency APVV 15-0435, KEGA 030TUKE-4/2020, VEGA 1/0141/20, VEGA 1/0355/18 and ITMS 26220220141.

Conflicts of Interest: The authors declare no conflict of interest. The funders had no role in the design of the study; in the collection, analyses, or interpretation of data; in the writing of the manuscript; or in the decision to publish the results.

Appendix A

1. Compute the combination strains p, q and t for the relieved strains ε_a, ε_b and ε_c measured by strain gauge rosette using:

$$\begin{aligned} p &= \frac{\varepsilon_c+\varepsilon_a}{2}, \\ q &= \frac{\varepsilon_c-\varepsilon_a}{2}, \\ t &= \frac{\varepsilon_c+\varepsilon_a-2\cdot\varepsilon_b}{2}. \end{aligned} \tag{A1}$$

2. Determine the calibration constants values $\bar{a}$, $\bar{b}$ corresponding to the hole diameter and type of strain gauge rosette used. Compute the three combination stresses P, Q and T corresponding to the three combination strains p, q and t using:

$$\begin{aligned} P &= \frac{\sigma_y+\sigma_x}{2} = -\frac{E\cdot p}{\bar{a}\cdot(1+\mu)}, \\ Q &= \frac{\sigma_y-\sigma_x}{2} = -\frac{E\cdot q}{\bar{b}}, \\ T &= \tau_{xy} = -\frac{E\cdot t}{\bar{b}}, \end{aligned} \tag{A2}$$

where P = isotropic (equi-biaxial) stress, Q = 45° shear stress, T = xy shear stress. The calibration parameters are determined either by experimental investigation of the calibration specimen or using numerical modeling.

3. Compute the principal stresses σ_{max} and σ_{min} using:

$$\sigma_{max}, \sigma_{min} = P \pm \sqrt{Q^2 + T^2}. \tag{A3}$$

Appendix B

The 3D measuring system used for the analysis consists of two Allied Stingray F504G cameras with a CCD sensor to capture images in maximum 5 Mpx resolution (2452 (H) × 2056 (V) px). The process of camera calibration, measurement as well as the evaluation was performed in Istra4D ver. 4.3.0 control software, provided with correlation systems Dantec Dynamics. Other technical parameters of the Q-400 correlation system are given in Table A1.

Table A1. Technical parameters of the Q-400 Dantec Dynamics system.

Measurement Area	Adjustable: from mm^2 to m^2
Measuring range	displacements: approximately 10^{-5} of the field of view, which corresponds to 1 μm at a field size of 100 mm^2, strains: from 0.01%—several 100%
Image resolution	max. 2452 × 2056 px up to 9 fps
Measurement results	the spatial contour of the object surface, 3D displacements, strains at each point
Control and monitoring electronics	portable laptop with Windows 7, 16-bit analog-to-digital converter (8 channels with a voltage range of ±0.05 V to ±10 V)
Lighting	halogen reflector with white light

Appendix C

The numerical determination of strains relieved is divided into several steps:

1. Creation of the analyzed specimen model (with dimensions according to Figure 9a) without the net cross-section in the specimen's analyzed area.

Note: the holes were created in the ending part of the model (at sufficient distance to the area of interest) for definition of boundary conditions, i.e., loading tension force of 250 N, and fixed support simulated the constrain of the specimen used in experimental analysis.

2. Convergence analysis of the solutions (Figure A1) performed by the stress analysis of the model with the net cross-section using different types of elements (e.g., CPS4R and CPS8R) with varying mesh refinement (e.g., 1/element size = 2, 1, 0.5, 0.25, 0.125, etc.). Selection of the proper element type (CPS8R) and sizing (0.5 mm) in the area of interest.

Note: the through-hole in the area of interest is created by removing of the finite elements which locations correspond to the hole with a diameter of 3.2 mm.

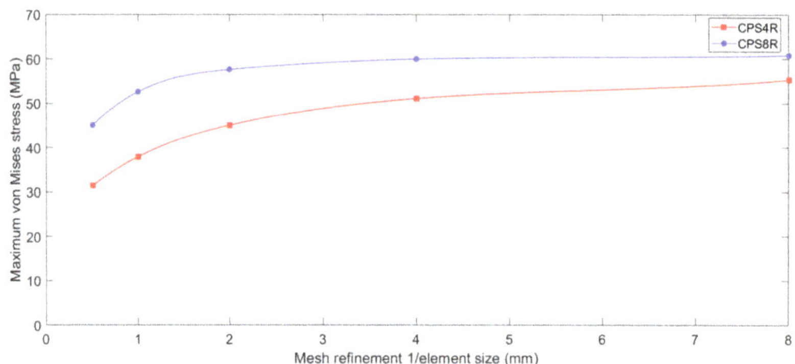

Figure A1. Convergence analysis of the solutions using two types of finite elements (CPS4R and CPS8R).

3. Creation of the finite element mesh using the linear quadrilateral elements of type CPS8R and its sizing (0.5 mm) defined in the analyzed area of interest A_0 (with the size of 25 (X) × 30 (Y) mm^2) with the total number of finite elements creating the area of interest 3560, and the number of nodes 10791 (Figure A2); definition of boundary conditions and material properties.

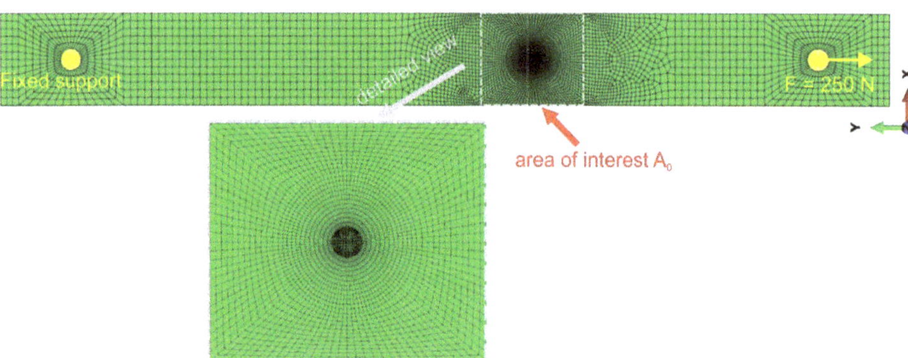

Figure A2. Meshed model and boundary conditions of the specimen without the net cross-section; detailed view on the finer mesh created in the area of interest.

4. Realization of the strain analysis on the specimen without the net cross-section; export data from the nodes located in the area of interest.
5. Realization of the strain analysis on the specimen with the cut through-hole (Figure A3); export data from the nodes located in the area of interest.

Note: the finite element model with the area of interest A1 (size of 25 (X) × 30 (Y) mm^2) was obtained, where the elements and the nodes have the same locations as in the case of the model without the net of the cross-section.

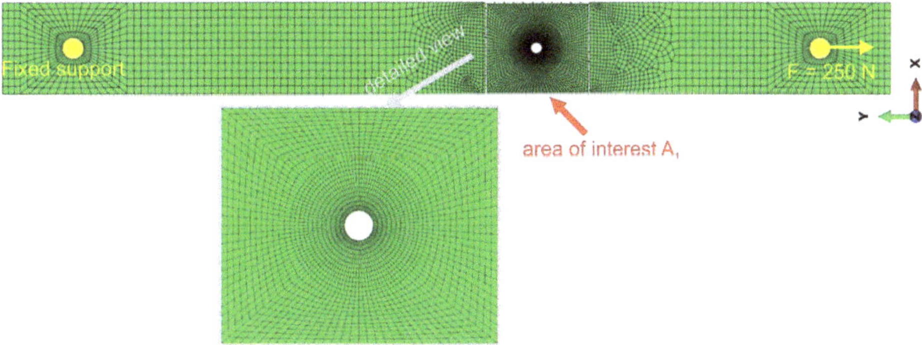

Figure A3. Meshed model and boundary conditions of the specimen with the through-hole created in the area of interest; detailed view on the finer mesh created in the area of interest.

6. Import data into the software Matlab R2020a (Mathworks, USA), the creation of a new mesh, assignment of the data values imported from the FE model nodes to the corresponding locations on the mesh created in Matlab.
7. Computation of the relieved strains as a difference in particular data (normal elastic strain X, normal elastic strain Y, tangential shear strain XY, strain intensity) values obtained in the areas of interest A1 and A0, respectively, their visualization for the squared areas A1M and A0M of 18×18 mm^2 (Figure A4).

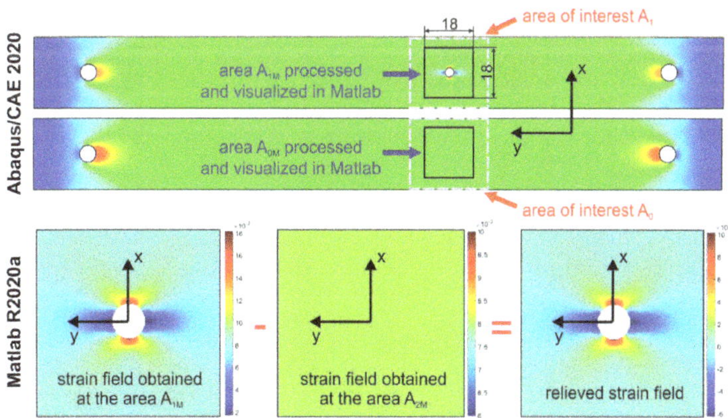

Figure A4. Schematic visualization of strains relieved determination using FEA performed in Abaqus/CAE 2020 and post-processing of the exported data in Matlab R2020a.

References

1. *ASTM E837-13a: Standard Test Method for Determining Residual Stresses by the Hole-Drilling Strain-Gage Method*; American Society for Testing and Materials: West Conshohocken, PA, USA, 2013.
2. Ya, M.; Miao, H.; Zhang, X.; Lu, J. Determination of residual stress by use of phase shifting moiré interferometry and hole-drilling method. *Opt. Laser Eng.* **2006**, *44*, 68–79. [CrossRef]
3. Shankar, K.; Xie, H.; Wei, R.; Asundi, A.; Boay, C.G. A study on residual stresses in polymer composites using moiré interferometry. *Adv. Compos. Mater.* **2004**, *13*, 237–253. [CrossRef]

4. Wu, Z.; Montay, G.; Lu, J. High sensitivity moiré interferometry and incremental hole-drilling method for residual stress measurement. *Comptes Rendus De L'académie Des Sci. Ser. IIB Mech.* **2001**, *329*, 585–593. [CrossRef]
5. Cárdenas-García, J.F.; Preidikman, S. Solution of the moiré hole drilling method using a finite-element-method-based approach. *Int. J. Solids Struct.* **2006**, *43*, 6751–6766. [CrossRef]
6. Ponslet, E.; Steinzig, M. Residual Stress Measurement Using the Hole Drilling Method and Laser Speckle Interferometry. Part II: Analysis Technique. *Exp. Tech.* **2003**, *27*, 17–21. [CrossRef]
7. Barile, C.; Casavola, C.; Pappalettera, G.; Pappalettere, C. Residual stress measurement by electronic speckle pattern interferometry: A study of the influence of analysis parameters. *Struct. Integr. Life* **2012**, *12*, 159–163.
8. Bingleman, L.; Schajer, G.S. ESPI Measurements in the Presence of Large Lateral Displacements. In *Application of Imaging Techniques to Mechanics of Materials and Structures, Volume 4*; Proulx, T., Ed.; Springer: New York, NY, USA, 2013; pp. 179–187.
9. Schajer, G.S.; Steinzig, M. Full-field calculation of hole drilling residual stresses from electronic speckle pattern interferometry data. *Exp. Mech.* **2005**, *45*, 526–532. [CrossRef]
10. Dolinko, A.E.; Kaufmann, G.H. A least-squares method to cancel rigid body displacements in a hole drilling and DSPI system for measuring residual stresses. *Opt. Laser Eng.* **2006**, *44*, 1336–1347. [CrossRef]
11. Viotti, M.R.; Dolinko, A.E.; Galizzi, G.E.; Kaufmann, G.H. A portable digital speckle pattern interferometry device to measure residual stresses using the hole drilling technique. *Opt. Laser Eng.* **2006**, *44*, 1052–1066. [CrossRef]
12. Lothhammer, L.R.; Viotti, M.R.; Albertazzi, A.; Veiga, C.L.N. Residual stress measurements in steel pipes using DSPI and the hole-drilling technique. *Int. J. Press. Vessel. Pip.* **2017**, *152*, 46–55. [CrossRef]
13. McGinnis, M.J.; Pessiki, S.; Turker, H. Application of three-dimensional digital image correlation to the core-drilling method. *Exp. Mech.* **2005**, *45*, 359–367. [CrossRef]
14. Nelson, D.V.; Makino, A.; Schmidt, T. Residual Stress Determination Using Hole Drilling and 3D Image Correlation. *Exp. Mech.* **2006**, *46*, 31–38. [CrossRef]
15. Lord, J.D.; Penn, D.; Whitehead, P. The Application of Digital Image Correlation for Measuring Residual Stress by Incremental Hole Drilling. *Appl. Mech. Mater.* **2008**, *13–14*, 65–73. [CrossRef]
16. Baldi, A.; Bertolino, F. A Low-Cost Residual Stress Measuring Instrument. In *Residual Stress, Thermomechanics & Infrared Imaging, Hybrid Techniques and Inverse Problems, Volume 9*; Quinn, S., Balandraud, X., Eds.; Springer International Publishing: Cham, Switzerland, 2017; pp. 113–119.
17. Baldi, A. Residual Stress Analysis of Orthotropic Materials Using Integrated Digital Image Correlation. *Exp. Mech.* **2014**, *54*, 1279–1292. [CrossRef]
18. Rief, T.; Hausmann, J.; Motsch, N. Development of a New Method for Residual Stress Analysis on Fiber Reinforced Plastics with Use of Digital Image Correlation. *Key Eng. Mat. Zur.* **2017**, *742*, 660–665. [CrossRef]
19. Brynk, T.; Krawczyńska, A.T.; Setman, D.; Pakieła, Z. 3D DIC-assisted residual stress measurement in 316 LVM steel processed by HE and HPT. *Arch. Civ. Mech. Eng.* **2020**, *20*, 65. [CrossRef]
20. Babaeeian, M.; Mohammadimehr, M. Investigation of the time elapsed effect on residual stress measurement in a composite plate by DIC method. *Opt. Laser Eng.* **2020**, *128*, 106002. [CrossRef]
21. Pástor, M.; Hagara, M.; Čarák, P. Residual stresses measurement around welds by optical methods. In *Proceedings of the 58th Conference on Experimental Stress Analysis, Zamecky resort Sobotin, Czech Republic, 19–22 October 2020*; VŠB—Technical University of Ostrava: Ostrava, Czech Republic, 2020; pp. 371–381.
22. Pástor, M.; Lengvarský, P.; Trebuňa, F.; Čarák, P. Prediction of failures in steam boiler using quantification of residual stresses. *Eng. Fail. Anal.* **2020**, *118*, 104808. [CrossRef]
23. Trebuňa, F.; Šimčák, F.; Bocko, J.; Pástor, M. Analysis of causes of casting pedestal failures and the measures for increasing its residual lifetime. *Eng. Fail. Anal.* **2013**, *29*, 27–37. [CrossRef]
24. Trebuňa, F.; Šimčák, F.; Bocko, J.; Trebuňa, P.; Pástor, M.; Šarga, P. Analysis of crack initiation in the press frame and innovation of the frame to ensure its further operation. *Eng. Fail. Anal.* **2011**, *18*, 244–255. [CrossRef]
25. Ajovalasit, A. Measurement of residual stresses by the hole-drilling method: Influence of hole eccentricity. *J. Strain Anal. Eng.* **1979**, *14*, 171–178. [CrossRef]
26. Beghini, M.; Bertini, L.; Mori, L.F. Evaluating Non-Uniform Residual Stress by the Hole-Drilling Method with Concentric and Eccentric Holes. Part I. Definition and Validation of the Influence Functions. *Strain* **2010**, *46*, 324–336. [CrossRef]
27. Barsanti, M.; Beghini, M.; Bertini, L.; Monelli, B.D.; Santus, C. First-order correction to counter the effect of eccentricity on the hole-drilling integral method with strain-gage rosettes. *J. Strain Anal. Eng.* **2016**, *51*, 431–443. [CrossRef]
28. Giri, A.; Pandey, C.; Mahapatra, M.M.; Sharma, K.; Singh, P.K. On the estimation of error in measuring the residual stress by strain gauge rosette. *Measurement* **2015**, *65*, 41–49. [CrossRef]
29. Post, D.; Zandman, F. Accuracy of birefringent-coating method for coatings of arbitrary thickness. *Exp. Mech.* **1961**, *1*, 21–32. [CrossRef]
30. Blum, A.E. The use and understanding of photoelastic coatings. *Strain* **1977**, *13*, 96–101. [CrossRef]
31. Herbst, C.; Splitthof, K.; Ettemeyer, A. *New Features in Digital Image Correlation Techniques*; SAE International: Warrendale, PA, USA, 2005.

32. Lu, P.; Liu, Q.; Guo, J. Camera Calibration Implementation Based on Zhang Zhengyou Plane Method. In Proceedings of the 2015 Chinese Intelligent Systems Conference, Yangzhou, China, 17–18 October 2015; Springer: Berlin/Heidelberg, Germany, 2016; pp. 29–40.
33. Hagara, M.; Trebuňa, F.; Pástor, M.; Huňady, R.; Lengvarský, P. Analysis of the aspects of residual stresses quantification performed by 3D DIC combined with standardized hole-drilling method. *Measurement* **2019**, *137*, 238–256. [CrossRef]
34. Hagara, M.; Pástor, M.; Delyová, I. Set-up of the standard 2D-DIC sysstem for quantification of residual stresses. In Proceedings of the 57th conference on experimental stress analysis, Luhacovice, Czech Republic, 3–6 June 2019; Czech Society for Mechanics: Prague, Czech Republic, 2019; pp. 106–114.
35. Mutafi, A.; Yidris, N.; Koloor, S.S.R.; Petrů, M. Numerical Prediction of Residual Stresses Distribution in Thin-Walled Press-Braked Stainless Steel Sections. *Materials* **2020**, *13*, 5378. [CrossRef]
36. Pástor, M.; Živčák, J.; Puškár, M.; Lengvarský, P.; Klačková, I. Application of Advanced Measuring Methods for Identification of Stresses and Deformations of Automotive Structures. *Appl. Sci.* **2020**, *10*, 7510. [CrossRef]
37. Pástor, M.; Bocko, J.; Lengvarský, P.; Sivák, P.; Šarga, P. Experimental and Numerical Analysis of 60-Year-Old Sluice Gate Affected by Long-Term Operation. *Materials* **2020**, *13*, 5201. [CrossRef]
38. Hagara, M.; Lengvarsky, P.; Bocko, J.; Pavelka, P. The Use of Digital Image Correlation Method in Contact Mechanics. *Am. J. Mech. Eng.* **2016**, *4*, 445–449. [CrossRef]
39. Pástor, M.; Frankovský, P.; Hagara, M.; Lengvarský, P. The Use of Optical Methods in the Analysis of the Areas with Stress Concentration. *Stroj. Časopis J. Mech. Eng.* **2018**, *68*, 61–76. [CrossRef]
40. Ramesh, K.; Sasikumar, S. Digital photoelasticity: Recent developments and diverse applications. *Opt. Laser Eng.* **2020**, *135*, 1–22. [CrossRef]
41. Makino, A.; Nelson, D. Residual-stress determination by single-axis holographic interferometry and hole drilling—Part I: Theory. *Exp. Mech.* **1994**, *34*, 66–78. [CrossRef]
42. Schajer, G.S. *Practical Residual Stress Measurement Methods*, 1st ed.; John Wiley and Sons, Ltd.: Chichester, UK, 2013; p. 320.

Article

Utilization of the Validated Windshield Material Model in Simulation of Tram to Pedestrian Collision

Stanislav Špirk [1,*], Jan Špička [1,2], Jan Vychytil [2], Michal Křížek [1] and Adam Stehlík [1]

[1] Regional Technological Institute, University of West Bohemia in Pilsen, Univerzitni 8, 306 14 Pilsen, Czech Republic; spicka@ntc.zcu.cz (J.Š.); krizek4@rti.zcu.cz (M.K.); stehlika@rti.zcu.cz (A.S.)

[2] New Technologies—Research Centre, University of West Bohemia in Pilsen, Univerzitni 8, 301 00 Pilsen, Czech Republic; jvychyti@ntc.zcu.cz

* Correspondence: spirks@rti.zcu.cz; Tel.: +420-37763-8728

Abstract: The rail industry has been significantly affected by the passive safety technology in the last few years. The tram front-end design must fulfill the new requirements for pedestrian passive safety performance in the near future. The requirements are connected with a newly prepared technical guide "Tramway front end design" prepared by Technical Agency for ropeways and Guided Transport Systems. This paper describes research connected with new tram front-end design safe for pedestrians. The brief description of collision scenario and used human-body model "Virthuman" is provided. The numerical simulations (from field of passive safety) are supported by experiments. The interesting part is the numerical model of the tram windshield experimentally validated here. The results of simulations are discussed at the end of paper.

Keywords: tram; pedestrian; crash; windshield model; HIC

Citation: Špirk, S.; Špička, J.; Vychytil, J.; Křížek, M.; Stehlík, A. Utilization of the Validated Windshield Material Model in Simulation of Tram to Pedestrian Collision. *Materials* **2021**, *14*, 265. https://doi.org/10.3390/ma14020265

Received: 21 October 2020
Accepted: 31 December 2020
Published: 7 January 2021

Publisher's Note: MDPI stays neutral with regard to jurisdictional claims in published maps and institutional affiliations.

Copyright: © 2021 by the authors. Licensee MDPI, Basel, Switzerland. This article is an open access article distributed under the terms and conditions of the Creative Commons Attribution (CC BY) license (https://creativecommons.org/licenses/by/4.0/).

1. Introduction

This paper is an extended research paper of the conference presentation EAN 2020 [1] prepared for the special issue "Selected Papers from Experimental Stress Analysis 2020" in Materials journal. Traffic injuries represent one the most significant cause of the death around the world [2]. Moreover, pedestrians are the most vulnerable road users and they are exposed to a high risk in the collisions with the vehicles. The statistical data show that pedestrians are still responsible for the second biggest number of fatalities and injuries on the road: about 2000 in the Czech Republic [3,4], 40,000 in the EU, and 1.25 million around the globe annually [2–6]. The recent studies in Europe indicate that the passenger cars are one of the most often involved in the collisions with the pedestrians. Figure 1 summarizes the distribution of the vehicle type participating in the pedestrian collisions in the Czech Republic for the years 2009 to 2014 [7]. However, the number of pedestrians involved in tram crashes is not insignificant (about 12% of total cases in CR between 2009 and 2014).

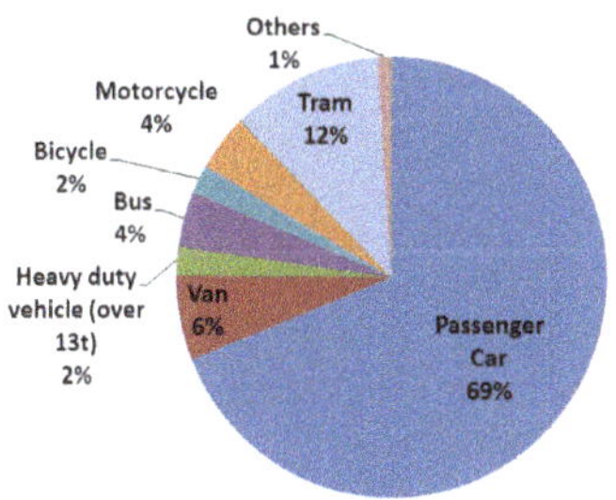

Figure 1. Distribution of the vehicle–pedestrian accidents in the Czech Republic in 2009–2014 [6].

The main aim of this paper is the utilization of the validated numerical model of human in the pedestrian–tram collision scenario. There is a new technical guide "Tramway front end design" currently under preparation by Technical Agency for ropeways and Guided Transport Systems [8]. This document is focused also on the guidance of the legislation of the new tram vehicles, with respect of the pedestrian safety and tram front-end design crashworthiness. The main aim is a condition of survivability of the pedestrian when the frontal crash occurs. The conditions of the collision are defined, as well as limits and threshold of the mechanical quantities describing the injury risk of the human. The legislative document is based on the automotive industry and its EuroNCAP regulation [9,10]. As the head is the more likely to be seriously injured in the collision with the tram, the main interest of the legislation is paid to its injury, monitored with the so-called Head Injury Criterion (HIC), where the peak of linear acceleration of the head center of gravity COG is to be the most responsible for the injury [11]. From the experience with tram development and from the experience with the automotive research, the authors assume that one of the most important part of the of the tram front end responsible for the level of head injury is the windshield [12–14]. This work is tightly connected with the research project focused on the development of the new tram, in the cooperation with worldwide tram developer acting in Czech Republic. As the design and the material need to be optimized, the mechanical behavior of all materials is required to build a suitable numerical model that will be used during the developing process. The main attention here is paid to windshield of tram as significant safety feature. The windshield consists of three layers (glass, PVB foil, and glass), and these must be described with the appropriate numerical model. The validated layered glass model from the automotive industry is used here and modify for the tram. In order to test its fidelity, the experimental test was done and compared with its numerical model.

The experimental test here is the pendulum test, where the rigid steel ball impacts the tram glass. The mass, diameter, and velocity of the impactor was defined in order to be close enough (similar kinetic energy) as a human head during the frontal collision with tram. The material specification, complete description of the test and level of the fidelity is described later.

The model of the windshield is then utilized in the full model of the tram (its front part) and together with the numerical model of the human Virthuman, the collision scenarios can be simulated. The Virthuman model is a model of the human body validated under specified conditions [15–19], and it has been successfully used in automotive research for the specified collision scenarios in correspondence with the tram regulation [8], to help in the process of the development of the new tram design. As this regulation defines the scenario, in which, the new vehicle is tested, the aim was to prepare and test this model, whether is a suitable tool and can help in the development of the new and pedestrian safety tram.

This paper is firstly focused on the material model of the windshield suitable for a Visual Performance Solution (VPS) [20]. There is an assumption, that the layered glass of a windshield is equivalent to the passenger vehicles, only the thickness of the glass and PVB foil has changed. Thus, the authors are using validated automobile windscreen model, modified it in order to represent a tram one, and if necessary, some material parameters optimization will be performed, to get the correct material models. In order to verify the numerical simulation, the experimental pendulum drop tests were done and results (especially crack propagation is being tested).

Second part of the paper uses the acquired windshield model in the full FE model of the tram required for the testing of the safety feature [8], where the pedestrian to vehicles collisions are defined. The aim is to prepare, to test and to verify this numerical model to be useful tool in the process of development of the tram front end design. The testing conditions as well as limits and threshold for the outputs are discussed. The new design and material of the tram, developed within the research group consisting of the research institute, tram producer, and testing institutions, is utilized here, and its description is provided here. Finally, the results of the tram vehicle, consisting of the validated windshield model, Virhuman model in the defined configuration [8] are presented.

The tram regulation defines the mechanical quantities on a human and its limits, to rate the safety of the vehicle, in case of collision with the pedestrian. In case of the human head injury, the regulation concerns only of the HIC criterion and the max threshold is 1000. For such purpose, the appropriate model of a human must be used, to get required results. The human model must have adequate level of biofidelity, must be validated for such purposes and must have the appropriate outputs. One of the suitable human model is a hybrid approach model Virthuman, developed for universal use with validation based on real human body behavior. Moreover, the authors have a great experience with the development and utilization of this model. The Virthuman model is briefly described in the next section and then, used in the defined crash scenarios. The list of the human body models can be found for instance in [21].

2. Material Model Description

The laminated glass generally composed of two outer layers of glass and one inner layer of polyvinyl butyral (PVB) [22,23]. The windshield modeling method here is based on the studies connected with the automotive industry, where the windshield has been studied and tested, and its material model was developed. The authors here defined an assumption: that the tram and passenger windshield are made of the aforementioned materials and only the thickness varies. Thus, the model from the automotive industry is further used as an initial material model for a tram modeling. The passenger car windshield is used here and further modified to better describe behavior of the tram windshield in the experiment done here. The laminated layered glass FEM is modeled as three layers of shell elements connected by tied contact (rigid node-segment link). The PVB foil is modeled as an isotropic nonlinear viscoelastic shell element of Maxwell type:

$$\sigma = k\left(1 - e^{-w\varepsilon}\right)\left(1 + h_1\varepsilon + h_2\varepsilon^2\right)\left(\frac{\dot{\varepsilon}}{\dot{\varepsilon}_{ref}}\right)^m \tag{1}$$

where $\dot{\varepsilon}$ is the plastic strain rate; $\dot{\varepsilon}_{ref}$ is the reference strain rate; and k, w, m, h_1, and h_2 are material constants. The glass is modeled as linear elastic material with a brittle failure criterion. For the fracture definition, the Rankine criterion is used [24]: fracture occurs when the maximum principal stress exceeds the critical value,

$$\sigma = \begin{bmatrix} (1-d_1)\sigma_{11} & (1-d_{max})\sigma_{12} & (1-d_{max})\sigma_{13} \\ (1-d_{max})\sigma_{12} & (1-d_2)\sigma_{22} & (1-d_{max})\sigma_{23} \\ (1-d_{max})\sigma_{13} & (1-d_{max})\sigma_{23} & 0 \end{bmatrix} \tag{2}$$

where σ is the damaged stress tensor, σ_{11},σ_{12},..,σ_{23} are components of undamaged tensor, d_1 and d_2 are damage values in two directions, and d_{max} is maximum of d_1 and d_2.

3. Experimental Testing and Validation of Windshield Numerical Model

The main idea here was to create an experimental setup that can be used for the validation and verification of the windshield model for the purposes of this work (i.e., impact of the human head, with the travelling speed of the tram). The standard process of head testing used the normalized head impactor [25,26], head FE model [27], or dummy model [28]. However, for our purposes, only the similar impact conditions (impactor stiffness, mass, velocity, and energy) must be satisfied, as the equivalent numerical model are created to compared with experimental results. The experimental testing with the safety laminated glass has its limitation. It is not possible to get tensile sample from producer (final glass cannot be cut), the brittle glass is not suitable for tensile test, PVB foil separated from glass has disrupted surface and the adhesion between glass and PVB foil is not known and cannot be experimentally tested (PVB material adheres to glass through hydrogen bridges [29]). In order to make the model simpler and to decrease the calculation time, the

2D elements are used here. This simplification has its basement also in automotive, aviation or civil engineering research, and thus can be used also here and should not significantly influence the final results [30–34].

In order to validate the material model for this purpose, the simple pendulum test and its numerical model are built, see Figure 2. The pendulum is made of steel S235 profile (50 × 30 mm^2, thickness 2 mm, length 2000 mm and 4.41 kg of mass) and the ball impactor (150 mm of diameter and 4.7 kg of mass). The circular section of the tram windshield is placed on the extruded polystyrene (with known properties) with diameter 300 mm. The steel impactor falls on the sample of the glass from the height of 2000 mm. The tested glass was a rea tram-layered glass (3 mm glass, 0.9 mm PVB foil, 3 mm glass). More than 10 tests were performed to get statistically significant results. In most of the tests, the ball was falling from full height (2000 mm). The friction between impactor and glass was 0.5 and the glass model was constrained to polystyrene basis via tied constraint (rigid node to node). The impactor in the model was constrained with revolute joint to the basic, and loaded with the initial velocity to COG. In order to speed up the calculation, the free fall of the pendulum was neglected, and the simulation starts when the impactor is just about to hit the glass. The initial velocity was calculated and verify with the data from experiment.

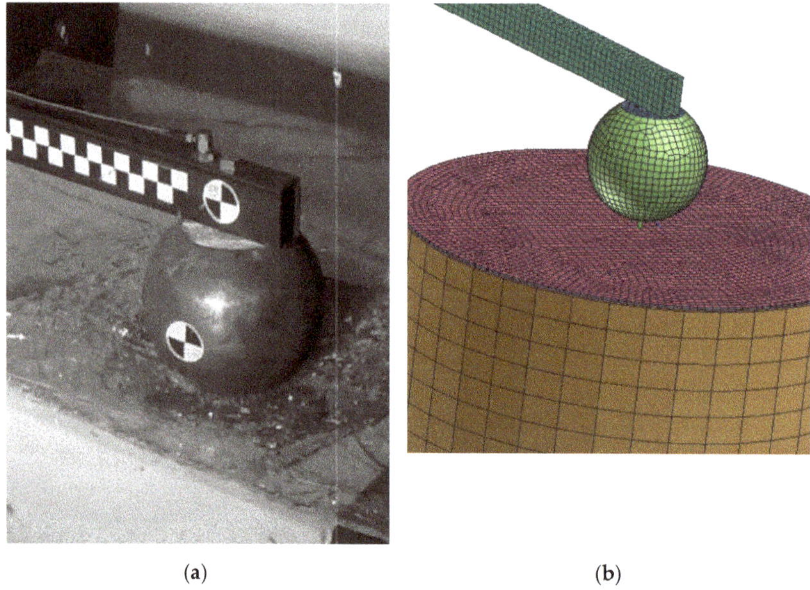

(**a**) (**b**)

Figure 2. Experimental test of windshield (**a**) and its numerical model (**b**).

Some tests were executed from a smaller height and also with initial crack on glass or with different shape and size of glass. These experimental results are not described here. However, the example of the experimental curves is presented in the Appendix A. The condition of smaller initial height allows us to validate the model also for range of initial velocities. The windshield model was validated only for full height with the impact velocity 6 m/s. The acceleration was recorded with Brüel & Kjær accelerometer (4533-B 10 mV/g, ±500 g) (Nærum, Denmark). Displacement of the impactor was recorded with high-speed camera (Photron fastcam SA X2 RV) (Tokyo, Japan) and reconstructed with the target focused method. In each photograph, the target (Figure 3) on the rigid impactor was selected and monitored. If the size of each pixel is known, than the displacement (and also its derivatives) can be calculated.

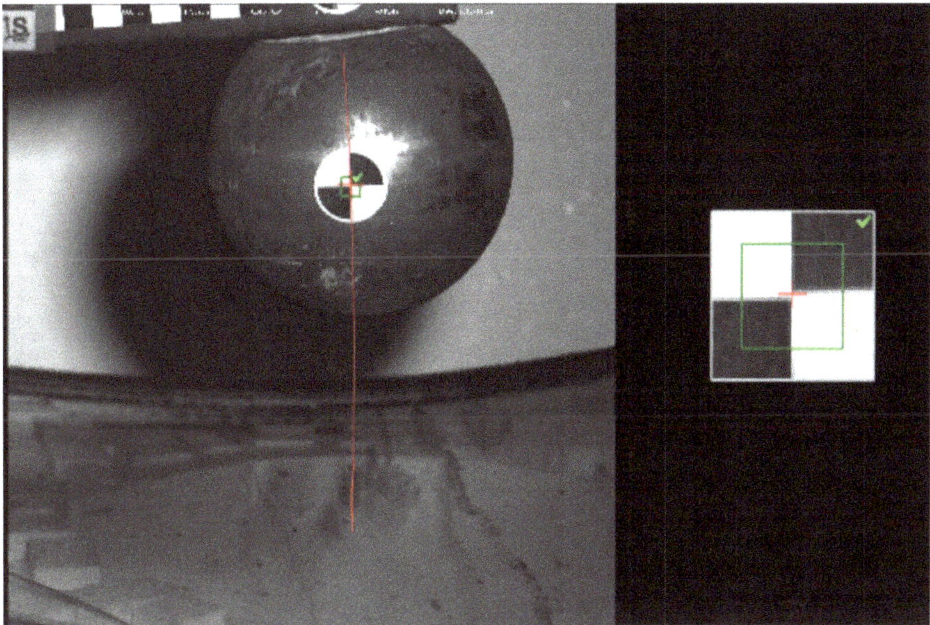

Figure 3. The impactor path reconstruction from target focusing.

This kinematics conditions were set up to be similar with pedestrian head impact during collision with tram (initial speed 20 km/h). The impactor is considered as a rigid body, as its purpose is to validate the glass model (not to evaluate head injury prediction). It is clear that this model cannot exactly predict the crack shape in detail, but it has been discovered that in repeated experiments and simulations the influence of crack shape difference is insignificant. In these phenomena, the crack shape has some general similarities in radial and circular direction, see Figures 4–7. Moreover, the acceleration results are influenced by the steel rod oscillations. This is one of the possible improvements for the further experiments. The glass rupture propagation is visible on the figures below.

The figures above (Figures 4 and 6) show the visible the rapture mechanism, crack propagation of the glass and the ultimate rupture during the maximum penetration. Similar behaviors were observed also in the simulation (Figure 7). Stress distribution and thickness of the PVB foil are shown in the Figures 8 and 9, respectively.

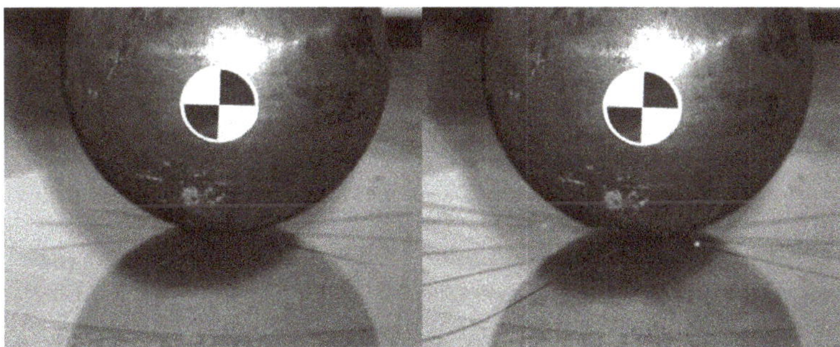

Figure 4. The radial crack propagation after the first contact of impactor.

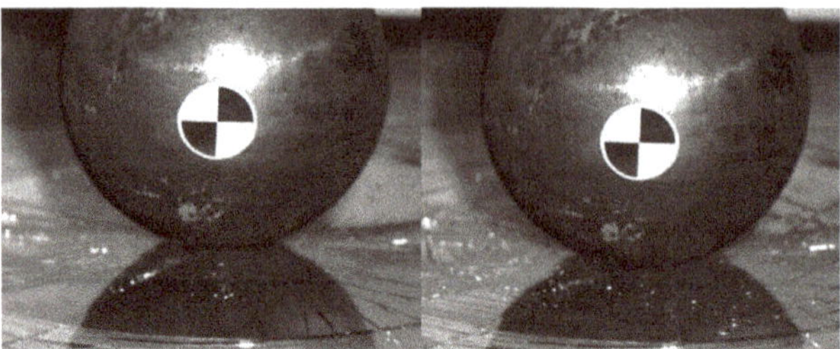

Figure 5. The concentric crack in the second state.

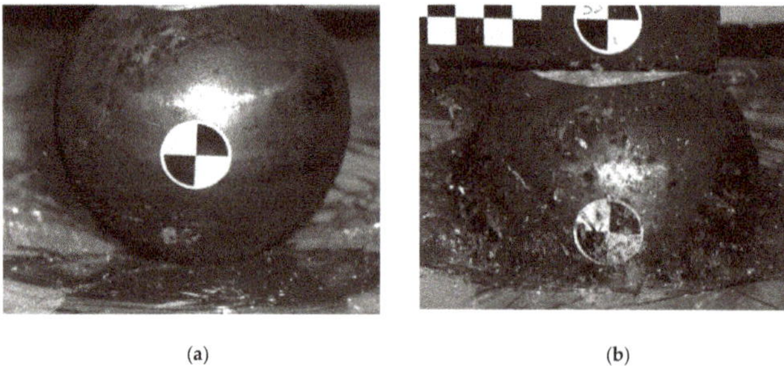

(**a**) (**b**)

Figure 6. The ultimate glass fracture (**a**); maximum penetration (**b**).

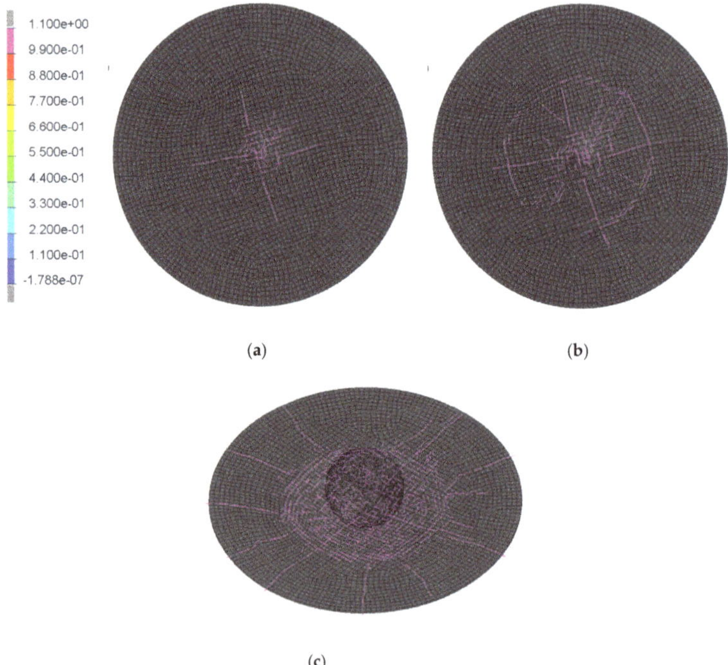

(**a**) (**b**)

(**c**)

Figure 7. The damage tensor directions; first radial cracks (**a**); secondary concentric crack (**b**); final ultimate fracture during the maximum impactor penetration (**c**).

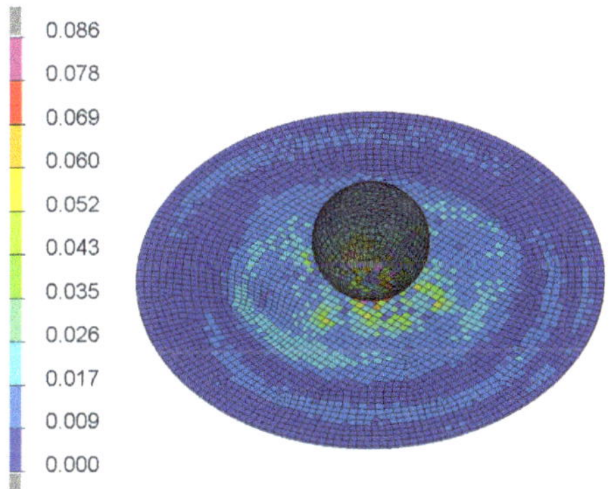

Figure 8. Discribution of Von Mieses stress of the glass after the first contact.

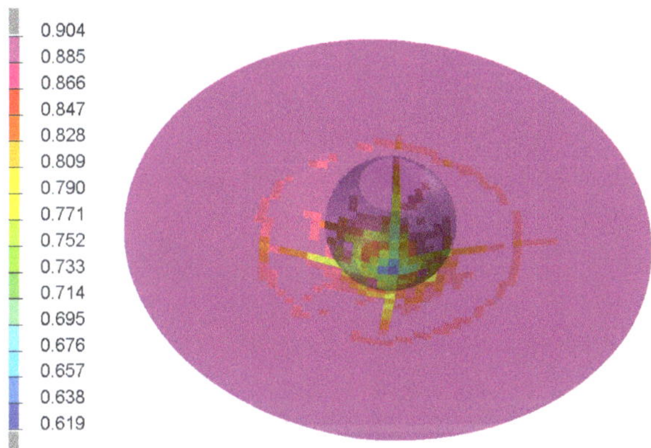

Figure 9. Thickness of the PVB foil at the maximum of penetration.

As was described above, the model of the passenger car windshield was used here as an initial guess. The thickness of the layers was modified based on the real tram data in order validate the model. First set simulations confirmed this theorem, as the simulated results were very close to the experimental ones. Thus, there were no requirements of numerical optimization (tuning of the material model parameters) to get the proper level of fidelity. Only minor modifications were done on the windshield model.

The comparison of the experimental data and numerical model (described above) shows a good correspondence. This coincidence is adequate and good enough for the purpose of safety simulations and for head impact injury predictions in case of windshield–head impact. The acceleration–time curves are very close to each other; see Figure 10, which indicates also similar result of HIC criterion. The deformation characteristics result in a good agreement of experimental and simulation curves Figure 11.

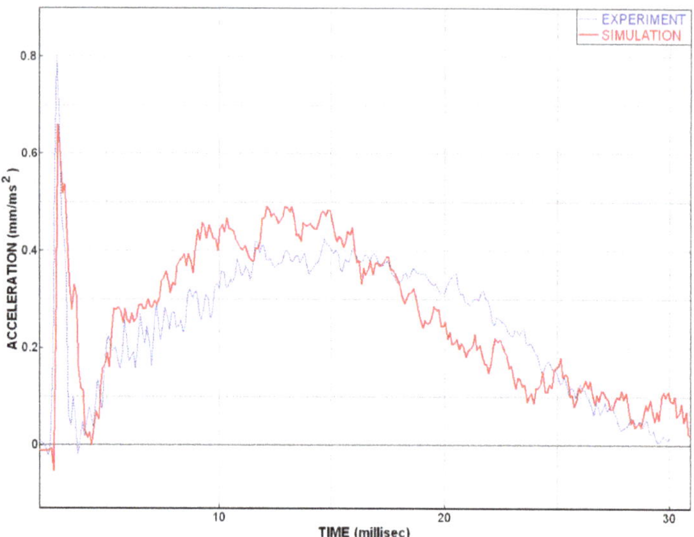

Figure 10. Plot of the acceleration vs. time.

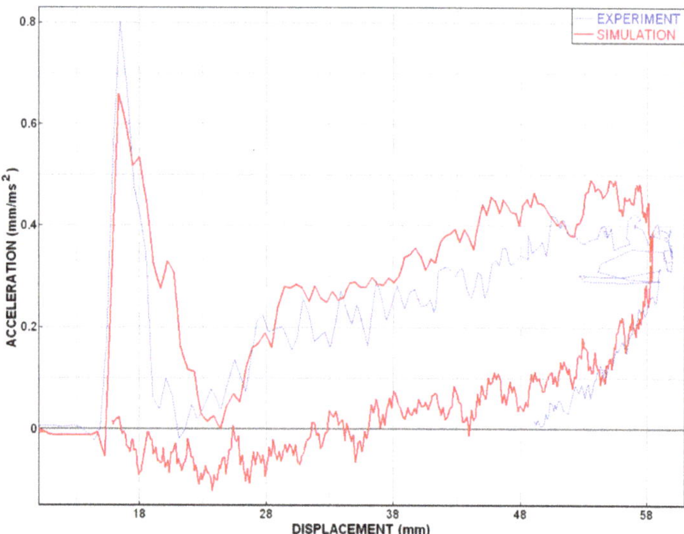

Figure 11. Plot of the acceleration vs. displacement.

4. Parameters of Validated Material Model

The description of material models of the glass and PVB foil, respectively, are presented below (units: mm kg ms), see Tables 1 and 2. These material parameters are used for tram windshield in simulation with three layers (3 mm outside glass, 0.9 mm PVB foil, 3 mm inside glass) connected with tied links. The presented material parameters are for the VPS software, where the entire model is built. The VPS software includes the Glass model, designed exactly for a modeling of the glass. The PVB foil is defined as a linear viscoelastic 2D material.

Table 1. Parameters of material model GLASS.

Mat Type	Density	Belyts.-Tsay Reduced Integration	Stiffness Elastic Hourglass	Quadratic Viscosity Multiply	
126	2.5×10^{-6}	0	0	1	
E	NU	Membrane Hourglass	Out of Plane Hourglass	Rotation Hourglass	Trans Share
70	0.2	0.01	0.01	0.001	0.8333
Sigmac	Time Filter	Stiff Damping			
0.031	0.01	0.1			

Table 2. Parameters of material model PVB-foil (nonlinear viscoelastic shell model).

Mat Type	Density	Belyts.-Tsay Reduced Integration	Stiffness Elastic Hourglass	Quadratic Viscosity Multiply	
121	1×10^{-6}	0	0	1	
E	NU	Membrane Hourglass	Out of Plane Hourglass	Rotation Hourglass	Trans Share
9	0.39	0.01	0.01	0.001	0.8333
G-Shell Param	K	M	H1	H2	W
	0.007	1.33	1.35	0	3

NOTE: The definition of all the defines parameters can be found in VPS manual [20]

5. Collision Scenario

The collision scenario defined in the technical guide is based on the statistical data of the tram to pedestrian collisions and also follows the automotive safety scenarios, defined in EuroNCAP [9,10]. The collision scenario is defined as a moving tram hitting the pedestrian (moving or standing) from his side. The pedestrian is moving perpendicular to the tram trajectory, in front of its front end. The technical report divides the impact into three phases, where the first phase is considered as an impact of the vehicle to the pedestrian. Second phase is an impact of the pedestrian onto the ground and the third impact phase deals with the scenario, where the pedestrian lays on the railway (ground) and can be overrun with the vehicle. The scope of the technical report is focused on the first and third phases. The second phase is connected mainly with urban and civil engineering and material of the surroundings (grass, concrete, pavement, asphalt, etc.) and thus it is not examined here.

- First collision scenario (type A): the pedestrians involved in the collision are specified to be mid-size male (175 cm, 78 kg—50th percentile) and 6 years old (YO) child (110 cm, 24 kg). The report also defines possible impact area and impact zones, with respect to the shape of the vehicle, for more specification, see the new regulation [8]. The collision scenario evaluating the first impact consider the tram moving with the initial velocity equals to 20 km/h and pedestrian standing still, left side to the vehicle, one step forward (not specified which leg to be forward) and the lateral position of the pedestrian relative to the vehicle has two specifications (H-point with respect to the tram):
    - 15% value of half of the tram width
    - 50% value of half of the tram width

The vehicle does not stop (not loaded with any deceleration pulse) only energy lost due to the impact. The pedestrian injury risk is monitored only with the Head Injury Criteria (HIC) [10], which should not exceed threshold of 1000. The second impact (pedestrian to the ground) is not included here, as it is connecter more with the urban engineering than traffic engineering.

- Second collision scenario (type B): (overrun of the pedestrian) is tested via four scenarios (each of them with the adult and child dummy). For this particular test, the dummies are specified to be the adult rescue dummy" (183 cm, 75 kg) and the child rescue dummy" (122 cm, 17 kg). The testing scenarios are defined as follows.
  - Test 1: transverse to the rail, centered
  - Test 2: transverse to the rail, off center (hip on the rail)
  - Test 3: lengthwise on the rail, centered (feet pointing towards the tram)
  - Test 4: lengthwise on the rail, off center (hip on the rail, feet pointing towards the tram)

This technical report also describes the protective technology to be used and how to be used, the distance between dummy and vehicle during test etc. The initial velocity of the tram in this collision scenario is 25 km/h and after reaching specified position, it starts to break within emergency breaking until it stops. The objective of this test is to verify capabilities of the vehicle during crash, with the following parameters.

- To stop any part of the rescue mannequin before the first wheel set
- Not to jam the rescue mannequin at its thighs, chest, or head
- Not to sever one of the rescue mannequin's limbs so that the rescue mannequin should remain intact
- To push the rescue mannequin away so that it does not come into contact with the wheels
- Not to trigger any debris or fracture on impact with the rescue mannequin (risk of aggravating injuries)

The full test protocol is available in the technical report, where all settings of the test conditions are specified. The conclusion of the test indicates whether it meets the objective or not. There is no threshold value specified to pass the tests. Only the position of the pedestrian with respect to the tram is monitored.

6. Human Body Model Virthuman

In order to represent a pedestrian in the collision scenario, the Virthuman model is considered. It is a virtual human body model where the skeleton is built based on the multi-body structure (MBS). The outer surface of the model consists of deformable segments that are connected to the skeleton via nonlinear springs and dampers to account for deformability of soft tissues. Individual rigid bodies of the MBS structure are interconnected via kinematics joints. Moreover, additional "breakable" joints are considered in lower extremities to include the possible fractures of both femur and tibia of the pedestrian in the collision scenario. The model has been validated extensively to ensure its boofidelity in the particular scenarios, connected mainly with the automotive industry [15–19]. The basic reference model (50th percentile male) can be scaled using the parameters of height, weight, age and gender. In this case, the 50th percentile male was used corresponding to the Hybrid III dummy (male, 172 cm, 78 kg). The scaling algorithm has its basement in the wide database of population in Czechoslovakia in the 1980s, where anthropometric dimension of more than 10,000 people were measured [35]. Thanks to the MBS structure of the model, the model is easy to position in any desired position respecting real human anatomy and physiology (range of motion of the real joint [36]). In this case, the positions defined in the chapter "Collision scenario" were considered for the crash scenario. Virthuman model includes an embedded algorithm in the model to evaluate standard injury criteria for individual body parts as defined by EuroNCAP testing procedures [10]. During the post processing, the particular injury criteria is being checked and the individual body segments are colored based the threshold of such criterion (Figure 12). In case of the head, the Head Injury Criterion is used in this study to predict injury sustained by the pedestrian in the collision with the tram [11].

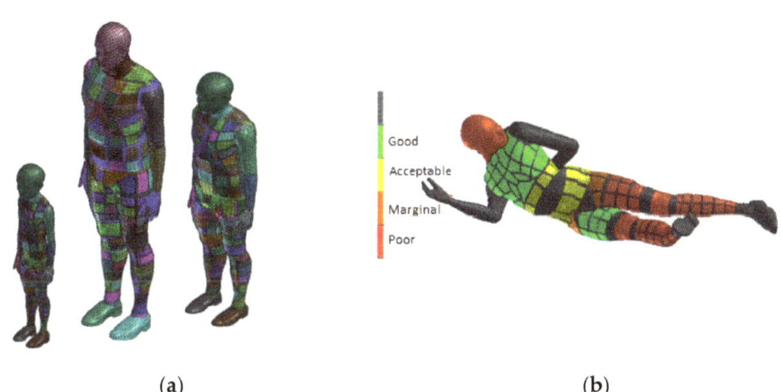

(**a**) (**b**)

Figure 12. Virthuman model: Scaled model. 6-year-old child, 110 cm, 17 kg; 40-year-old male, 190 cm, 100 kg; 25-years woman, 50 kg (**a**); Injury risk rating (**b**).

7. Numerical Model of Collision with Windshield Material Model

For the dynamic structural analysis (with significant nonlinearities), the explicit integration method is used. The tram to pedestrian collision is modeled with numerical software Visual Performance Solution (VPS) [20] and human body model Virthuman described above is used. The model of the vehicle is created mainly with quad and brick elements with one gauss integration point. The model was discretized into 1.6 million elements with the smallest element characteristic length 2 mm (leading to a time step 5×10^{-6} ms). The contacts and links (node to segment connection) are realized by penalty algorithm. The simulation time of the defined scenario is 390 ms. The mechanical properties of steel S235 [37] and steel 1.4301 are known. The top shell cover is made from polymer (acrylonitrile butadiene styrene) with acceptable fire protection and recycling possibilities. Unfortunately, mechanical properties of this material used in simulation are confidential (courtesy of the company) and cannot be provided. The Virhuman model is loaded only with gravity, while the vehicle is loaded with the specified velocity (constant during entire simulation—type A; constant until specified time—type B). Contact between the human and the tram is defined as an asymmetric node-to-segment contact with the friction coefficient equals to 0.3. The connection between the structural compartment of the vehicle is realized with the rigid (tied, rigid spot-weld, rigid constraint) or deformable (glued) connection. For instance, the glass is connected to the frame by glue, modeled as deformable solid, connected to frame.

There are two different setups of the simulations: The first type is a collision with standing pedestrian. The second one is a collision with pedestrian lying on the ground. All simulations are performed with the same human and tram model, respectively. In previous section, the windshield model was tested with the experimental set up, to proof its fidelity for this particular scenario. The main aim here is to apply the simulations from the field to provide passive safety to the pedestrian. The crashworthiness of the newly developed tram design is evaluated here, with respect to defined regulation for tram safety [8]. The structure of the Virthuman model is based on hybrid approach (MBS basic skeleton and rigid super elements connected via springs and dampers to the basic structures) to account for the local deformations. Consequently, the particular body segments do not catch the full local deformation, such as full FE model (GHBMC or THUMS [21,38]). However, the deformation is represented with the deformation of the spring and dampers, and the motion (displacement, velocity, acceleration) of the COG is close to real motion (validated model). Thus, the user cannot see the deformation via visual observation, but the acceleration curves and HIC value are adequate.

8. Results

The sequence Figures 13 and 14 in the time shows the detail of the head impact to the tram front-end. This part of collision is the most important, as the head injury connected with significant severity occurs here. It is clearly visible that the head impact occurs directly to the windshield. However, the head acceleration does not exceed limit (Figure 15) of 0.8 m/s^2 and the HIC criterion is only 234 (below the threshold of 1000). This indicates low or acceptable injury risk of the head. The peak in the curve is results from the rapid change in acceleration and velocity (in accordance with the crash impact theory: large magnitude in a short time). The velocity of the head during the impact does not have any threshold value (no specified criterion or limit); however, this value can help quantified the injury risk. In this particular scenario, the velocity of the head COG was about 7 m/s in magnitude, see Figure A4.

Figures 16 and 17 show the stress distribution of the layered glass under the impact of the body (shoulder and head, respectively).

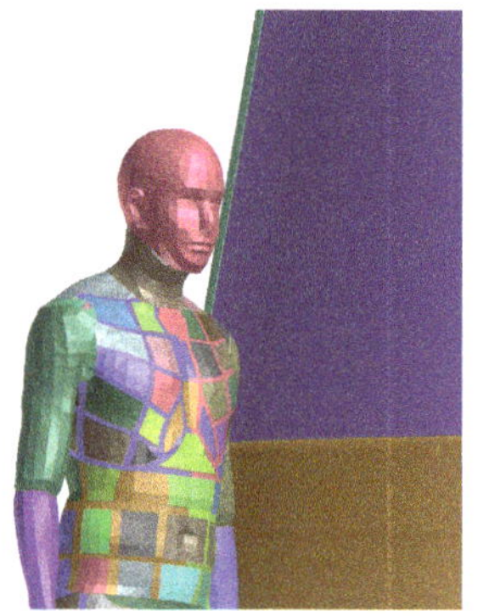
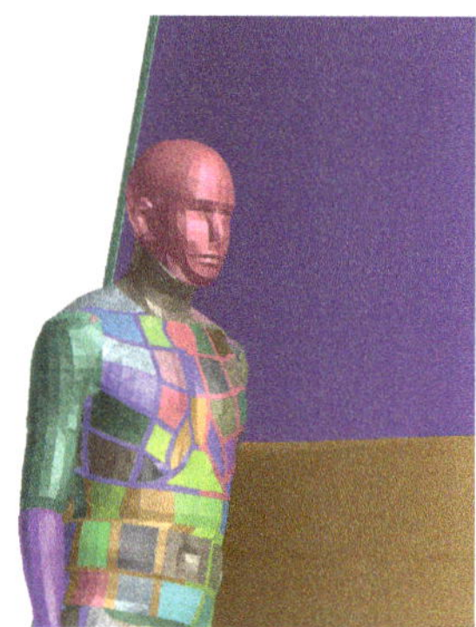

Figure 13. Results of simulation (in time 0 and 25 ms) where the head impacts the windshield.

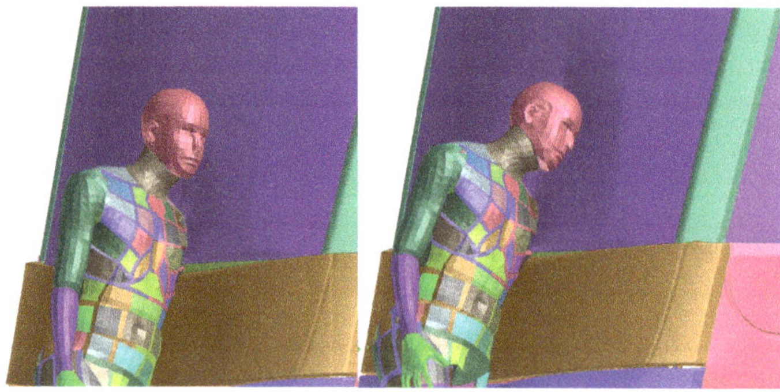

Figure 14. Results of simulation (in time 50 and 75 ms) where the head impacts the windshield.

The presented results of the pedestrian–tram collision are only an example of the calculated results. The aim of this paper was to validate the windshield model for this scenario, and to use this model in the full model of the tram and pedestrian collision. Moreover, this work is tightly connected with the research project developing a brand new tram (in cooperation with the worldwide producer) [14] and is based on the safety requirement [8]. Thus, only the crash configuration defined in this regulation is analyzed. However, from the authors' experience, we can expect a different dynamic an injury risk of the pedestrian for different initial condition (lateral position of the pedestrian with respect to the vehicle or rotation of the body) [13]. Moreover, the location of the first contact can significantly change the results (effect of the shoulder and arm or elbow respectively). All these aspects are considered in the tram design development, however not presented here. The paper here is testing the crashworthiness of the new vehicle in the conditions specified in the European regulation [8].

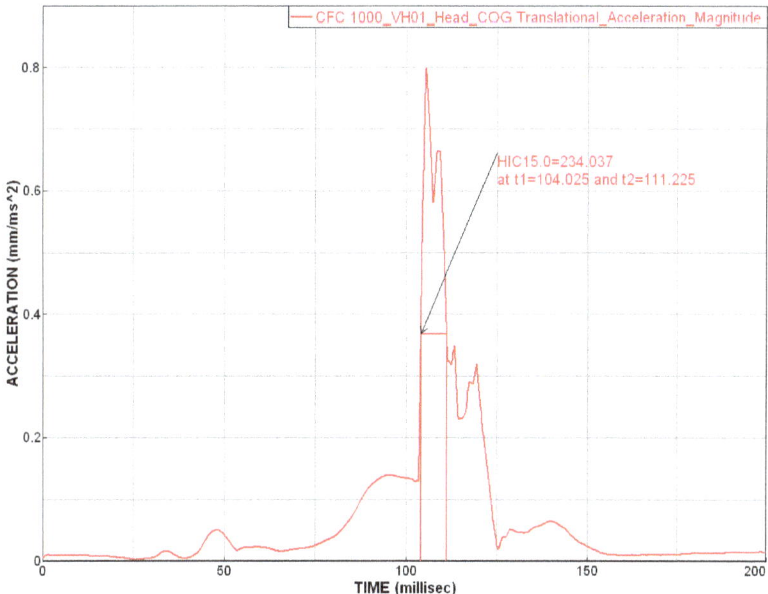

Figure 15. Acceleration of the COG (center of gravity) filtered by CFC1000 filter with HIC15 injury criterion assessment.

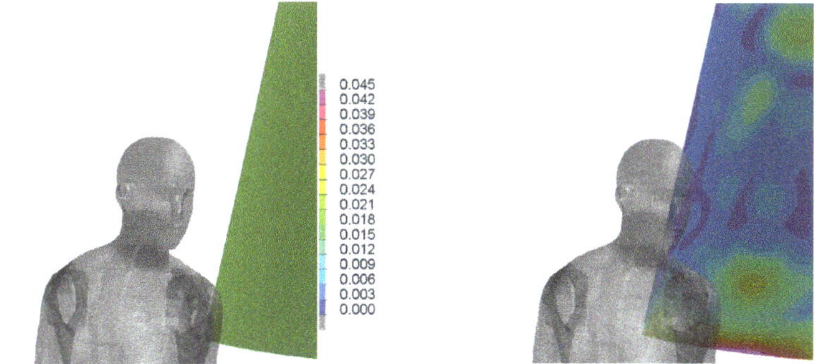

Figure 16. Results of simulation (in time 0, 40 ms) where the stress distribution of the glass is shown.

The collision of tram and laying pedestrian shows how the requirements for the pedestrian anti-crush mechanism are met (Figure 18). With the advantages of the simulations, the effect of mechanism with correct clearance is visible. It can be said that the injury of laying pedestrian during collision with tram without pedestrian anti-crush mechanism are crucial (the most of trams has no pedestrian anti-crush mechanism see Figure 19). The design of a new tram can save many lives and significantly reduce number and severity of injuries. Based on the calculated simulations, the new tram is passing the requirements.

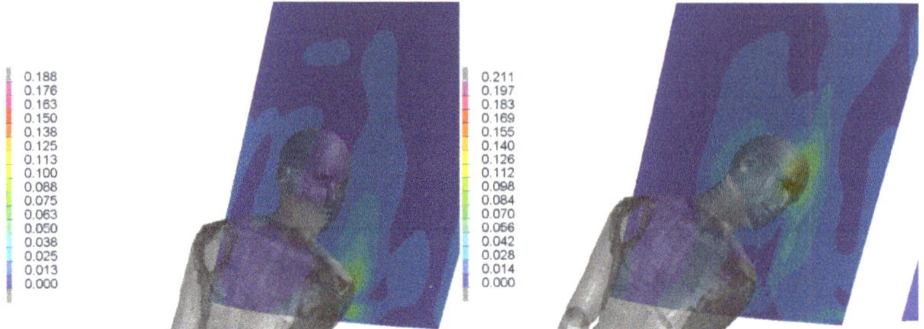

Figure 17. Results of simulation (in time 80, 120 ms) where the stress distribution of the glass is shown.

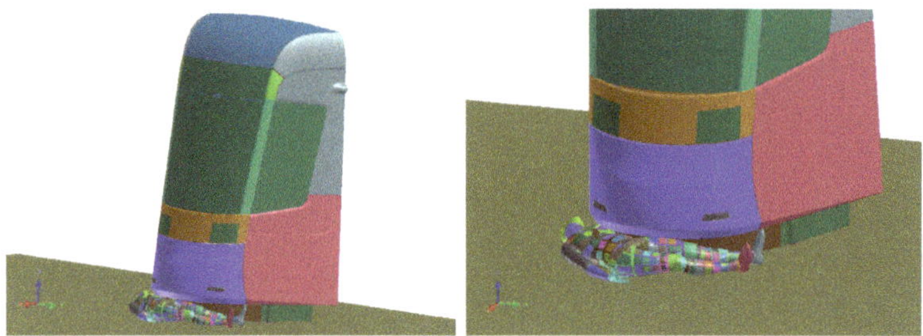

Figure 18. Crash scenario B, overriding with adult rescue dummy, transverse to the rail, centered.

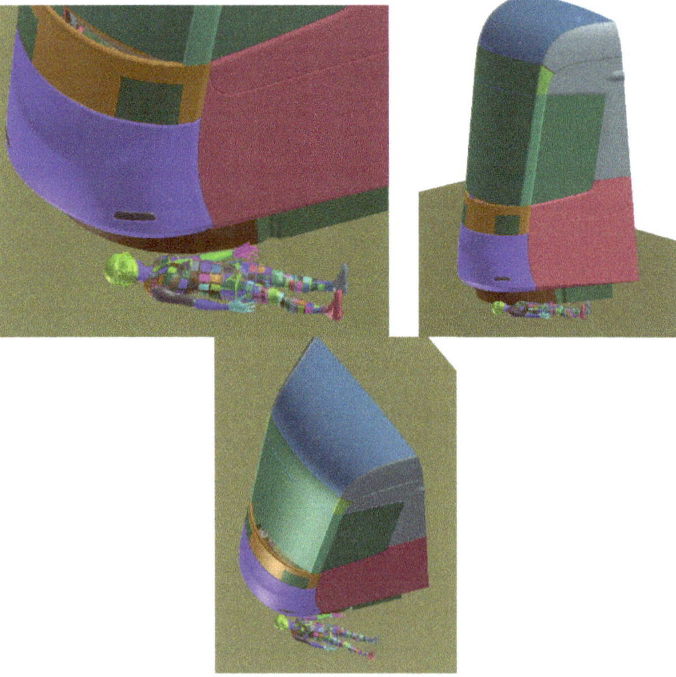

Figure 19. Crash scenario B, overriding with child rescue dummy, transverse to the rail, off-centered.

9. Conclusions

This paper contains a description of experimental material testing and its validation for the purpose of pedestrian collision. As the numerical model of the passenger car is available, the assumption of its similarity with the tram glass was defined. We assume that the materials are equivalent and only the thickness of the particular layers (inner and outer glass and PVB foil) varies. The experimental test of the sphere impact to the tram glass was performed and the results were utilized in the material testing and verification. Since only the minor modification of the car windshield was necessary, this statement was confirmed and the obtained model of the glass can be further used in the full tram model, for the scenarios of pedestrian impact. Furthermore, the impact of the head and its dynamic was the main concern, and thus, this head-like impactor test was appropriate. If the glass model would be used in other impact conditions (especially under different impact condition: change in the mass, shape or velocity of the impactor) a new validation would be necessary. The final material model of the glass and PVB foil for the VPS software is presented (material parameters table). The best comparison of simulation and experiment (for the mode of impact loading) is in macroscopic characteristics (like deformation characteristic). The simple pendulum test has some inaccuracies (see above), which can be further improved. The validated material model shows a good coincidence with the experiment.

The windshield model is further used in the tram model. The collision of tram to pedestrian is described in second half of paper, with the respect to the newly defined certification conditions. The collision scenario was simulated with the advantage of the virtual human body model "Virthuman". The injury of the pedestrian head (HIC) is highly influenced by the windshield behavior and thus the windshield was the main concern of this paper. The simulation (of pedestrian collision with full-scale tram face) indicates that the defined tram to pedestrian crash scenario results in the HIC value (234) significantly smaller than the threshold limit of 1000. The Virthuman model is not a full FE model of the human body, and of course it has some limitations. However, in the case of traffic accident and human injury, especially injury of the head, it results in a good level of boofidelity. Thus, the results of the tram to pedestrian collision are reliable for this particular research. The first results of the experiments suggest very similar material behavior of tram and road vehicle windshield. Therefore, it is possible to use the material model of glass and PVB foil from an automotive industry with modified thickness and minor modification of the material parameters also in the tram model. The results of the simulation (with experimentally validated material) indicate that the windshield is feature with good crashworthiness.

The new design of tram with low height of bottom windshield show the good behavior with respect to the pedestrian injury both in the frontal crash (scenario A) and overriding of the laying pedestrian (scenario B). However, the further research is still an ongoing task, and full vehicle testing must be done.

Author Contributions: Conceptualization, S.Š. and J.Š.; software, S.Š, J.Š, and J.V.; validation; S.Š; formal analysis, S.Š. and J.Š.; investigation, S.Š.; resources, S.Š. and M.K.; data curation, S.Š., M.K. and A.S.; writing, S.Š, J.Š, and J.V.; visualization, S.Š. and M.K.; supervision, S.Š. All authors have read and agreed to the published version of the manuscript.

Funding: This work has been supported by the project TRIO FV20441 "Research and development of safe tram face" provided by the Ministry of Industry and Trade of Czech Republic).

Institutional Review Board Statement: Not applicable.

Informed Consent Statement: Not applicable.

Conflicts of Interest: The authors declare no conflict of interest.

Appendix A

The experimental results of the spherical metal impactor to the layered glass (windshield) are presenter here.

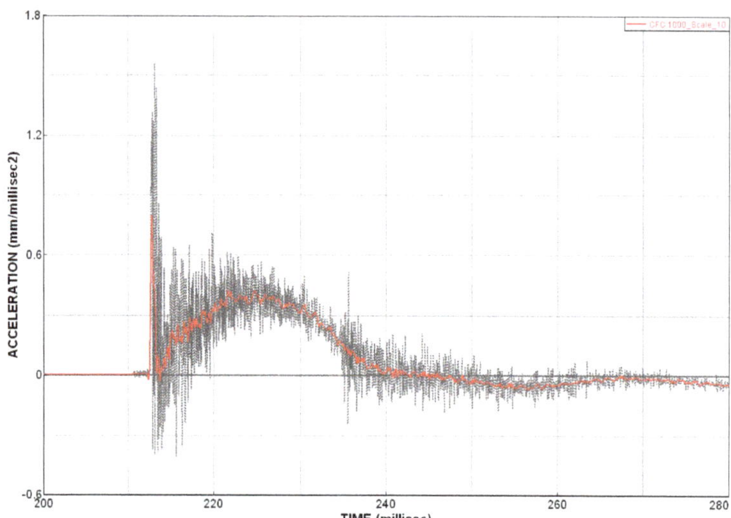

Figure A1. Original and filtered acceleration curve (sampling rate 48 kHz filtered with the CFC 1000 filter).

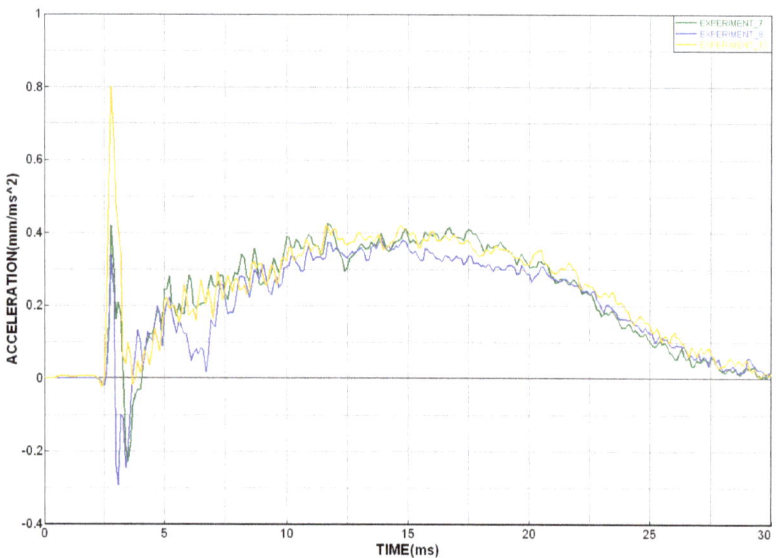

Figure A2. Examples of the experimental acceleration curves.

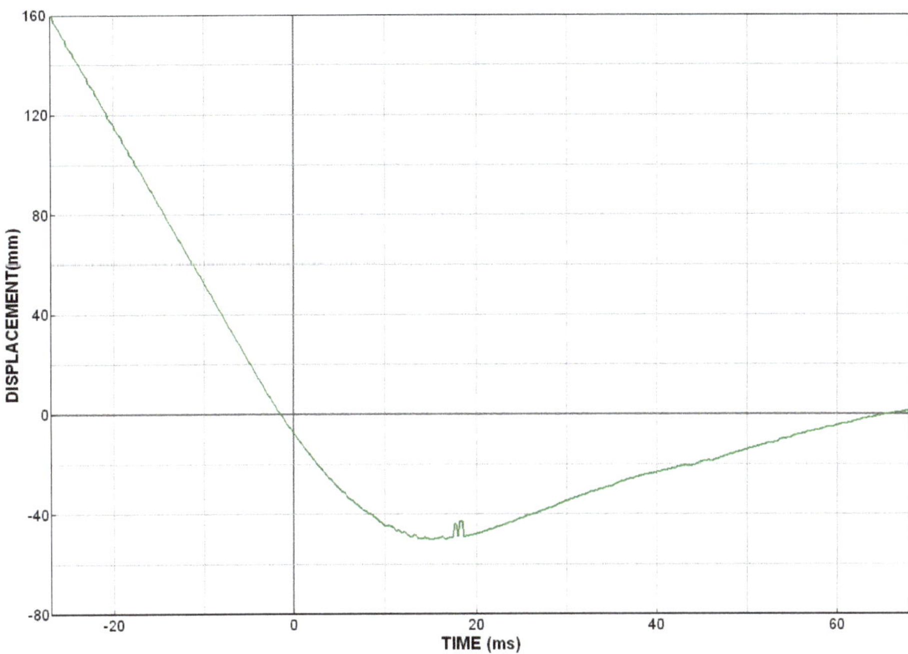

Figure A3. Displacement of the impactor COG, data from high-speed camera.

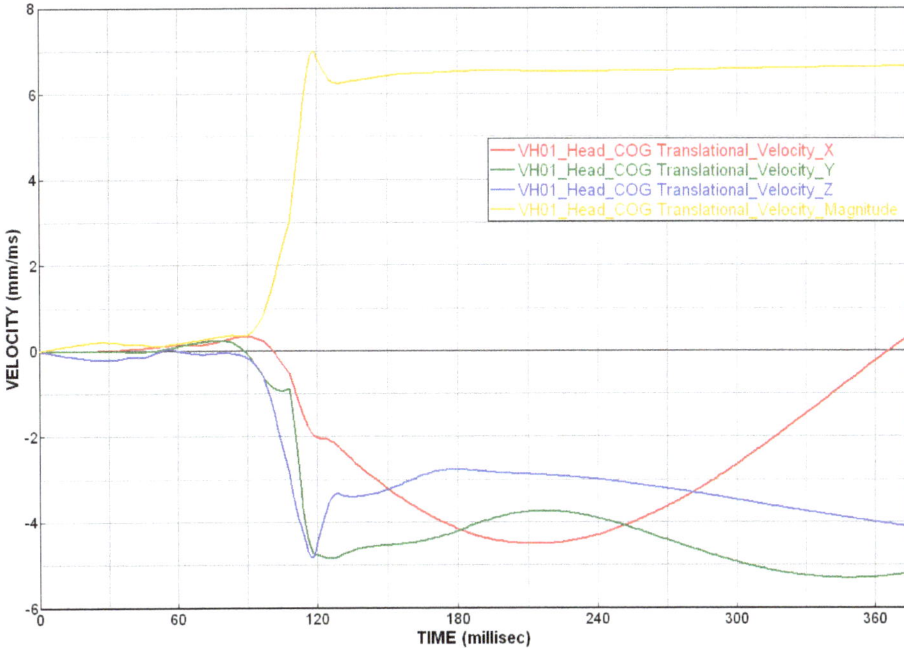

Figure A4. Velocity of the head in the simulation of collision scenario A.

References

1. Špirk, S.; Špička, J.; Vychytil, J. Simulation of Tram-Pedestrian Collision with Validated Windshield Material Model. In Proceedings of the 58th International Conference on Experimental Stress Analysis-EAN2020, Sobotín, Czech Republic, 19–22 October 2020.
2. WHO. *Global Status Report on Road Safety in 2015*; World Health Organisation: Geneva, Switzerland, 2015; Available online: https://www.who.int/violence_injury_prevention/road_safety_status/2015/en/ (accessed on 13 June 2015).
3. Besip. *Overview of the Road Accident in the EU [Online]*; Ministry of Transportation: Toronto, Canada, 2016; Available online: http://www.ibesip.cz/cz/statistiky/statistiky-nehodovosti-v-evrope/prehled-vyvojedopravnich-nehod-v-eu (accessed on 6 December 2016).
4. Policie, Č.R. *Overview of Road Accident in the Czech Republic in the 2014*; Directorate of Transport Police; Police Presidium of the Czech Republic: Praha, Czech Republic, 2015; Available online: http://www.policie.cz/clanek/statistika-nehodovosti-900835.aspx?Q=y2hudw09mw%3d%3d (accessed on 3 February 2015).
5. Ramamurthy, P.; Blundell, M.V.; Bastien, C.; Zhang, Y. Computer simulation of real-World vehicle–Pedestrian impacts. *Int. J. Crashworthiness* **2011**, *16*, 351–363. [CrossRef]
6. Yang, J.; Yao, J.; Otte, D. Correlation of different impact conditions to the injury severity of pedestrians in real world accidents. In Proceedings of the 19th International Technical Conf. Enhanced Safety of Vehicle, Washington, DC, USA, 6–9 June 2005.
7. TAČR TA04030689. *Development of Active Car Bonnet with Respect to the Diversity of the Human Population and Implementation of the Biomechanical Model of Human Body*; Intermediate Report: Pilsen, Czech Republic, 2015.
8. Guesset, A. Technical Agency for Ropeway and Guided Transport System. In *Tramway Front End Design STRMTG*; Ver. 01; The Technical Guides: Grenoble, France, 2016.
9. EuroNCAP. *Assessment Protocol–Vulnerable Road User Protection*; Ver. 10.0.2; EuroNCAP: Leuven, Belgium, 2019; Available online: https://www.euroncap.com/en/for-engineers/protocols/vulnerable-road-user-vru-protection/ (accessed on 11 December 2019).
10. EuroNCAP. *Assessment Protocol–Adult Occupant Protection*; Ver. 6.0; EuroNCAP: Leuven, Belgium, 2013; Available online: http://www.globalncap.org/wp-content/uploads/2013/06/assessment-protocol-Adult.pdf (accessed on 20 September 2013).
11. Schmitt, K.U.; Niederer, P.; Muser, M.; Walz, F. *Trauma Biomechanics*; Springer: Berlin/Heidelberg, Germany, 2010; pp. 143–152.
12. Špička, J.; Vychytil, J.; Maňas, J.; Pavlata, P.; Motl, J. Modelling of Real Car-To-Pedestrian Accident: Comparison of Various Approaches in the Car Bonnet Modelling. In Proceedings of the ECCOMAS Thematic Conference on Multibody Dynamics, Prague, Czech Republic, 2 February 2017.
13. Špička, J.; Vychytil, J.; Hynčík, L. Numerical Analysis of a Pedestrian to Car Collision: Effect of Variations in Walk. In *Applied and Computational Mechanics*; The University of West Bohemia: Pilsen, Czech Republic, 2017; Volume 10.
14. TRIO FV20441. *Research and Development of Safe Tram Face—Parametric and Sensitivity Studies of Alternative Solutions for Tram Face Shaping*; Intermediate Report: Pilsen, Czech Republic, 2019.
15. Vychytil, J.; Maňas, J.; Čechová, H.; Špirk, S.; Hynčík, L.; Kovář, L. *Scalable Multi-Purpose Virtual Human Model for Future Safety Assessment (No. 2014-01-0534)*; SAE Technical Paper; The University of West Bohemia: Pilsen, Czech Republic, 2014.
16. Maňas, J.; Kovář, L.; Petřík, J.; Čechová, H.; Špirk, S. Validation of human body model VIRTHUMAN and its implementation in crash scenarios. In *Advances in Mechanisms Design*; Springer: Berlin/Heidelberg, Germany, 2012; pp. 351–356.
17. Vychytil, J.; Hynčík, L.; Maňas, J.; Pavlata, P.; Striegler, R.; Moser, T.; Valášek, R. *Prediction of Injury Risk in Pedestrian Accidents Using VIRTual Human Model VIRTHUMAN: Real Case and Parametric Study (No. 2016-01-1511)*; SAE Technical Paper; The University of West Bohemia: Pilsen, Czech Republic, 2016.
18. Hynčík, L.; Špička, J.; Maňas, J.; Vychytil, J. *Stature Based Approach towards Vehicle Safety, (No. 2015-26-0209)*; SAE Technical Paper; The University of West Bohemia: Pilsen, Czech Republic, 2005.
19. Hynčík, L.; Bońkowski, T.; Vychytil, J. Virtual hybrid human body model for PTW safety assessment. *Appl. Comput. Mech.* **2017**, *11*, 2017137. [CrossRef]
20. ESI. *Virtual Perfomance Solution*; Rev.2; VPS Explicit MBS Model; Mecas ESI: Plzeň, Czech Republic, 2019.
21. Scataglini, S.; Gunther, P. *DHM and Posturography*; Academic Press: New York, NY, USA, 2019.
22. Kosiński, P.; Osiński, J. Laminated Windshield Breakage Modelling in the Context of Headform Impact Homologation Tests. *Int. J. Appl. Mech. Eng.* **2015**, *20*, 87–96. [CrossRef]
23. Peng, Y.; Yang, J.; Deck, C.; Willinger, R. Finite element modeling of crash test behavior for windshield laminated glass. *Int. J. Impact Eng.* **2013**, *57*, 27–35.
24. Pyttel, T.; Liebertz, H.; Cai, J. Failure criterion for laminated glass under impact loading and its application in finite element simulation. *Int. J. Impact Eng.* **2011**, *38*, 252–263.
25. Barbat, S.D.; Jeong, H.Y.; Prasad, P. *Finite Element Modeling and Development of the Deformable Featureless Headform and Its Applications to Vehicle Interior Head Impact Testing*; SAE Transactions SAE: Warrendale, PA, USA, 1996.
26. Kamalakkannan, S.B.; Guenther, D.A.; Wiechel, J.F.; Stammen, J. MADYMO Modeling of the IHRA Head-Form Impactor. In Proceedings of the SAE Digital Human Modeling for Design and Engineering Conference, Warrendale, PA, USA, 4–6 July 2005.
27. Barbosa, A.; Fernandes, F.A.O.; Sousa, A.d.R.J.; Ptak, M.; Wilhelm, J. Computational Modeling of Skull Bone Structures and Simulation of Skull Fractures Using the YEAHM Head Model. *Biology* **2020**, *9*, 267. [CrossRef] [PubMed]
28. Mohan, P.; Marzougui, D.; Kan, C.-D. Development and Validation of Hybrid III Crash Test Dummy. In Proceedings of the SAE World Congress & Exhibition, Detroit, MI, USA; 2009.

29. Keller, U.H.; Mortelmans, H. *Adhesion in Laminated Glass; What Makes it Work?* Glass Performance Days: Tampere, Finland, 1999; pp. 353–356.
30. Santarsiero, M.; Louter, C.; Nussbaumer, A. Laminated connections for structural glass applications under shear loading at different temperatures and strain rates. *Constr. Build. Mater.* **2016**, *128*, 214–237. [CrossRef]
31. Pelfrene, J.; Dam, S.; Spronk, S.; Paepegem, V.W. Experimental Characterization and Finite Element Modelling of Strain-rate Dependent Hyperelastic Properties of PVB Interlayers. In Proceedings of the Challenging Glass Conference 6At, Delft, The Netherlands, 17–18 May 2018.
32. Prasongngen, J.; Putra, I.P.A.; Koetniyom, S.; Carmai, J. Improvement of windshield laminated glass model for finite element simulation of head-To-Windshield impacts. In Proceedings of the 2019 IOP Conference Series: Materials Science and Engineering, Kunming, China, 25–29 May 2019.
33. Warsiyanto, B.A. Bird Strike Analysis on 19 Passenger Aircraft Windshield with Different Thickness and Impact Velocity. *J. Teknol. Kedirgant.* **2020**, *5*. [CrossRef]
34. Schuster, M.; Schneider, J.; Nguyen, T.A. Investigations on the execution and evaluation of the Pummel test for polyvinyl butyral based interlayers. *Glas. Struct. Eng.* **2020**, *5*, 371–396. [CrossRef]
35. Bláha, P.; Šedivý, V.; Čechovský, K.; Kosová, A. *Anthropometric Studies of the Czechoslovak Population from 6 to 55 Years*; part 2; Czechoslovak spartakiade: Praha, Czech Republic, 1985; Volume 1.
36. Robbins, D.H. *Anthropometric Specifications for Mid-Sized Male Dummy*; University of Michigan Transportation Research Institute (UMTRI): Pilsen, Czech Republic, 1983.
37. Kossakowski, P. Influence of Initial Porosity on Strength Properties of S235JR Steel at Low Stress Triaxiality. *Arch. Civ. Eng.* **2012**, *58*, 293–308.
38. Duffy, V.; Salvendy, G. *Handbook of Digital Human Modeling: Research for Applied Ergonomics and Human Factors Engineering*; CRC Press: Boca Raton, FL, USA, 2008.

 materials

Article

Modeling and Testing of Flexible Structures with Selected Planar Patterns Used in Biomedical Applications

Pavel Marsalek [1,*], Martin Sotola [1], David Rybansky [1], Vojtech Repa [1], Radim Halama [1], Martin Fusek [1] and Jiri Prokop [2,3]

1 Deparment of Applied Mechanics, Faculty of Mechanical Engineering, VŠB—Technical University of Ostrava, 17. listopadu 2172/15, 708 00 Ostrava, Czech Republic; martin.sotola@vsb.cz (M.S.); david.rybansky@vsb.cz (D.R.); vojtech.repa.st@vsb.cz (V.R.); radim.halama@vsb.cz (R.H.); martin.fusek@vsb.cz (M.F.)

2 Department of Surgical Studies, Faculty of Medicine, University of Ostrava, Dvorakova 7, 701 03 Ostrava, Czech Republic; jiri.prokop@fno.cz

3 Department of Surgery, University Hospital Ostrava, 17. listopadu 1790/5, 708 00 Ostrava, Czech Republic

* Correspondence: pavel.marsalek@vsb.cz

Abstract: Flexible structures (FS) are thin shells with a pattern of holes. The stiffness of the structure in the normal direction is reduced by the shape of gaps rather than by the choice of the material based on mechanical properties such as Young's modulus. This paper presents virtual prototyping of 3D printed flexible structures with selected planar patterns using laboratory testing and computer modeling. The objective of this work is to develop a non-linear computational model evaluating the structure's stiffness and its experimental verification; in addition, we aimed to identify the best of the proposed patterns with respect to its stiffness: load-bearing capacity ratio. Following validation, the validated computational model is used for a parametric study of selected patterns. Nylon—Polyamide 12—was chosen for the purposes of this study as an appropriate flexible material suitable for 3D printing. At the end of the work, a computational model of the selected structure with modeling of load-bearing capacity is presented. The obtained results can be used in the design of external biomedical applications such as orthoses, prostheses, cranial remoulding helmets padding, or a new type of adaptive cushions. This paper is an extension of the conference paper: "Modeling and Testing of 3D Printed Flexible Structures with Three-pointed Star Pattern Used in Biomedical Applications" by authors Repa et al.

Keywords: wearable; flexible; structure; stiffness; biomedical; mechanics; simulation; pattern; 3D print; PA12

Citation: Marsalek, P.; Sotola, M.; Rybansky, D.; Repa, V.; Halama, R.; Fusek, M.; Prokop, J. Modeling and Testing of Flexible Structures with Selected Planar Patterns Used in Biomedical Applications. *Materials* **2021**, *14*, 140. https://doi.org/10.3390/ma14010140

Received: 19 November 2020
Accepted: 28 December 2020
Published: 30 December 2020

Publisher's Note: MDPI stays neutral with regard to jurisdictional claims in published maps and institutional affiliations.

Copyright: © 2020 by the authors. Licensee MDPI, Basel, Switzerland. This article is an open access article distributed under the terms and conditions of the Creative Commons Attribution (CC BY) license (https://creativecommons.org/licenses/by/4.0/).

1. Introduction

Simple shapes have been connected to form patterns since ancient times. Mostly, such tasks had only an aesthetic function, such as the decoration of exterior surfaces. The process of covering the surface with a pattern is called tesselation and the geometric shapes forming the pattern are called tiles. Tesselation in two dimensions (2D), also called planar tiling, is a process of arranging the tiles to fill a surface according to predefined rules. Periodic tiling has a repeating pattern. Patterns are common also in nature—for instance, an almost perfect pattern of hexagonal cells can be found in honeycombs [1,2]. Recent studies showed the potential of patterns in designing structures as an emerging solution for the reduction of the product weight, consumption of energy during production, and of manufacturing time [3,4]. For example, a lattice structure is often recommended to maintain the stiffness of structures while reducing weight (and, thus, material consumption) [5,6]. An application of an aluminum honeycomb structure can be found in the article by Phu et al. [7] investigating the impact properties of such a solution in impact attenuators, requiring a strictly defined crushing behavior.

The requirements for controlling the stiffness of structures are becoming more common in biomedical applications. Therefore, the authors of this work focused on the design of flexible structures (FS) that can be produced by means of additive manufacturing, especially by 3D printing. The term FS typically describes a thin shell with a pattern of holes decreasing the structure's stiffness in the normal direction. This approach allows the stiffness and other mechanical properties of the part to be controlled by the shape of the empty areas (holes) such as the size of the gaps, structure thickness, curvature, etc., rather than by the choice of material. Typical applications of FS include the flexible parts of orthoses or prostheses that are in direct contact with the patient's body [8], cranial remoulding helmets padding [4], new types of adaptive cushions and others [9]. In this paper, authors focused mainly on external biomedical application (for example, already mention helmet padding). However, it should be noted that some FS are used in vivo, such as flexible micro-LEDs for optogenetic neuromodulation [10] or flexible films (patches) for energy harvesting [11].

Parts with such a flexible structure can be designed with desired deformation behavior. For example, Schumacher et al. [12] were studying mechanical properties of structured sheet materials. They managed to establish a link between the complex deformation behavior and elastic properties through numerical methods (homogenization). Doing so, they calculated various structures and evaluated their Young's modulus, Poisson's ratio and bending stiffness, which helped them determine if the structures had isotropic or anisotropic behavior.

Bickel et al. [13] also studied the deformation behavior of flexible structures. They aimed to optimize the base material structure to achieve desired deformation behavior. For this purpose, they used the finite element method (FEM) as the design tool, combined with additive manufacturing. Lastly, it is possible to design aesthetic modifications of parts that also preserve their structural integrity; see e.g., various papers by Schumacher et al. [14]. Those authors presented a novel method to design shells factoring at the same time the aesthetics of the product, its stability and material efficiency into the design through structural optimization.

Wearable FS applications must strictly meet the stiffness requirements established through medical examination because they must be adapted to the user body. One of the aims of the use of flexible structures is to reduce swelling and to allow a proper flow of body fluids. The use of holed design also facilitates breathing through the skin. Due to the enormous advantages of 3D printing methods such as Selective Laser Sintering (SLS) or Multi-Jet Fusion (MJF), the complex shape of FS can be printed without supports, which allows the production of the final shape without additional machining. SLS is based on gradual sintering of powder layers using a laser beam, offering high accuracy even with complex shapes [15]. The procedure is simple: A thin layer of powder is applied on the working surface, heating the material slightly below the sintering temperature. Next, the laser beam heats the exact locations of the area, which leads to sintering. After sintering, the base plate is moved down by the thickness of a single layer and the whole process is repeated. The base plate and work chamber are heated, which prevents the part from warping or shrinking during printing. MJF is similar to SLS but after the powder layer is applied, an additional fusion agent must be jetted with precision on the exact locations to be sintered. Then, a high-power infrared (IR) light passes over the bed and the areas containing the fusion agent reach the sintering temperature while other areas remain untouched. This method is significantly faster and, in some cases, more accurate than SLS [16]. However, the color of the product, which is usually grey to black due to the use of IR light, represents (for some applications, at least), a possible disadvantage.

The objective of this work was to prepare a method for determining the stiffness of an isohedral flexible structure using planar patterns. However, as performing all experiments with custom-printed structures would be extremely expensive, a computer model using the Finite element method (FEM) was created for evaluation of the structures' stiffness. This approach allowed a great improvement in cost-effectiveness as only a tensile test of the material used for printing was necessary to acquire the input for the model. The stiffness derived from the model was then compared with experimental results. This approach does not aim to fully remove the experiments but to significantly decrease their number. After preparing the computer model, an additional parametric study was performed to demonstrate the advantages of this approach. The preliminary results of the work using a linear computational model were presented at the EAN 2020 conference and published in the book of extended abstracts by authors Repa et al. [17].

2. Materials and Methods

The proposed methods are divided into four stages. The first stage describes the design of the experiment, aiming at the selection of patterns suitable for the specimens. Moreover, the choice of the material of the specimens is discussed in the first stage. Necessary experiments evaluating the stiffness of the structure are also performed at this stage. The remaining stages focus on the design of the test specimens and testing device. In the second stage, the design of a simplified non-linear computational model replacing extensive experiments is described. The result of the second stage is a validated model that sufficiently describes the experiment in terms of structural stiffness. In the third stage, the geometric properties of the selected patterns are analyzed using a parametric study. Since the designed specimens are sensitive to failure when subject to large deformations, a full non-linear computer model of the selected pattern is designed in the last stage, used for comparison of load-bearing capacity. In this way, it is possible to assess not only the stiffness of the structure but also its load-bearing capacity. Figure 1 shows a diagram summarizing the workflow described in the previous paragraph.

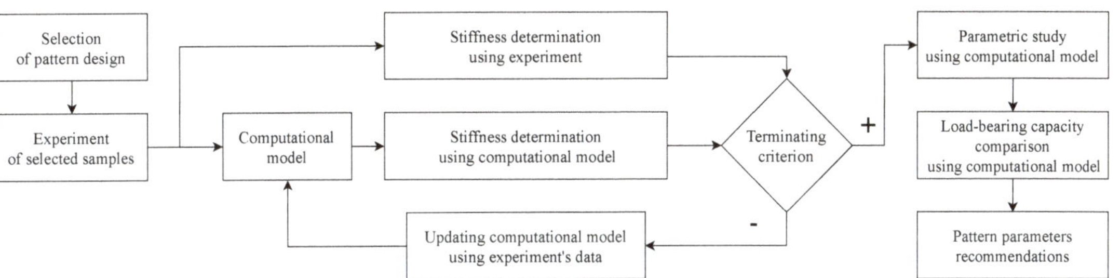

Figure 1. A diagram showing the research workflow of modeling and testing of flexible structures.

2.1. First Stage: Experimental Design

2.1.1. Selection of Pattern Design

In this work, the starting patterns were selected based on the authors' experience with designing biomedical applications.

Three types of FS pattern were used, namely the three-pointed star pattern (A), modified three-pointed star pattern (B) and tri-hexagonal pattern (C). All patterns were created by isohedral tesselation and prepared based on custom-selected input parameters, which are detailed in Figure 2. The three-pointed star (A) and tri-hexagonal (C) patterns have a three-axis reflection symmetry while the modified three-pointed star pattern (B) has a three-fold rotation symmetry; this, according to Schumacher et al., [12] means that their behavior during loading should be isotropic.

2.1.2. Material Selection

The complex geometry of FS is difficult to be manufactured using conservative ways, see Figure 3. Additive manufacturing (AM), on the other hand, offers freedom in designing complex geometries. Nevertheless, not all AM methods are suitable for FS printing. The methods that can be used for their production should offer support-free printing. Suitable methods are mentioned in the Introduction.

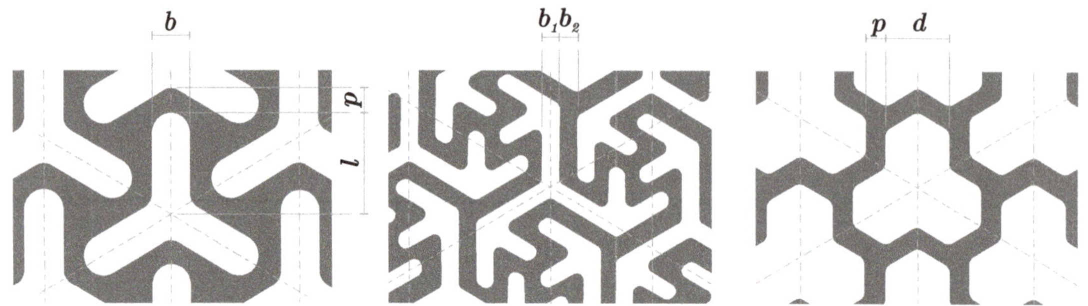

Figure 2. Selected pattern designs for flexible structures—three-pointed star pattern A (**left**), modified three-pointed star pattern B (**middle**) and tri-hexagonal pattern C (**right**).

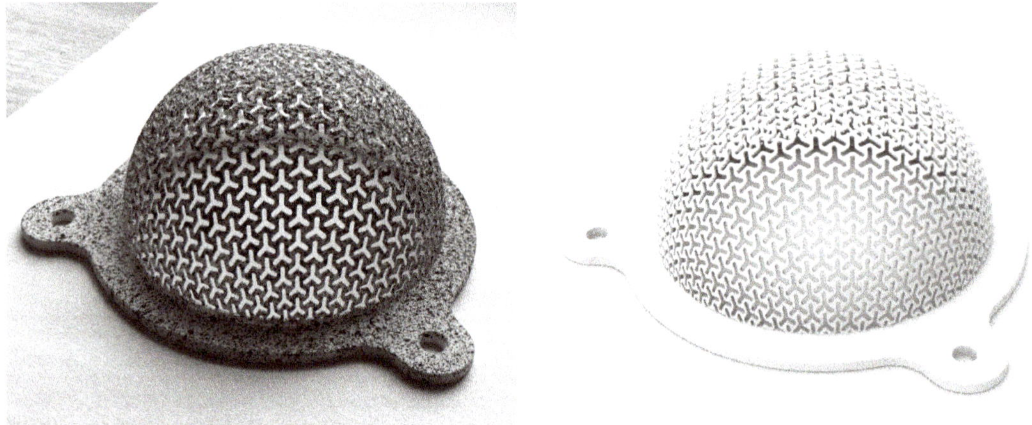

Figure 3. Printed specimen A with a spot pattern for DIC method (**left**) and visualization of geometric model (**right**).

For the purposes of this study, Nylon (Polyamide 12, PA12), was chosen as an appropriate material suitable for 3D printing due to its material properties described in the article by Marsalek et al. [4]. PA12 is a thermoplastic material offering good chemical and wear resistance and high toughness. It has a great strength to weight ratio and is also recommended for biomedical applications [18]. 3D printed products usually have anisotropic behavior dependent on the orientation of printing. However, the SLS printing method used in our case study allowed us to neglect the variation of Young's modulus [19]. The initial elastic material properties of PA12 are detailed in Table 1; they were obtained from the article by Stoia et al. [20] (in that article, PA12 is labeled as PA2200). Obtained values were for printed samples with the orientation angle of 90 degrees. The table also contains the yield strength and tangent modulus of the PA12 material to describe the plastic behavior of the specimen for computer modeling of designed experiment.

2.1.3. Description of Testing Methodology

The experimental verification of the shell stiffness was inspired by the laboratory testing of cranial orthoses published by Marsalek et al. [4].

Table 1. Mechanical properties of PA12.

Property	Symbol	Value	Unit
Young's modulus	E	1224	MPa
Poisson ratio	μ	0.39	–
Density	ρ	1010	kgm^{-3}
Yield stress	σ_Y	21	MPa
Tangent modulus	E_T	334	MPa

The authors of the article used a plunger to gradually deform the structure and analyze the displacement. Also, the authors used a segment of a large radius sphere as a testing specimen in their paper. In the new testing methodology proposed in the presented paper, the specimen representing the FS is a hemisphere with a fixed inner radius of $R_1 = 65.0$ mm, see the scheme in Figure 4. We expect the new specimen shape to be suitable for a wider range of applications due to type of loading. The hemisphere shape was subjected to two types of loading. The area near the plunger was subjected mainly to bending while the area near the rim was subjected to a combination of bending and tensile load. The curvature of the hemisphere was also closer to applications such as helmet padding (almost hemispheric shape) or orthoses and prostheses (almost cylindrical shape). A fastening system was designed for clamping specimens and a plunger with dimensions of $R_2 = 65.0$ mm and $h = 18.0$ mm was prepared for loading the specimens. The continuous load was generated and measured using a universal testing machine TESTOMETRIC 50050CT (Testometic, Rochdale, UK). The fastening system was designed using the CAD software (SW) Autodesk Inventor (Autodesk, San Rafael, CA, USA). Every component of the testing device, including samples, was made by additive manufacturing. Most of them were printed using the SLS method and PA12 material. Only the plunger was printed using the Fused Filament Fabrication (FFF) method and PLA (polylactic acid) material. It must be emphasized that specimens were more compliant than the fastening system and plunger. The fastening system was designed with robust legs with stiffeners. Also, both the fastening system and plunger were printed as full material (i.e., without using lattice infilling), which further increased their stiffness.

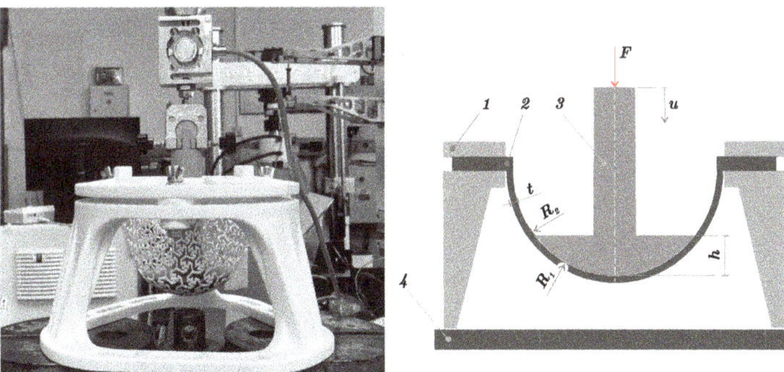

Figure 4. The experimental equipment (**left**) and its scheme (**right**): 1—fastening system, 2—specimen, 3—plunger, 4—frame.

Samples were designed using a combination of two CAD SW, Rhinoceros (Robert McNeel & Associates, Seattle, WA, USA, RHN) and Ansys Spaceclaim (Ansys, Canonsburg, PA, USA, SCDM). Most of the common CAD software solutions tend to deform the shape of the structure when applied on curved surfaces. However, CAD RHN offers robust modeling of such complex structures even on non-planar surfaces. SCDM supports the use of custom-made scripts for creating such structures on the reference surface. At the same time, SCDM supports easy export of the geometric model representation (CAD) into ANSYS Workbench

(Ansys, Canonsburg, United States of America, AWB). The designed specimens (3D printed and virtual one) using pattern A are shown in Figure 3.

Altogether, four specimens were printed for the purposes of the experiment, namely two specimens marked A1 with a three-pointed star pattern employing parameters $b = 2.00$ mm, $l = 5.00$ mm, and $p = 2.35$ mm, and two specimens with a modified three-pointed star pattern marked B1 with parameters of $b_1 = 1.75$ mm, and $b_2 = 1.75$ mm, see Figure 2. Both A1 and B1 specimens were printed with two thicknesses ($t = 1.00$ and 1.25 mm, respectively). The parameters of these patterns are listed in Table 2.

Table 2. Parameters of tested structures.

Specimen	Parameter of Pattern			
Specimen A1	b [mm]	l [mm]	p [mm]	t [mm]
	2.00	5.00	2.35	1.00 1.25
Specimen B1	b [mm]	l [mm]	-	t [mm]
	1.75	1.75	-	1.00 1.25

2.2. Second Stage: Virtual Modeling of Structural Stiffness

To be able to replace the FS experimental measurements with a virtual one, two different computational models were designed. These static computational models were solved using FEM in SW AWB. The spatial geometry of the samples was converted into a shell structure and the specimen rim was removed. A surface body for discretization was obtained as the inner surface of the specimen. Shell discretization (elements SHELL181) was used to describe the behavior of the structure. The thickness of elements was set to be outwards. The basic characteristics of the shell elements in the application are described in [21]. The number of finite elements and nodes used for models is described in the Table 3. The discretization is shown in Figure 5.

Table 3. Finite element mesh statistics.

	Number of Elements	Number of Nodes
Modeling of structures stiffness	81,500	93,200
Modeling of load-bearing capacity	158,000	177,000

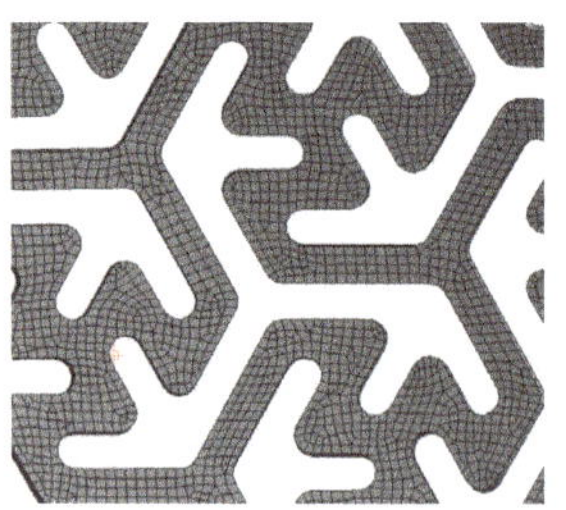
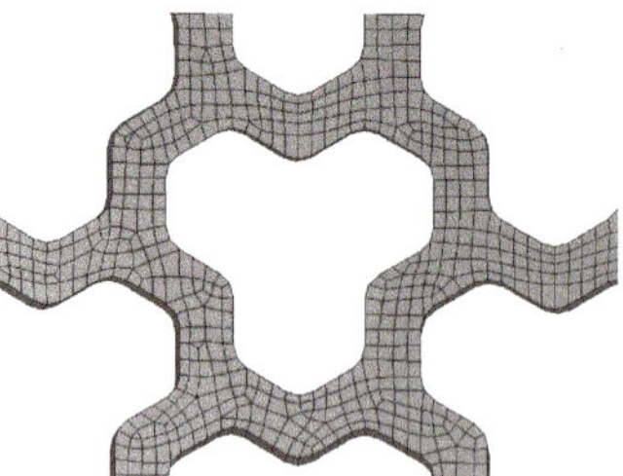

Figure 5. Visualization of the detail of the finite element mesh (specimen A—**left**, specimen B—**middle** and specimen C—**right**).

The first simplified non-linear model was used for determining structural stiffness. The simplified model did not contain plunger. The second full non-linear model was used for comparison of the load-bearing capacity. Analyses of both models took into account large deformation and displacement. In both models, a bilinear PA12 material elastic-plastic was used for describing plastic behavior. The mechanical properties for both material models are listed in Table 1. It is necessary to mention that modeling the load-bearing capacity of structures is very difficult as it is necessary to consider the complexity of problem using an iterative solver with an unsymmetric Newton Raphson option. This topic was studied previously by the group of Horyl and Marsalek [22,23].

The simplified non-linear computational model for prediction of the structure stiffness used only a simplified structure of the individual specimens A, B and C (without the rim), see Figure 6. Since the plunger was much stiffer than the tested structure, the boundary condition corresponding to the specimen loading (by plunger) was defined by using an ANSYS function Remote force with deformable behavior (red color). Remote force is placing remote (pilot) node that is connected to the part using multi-point connection, which can be set as deformable or rigid behavior. Rigid behavior of the remote force is not appropriate due to significant alteration of stiffness in the area of application. Using this function, the force acting on the remote point is distributed to the applied area. The static load was gradually increased. The outer part of the structure (where the rim was removed) was fixed (blue color), all degrees of freedom (three displacement and three rotations) were set to 0 mm, respectively 0 rad. The friction between the plunger and the specimen was disregarded. The load on the structure was gradually increased until the established value of the loading force was reached. In the case of specimen A, the maximal load was $F = 250$ N, in the case of specimen B, the maximal load was $F = 50$ N. Specified values of the loading forces correspond to the experiment, see Figure 7. For the last type of specimens (specimen C), the maximal load was $F = 150$ N.

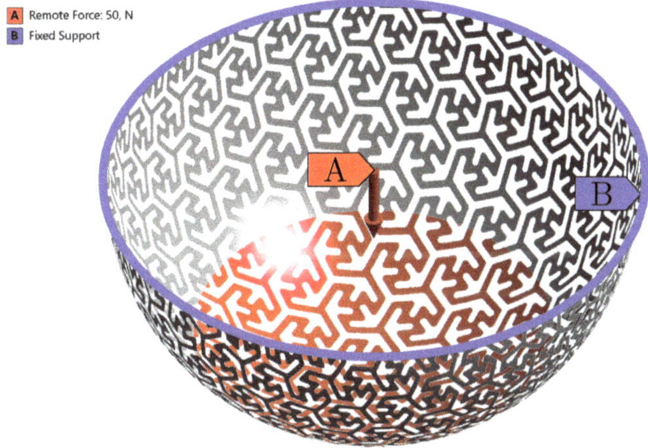

Figure 6. Boundary conditions for a simplified non-linear model analyses of structure stiffness.

Due to the 3D printer capabilities and printing tolerances, the resulting thicknesses of the printed specimens were lower than required. As a result, we had to adjust the models to the real thicknesses, i.e., the t_r values were reduced by 0.20 mm during stiffness modeling.

2.3. Third Stage: Parametric Study

All selected patterns were used for the parametric study of the FS stiffness employing the validated simplified computer model. In all, the stiffness was evaluated for 54 combinations of the pattern and real thickness. For each of the three pattern types (3-point star, modified three-point star, tri-hexagonal), three variants of the pattern geometry were analyzed, see Table 4 for individual parameters. In addition, five thicknesses for each of these pattern geometries were evaluated, ranging from 0.80 mm to 2.05 mm with the step size of 0.25 mm.

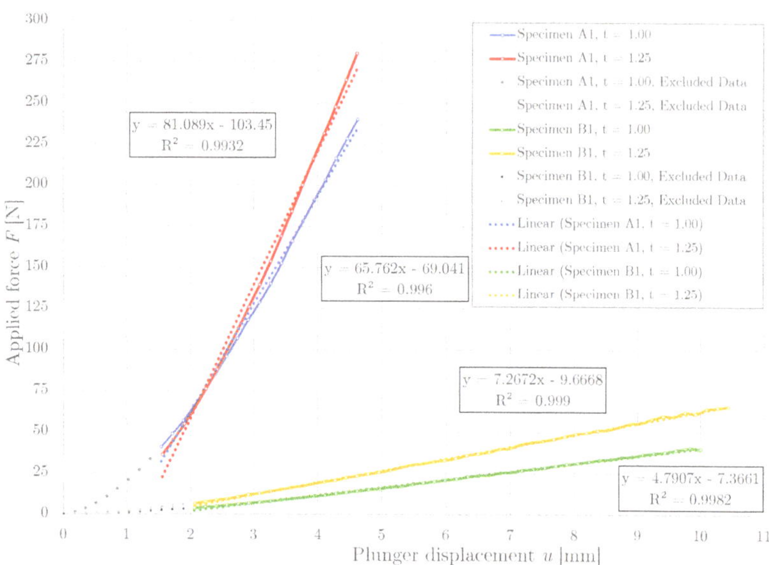

Figure 7. The relationship between the applied force F and measured plunger displacement u.

Table 4. Used specimen pattern parameters.

Pattern Parameters	Opening Parameter b [mm]	Opening Parameter l [mm]	Width between Pattern Openings p [mm]
Specimen A1	2.00	5.00	2.35
Specimen A2	2.00	5.00	2.60
Specimen A3	2.25	5.25	2.35
	Opening Parameter b_1 [mm]	**Width between Pattern Openings b_2 [mm]**	-
Specimen B1	1.75	1.75	-
Specimen B2	1.75	2.00	-
Specimen B3	2.00	1.75	-
	Opening Parameter d [mm]	**Width between Pattern Openings p [mm]**	-
Specimen C1	5.00	1.75	-
Specimen C2	5.00	2.00	-
Specimen C3	5.25	1.75	-

2.4. Fourth Stage: Modeling of the Load-Bearing Capacity

To capture the physical behavior of the specimen a full non-linear computational model was created. The computational model was based on a parametric geometrical model of the specimen with a three-point star pattern, just like in the case of simulation for stiffness determination. In this case, however, the computational model was complemented by a spatial geometrical model of the plunger. The plunger was considered rigid, as its stiffness is significantly higher than that of the specimens. Also, the Augmented Lagrange contact algorithm using friction between the specimen and plunger was considered. The friction coefficient f_s = 0.30 was obtained from the article by Bai et al. [24] focusing on tribological and mechanical properties of PA12. The gradually increasing displacement was applied on the plunger, up to the value of u_{max} = 4.00 mm. The boundary conditions for the non-linear model are depicted in Figure 8.

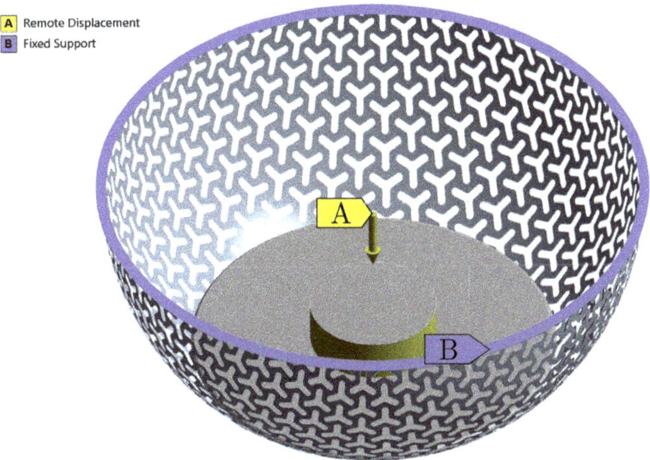

Figure 8. Boundary conditions for a full non-linear model analyses of load-bearing capacity.

3. Results

Experiments were performed on 4 specimens (A1 and B1 in two thickness variants) and their results were used for validation of the computational model. This model was in turn used for a parametric stiffness study that was performed for various parameters of patterns A, B, and C. Based on the results, the pattern type combining the best properties (best stiffness:load-bearing capacity), namely pattern A, was selected for the load-bearing capacity modeling.

3.1. Stiffness Determination Using Experiment

The measured relationship between the applied force F and the plunger displacement u is depicted in Figure 7. It is apparent that the relation between the force F and displacement is non-linear at the beginning. This was caused by adjusting the backlash (gap) between the plunger and the specimen. After adjustment, it can be viewed as an almost linear dependence. This meant that the beginning had to be excluded from the dataset. The aim of the data exclusion was to obtain a linear regression with a coefficient of determination R^2 of > 0.99. The resulting stiffness values are listed in Table 5.

Table 5. Values of determined stiffness k from experiment.

Specimen	Designed Thickness t [mm]	Experiment k [N/mm]
A1	1.00	65.76
	1.25	81.09
B1	1.00	4.79
	1.25	7.27

3.2. Stiffness Determination Using the Computational Model

The validated computational model calculated the relationship between the applied force F and maximal displacement u, from which stiffness was determined by linear approximation (in the same way as in the case of experimental measurement). A high agreement between the experiment and simulation was achieved (see Table 6). The vector displacement field and Von Mises stress field corresponding to the loading forces of F = 250 N and F = 50 N, respectively, are shown in Figures 9 and 10. It can be seen that Von Mises Stress is higher than the considered yield strength σ_Y = 21 MPa at local areas (near the notch, depicted by red color). Plastic deformation occurs in these areas.

Table 6. Comparison of determined stiffness k from experiment and simulation.

Specimen	Designed Thickness t [mm]	Real thickness t_r [mm]	Experiment k [N/mm]	Simulation k [N/mm]	Difference [%]
Specimen A1	1.00	0.80	65.76	64.17	2.42
	1.25	1.05	81.09	86.28	6.40
Specimen B1	1.00	0.80	4.79	4.80	0.21
	1.25	1.05	7.27	6.37	12.38

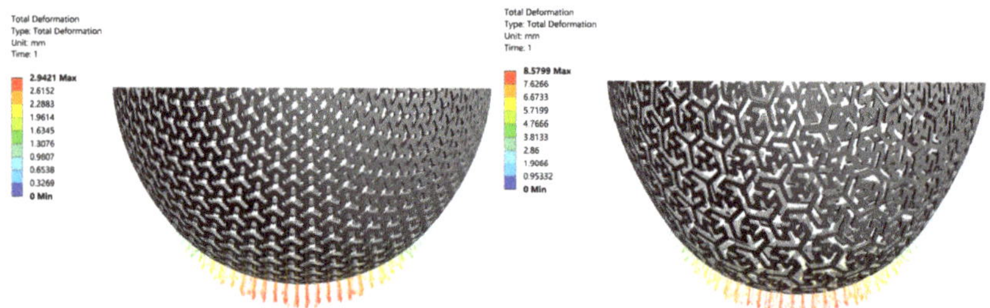

Figure 9. The vectordisplacement fields corresponding to the applied force F = 250 N for Specimen A1 (**left**), respectively the applied force F = 50 N for specimen B1 (**right**), real thickness t_r = 1.05 mm.

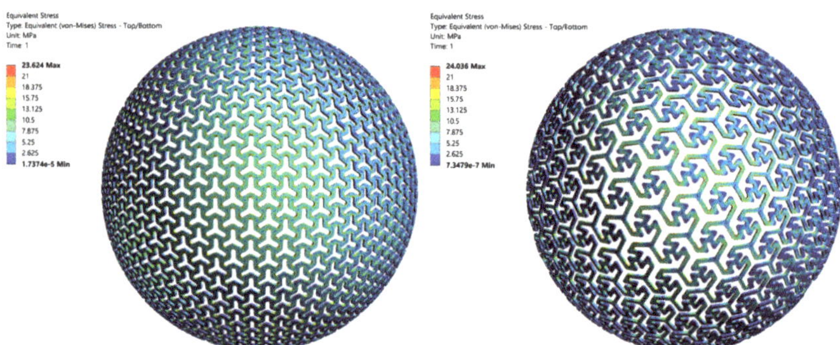

Figure 10. Von Mises stress fields corresponding to the applied force F = 250 N for Specimen A1 (**left**), respectively the applied force F = 50 N for specimen B1 (**right**), real thickness t_r = 1.05 mm.

3.3. *Stiffness Determination Using a Parametric Study*

Obtained results are listed in Table 7. It shows that the influence of the stiffness k grows with increasing the real thickness t_r. Figure 11 displays mentioned results in the graphical form. Results show that stiffness can be controlled by parameters of the gaps. For example, increasing gaps parameters b and l by 0.25 mm (this was difference between specimen A1 and A3) reduced stiffness approximately by 10%.

Table 8 shows the influence of the thickness demonstrated on the pattern A. It is apparent that the FS stiffness almost linearly grew with the real thickness in the range of t_r = 0.80–2.05 mm. Thanks to this growth, which can be described by linear regression with coefficient of determination R^2 > 0.99, see Figure 11, stiffness can be easily controlled by changing thickness.

Table 7. Evaluation of stiffness from parametric study.

Real thickness t_r [mm]	0.80	1.05	1.30	1.55	1.80	2.05
Flexible Structure	**Determined Stiffness k [N/mm]**					
Specimen A1	64.17	86.28	110.67	132.58	152.38	170.96
Specimen A2	81.69	109.08	137.04	162.00	185.30	208.22
Specimen A3	57.52	76.40	99.11	119.04	137.04	153.74
Specimen B1	4.80	6.37	7.69	8.90	10.05	11.17
Specimen B2	5.36	7.10	8.53	9.89	11.19	12.46
Specimen B3	3.98	5.52	6.76	7.85	8.87	9.85
Specimen C1	54.21	69.93	84.87	99.66	114.30	128.80
Specimen C2	70.65	90.91	110.96	130.80	150.38	169.74
Specimen C3	47.72	62.00	75.14	88.06	100.85	113.53

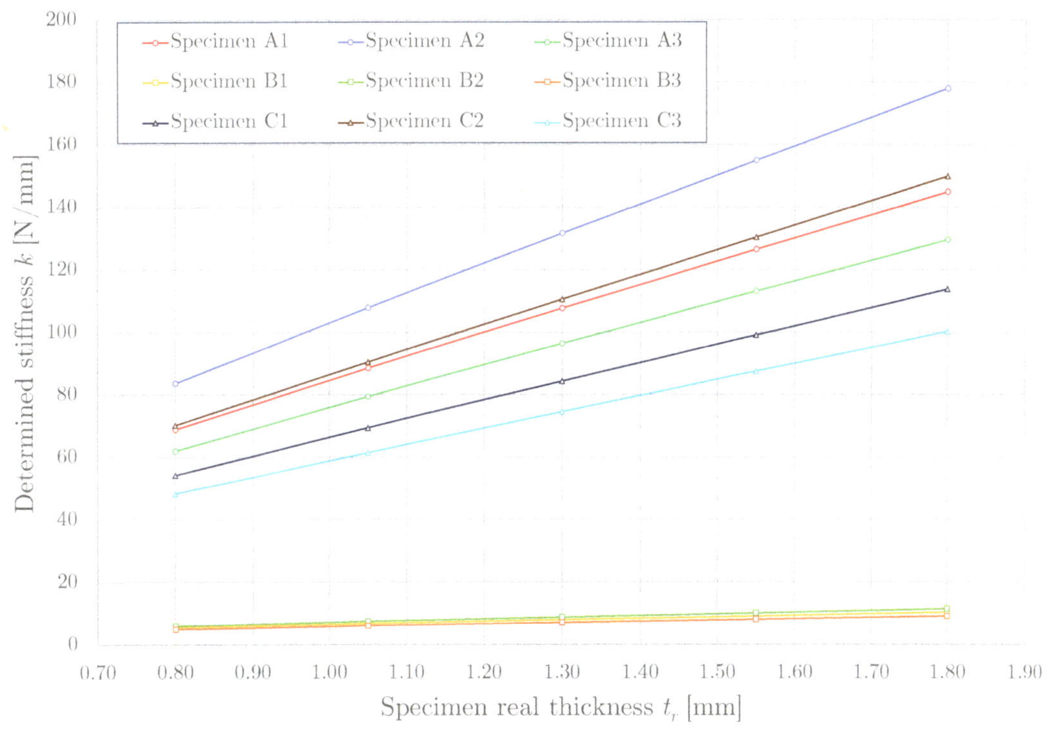

Figure 11. The relationship between the determined stiffness k and specimen real thickness t_r.

Table 8. Percentage stiffness of A-series specimens.

Real Thickness t_r [mm]	0.80	1.05	1.30	1.55	1.80	2.05
Flexible Structure	**Percentage Stiffness [%]**					
Specimen A1	100	134.5	172.5	206.6	237.5	266.4
Specimen A2	100	133.5	167.8	198.3	226.8	254.9
Specimen A3	100	132.8	172.3	206.9	238.2	267.3

3.4. Comparison of the Load-Bearing Capacity

To design the computational model, we destroyed the specimen A1 with a real thickness of t_r = 0.80 mm (after evaluating stiffness data) to obtain the value of the load-bearing capacity. The maximal failure load of the FS $F_{max} = 276$ N was measured in the experiment. This force was important for estimating the value of the plastic strain ε_{pb} = 0.0339, which was obtained using full non-linear model (with contact) after subjecting specimen A1 to the maximal failure load F_{max}, see Table 9. The plastic strain field corresponding to the failure load force F_{max} N with details on the critical area is shown in Figure 12. Plastic deformation occurred locally on the borders of the pattern empty area. A critical area with the highest plastic strain was near the edge of the plunger, where additional bending of FS acted. This is also the area of the breakage of the specimen during the initial experiment.

After calculating the value of plastic strain and locations, authors were able to compare the pattern types A2 and A3 with A1. A failure load was also calculated for specimens A2 and A3 with the real thickness of t_r = 0.80 mm. The highest plastic strain on specimen A2 and A3 occurred in the area similar to that of the specimen A1. The load-bearing capacity values corresponding to all calculated failure loads are listed in Table 10. It should be noted that only one specimen was destroyed and values in Table 10 cannot be viewed as statistically correct; still, it gave us some approximate values of the load-bearing capacity of all geometries.

Table 9. Comparison of failure load F_{max} from experiment and simulation.

Pattern Type	Real Thickness t_r [mm]	Experiment F_{max} [N]	Simulation F_{max} [N]	Failure Plastic Strain ε_{pb} [%]
Specimen A1	0.80	276	276	3.39

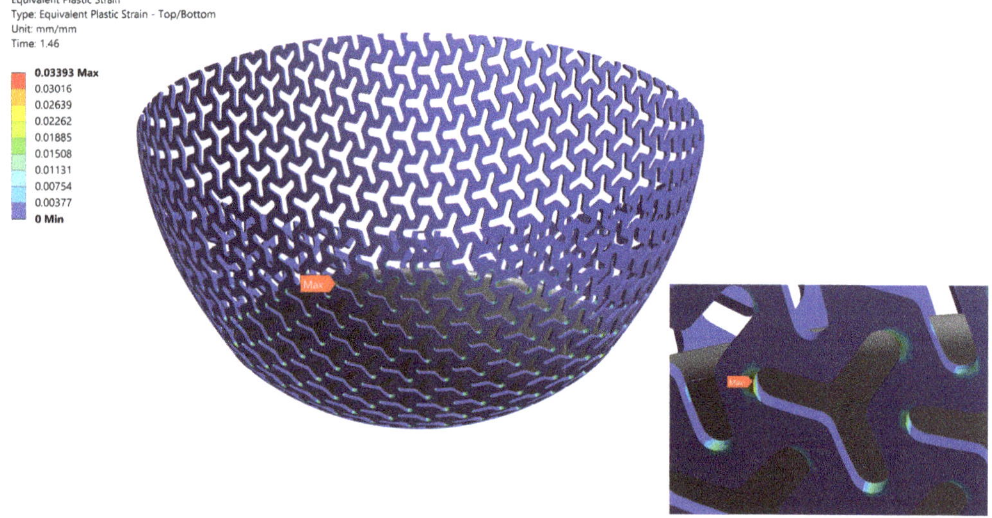

Figure 12. The plastic strain field corresponding to the failure load F_{max} = 276 N for specimen A1, real thickness t_r = 0.80 mm.

Table 10. Calculated values of load-bearing capacity.

Type of Pattern	Real Thickness t_r [mm]	Simulation F_{max} [N]
Specimen A1	0.80	276
Specimen A2	0.80	318
Specimen A3	0.80	258

4. Discussion

In this paper, two ways of determining the stiffness of 3D printed flexible structures (FS) are presented. The first way was the construction of a novel testing device replacing the traditional tensile and flexural testing techniques used for testing of internal structure properties [25,26]. The second way is a custom-developed validated computational model based on the Finite element method (FEM). It was also demonstrated that the computational model is suitable for the description of the FS behavior and, as such, could be used to evaluate the FS stiffness. The agreement between the initial experiment and simulation was achieved and the model was subsequently used for a parametric study with three different FS pattern types, each of them with three variations of pattern geometries and five thickness variations (Figure 11).

Even though all testing devices were 3D printed, they were stiffer than tested specimens. This influence was analyzed by finite element analysis of the fastening system. Maximal deformation of the fastening system (in the direction of force) was only $u = 0.11$ mm corresponds to load force $F = 250$ N. Deformation of the fastening system did not significantly affect performed experiments (compared to deformations of specimens). Therefore, the influence of the fastening system stiffness could be neglected. For measuring the specimens with higher thickness, a new robust fastening system is necessary to design.

The proposed experimental technique was found to be an important complementary tool in the design of the 3D printed FS with a three-point star pattern by non-linear computational modeling. Moreover, a computational model for analyzing the FS load-bearing capacity was presented, yielding results with satisfactory accuracy compared to the real, experimentally verified, failure load. Such a non-linear computational model can be used for prediction of the FS load-bearing capacity and for prevention of specimen breakage during testing. One should not forget the possibility of plastic strains occurring in small curvatures at higher load forces. It should be noted that comparing linear and non-linear computational models, the latter one describing laboratory experiment closely.

This paper presents two different computational models, both solving a static structural problem with elasto-plastic behavior defined by bilinear material model. In the first model for stiffness determination, we used a simplified non-linear model without a plunger. For precise evaluation of the load-bearing capacity, however, a non-linear model with a plunger had to be constructed. Although the material model provided satisfactory results, it could be substituted with an even better one, such as Chaboch material model, if even more accurate results were necessary. Nevertheless, the main advantage of the presented non-linear material model is its simplicity, needing only two input variables.

It should be noted that the thicknesses of the computational models were corrected to fit the real printed ones and, therefore, they were comparable with experiments. The deviation from the planned specimen thickness occurred during printing. This might have been caused by the low thickness of the shell to which a lower than required amount of the Nylon powder may have adhered. Measurement was performed on various places of the specimen using the vernier caliper. In some areas, the thickness was reduced by 0.18 mm, sometimes by 0.22 mm. The thickness of the computational model was reduced on average by 0.20 mm. To increase the accuracy of the thickness, it would be necessary to use a very accurate 3D scanner and to subsequently insert the acquired point cloud into the calculation.

Assuming the isotropic material behavior could be viewed as an incorrect premise since printed materials do not behave the same in all directions. The material model should be constructed for orthotropic or, even better, anisotropic behavior. Some results from the measurement suggest a possibility of creep and viscoelastic behavior under load.

During the experiment, strain and displacement were also measured by the DIC method; however, obtained data were not satisfying. The measurement was limited by the used template size, which had to be set to "large" to obtain some data. Another complication was the presence of a "holed" curved surface, which caused difficulties in detecting the pattern on the outer surface. All these problems meant that although it was possible to correctly assess the displacement, the strain/stress measurement could not be properly evaluated. In future experiments, therefore, the use of the DIC method should be redesigned to prevent the mentioned problems. Successful use of this method on a curved surface is described in a paper by Halama et al. [27].

Although the measurement using the DIC method was not successful, it sparked a few ideas on how to improve the procedure. First, background should be added to the inner surface to create contrast with the outer pattern. Second, some post-manufacturing process should be applied to ensure that the outer surface does not remain as porous as it is after printing. Third, the measuring software should be set up in a better way for this analysis; in particular, a better template size should be chosen. In our experimental setup, unfortunately, only the displacement could be evaluated without a significant error. Fourth, cameras should be better positioned to ensure measurement of displacement in the part of the sphere furthest from the rim. The main reason for preparing a better DIC measurement is obtaining a strain field from which one could derive the limit value of plastic strain (currently, the limit value is obtained from the computational model).

5. Conclusions

In this paper, we present that using a non-linear computational model, the stiffness can be determined with satisfactory results. The difference between modeling results and experimental data is less than 15%, which means that our simplified non-linear computational model was suitable for evaluating the stiffness properties of FS. As far as the load-bearing capacity is concerned, a full computational non-linear model was calibrated for force F_{max} = 276 N which gave us limit value of plastic strain. This being said, the simplified model had an overwhelming advantage in solving time—it was approximately 160% faster than the full model; therefore, the simplified model was more useful in the parametric study. For example, the computing time for a simplified non-linear model (Specimen A1, t_r = 0.80 mm) was 35 min. (84 iterations) on a standard machine (workstation Intel i7-8700K, 12 cores, 16 GB RAM, SSD). Approximately 25 s. was needed for each iteration. If we used a full non-linear model, the time for each iteration would be multiplied by more than 2 (i.e., 53 s. per iteration, summary 103 iterations, total time 91 min.). It must be also emphasized that the full non-linear model was not ideally converging to a solution every time. Additional tuning of the contact pair was necessary and better boundary condition needed to be set (in this case, the displacement boundary condition converged better than the force boundary condition). However, this model is highly recommended for evaluating load-bearing capacity.

Out of the three virtually tested pattern types, the pattern A performed best, mainly thanks to its high load-bearing capacity. It offered a high stiffness, which helped prevent the specimen A from changing shape under the applied load (see Figure 9). This was also confirmed in the analysis calculating the failure load. Another advantage was that the stiffness of specimen A scaled almost linearly with thickness with thickness, making it easy to design a specific stiffness value. Specimen A2 was the stiffest of the patterns A. Also, it had the highest load-bearing capacity. These properties were caused by having smaller gaps than the specimen A1. On the other hand, Specimen A3 was, compared to the other two geometries compliant, which also meant a low failure force. It was thanks to wider gaps than specimen A1. Other patterns did not offer such mechanical properties (such as stiffness and load-bearing capacity).

Our paper studied a shell, considering 2D pattern only. However, if an application needs a compliant (or rather even more compliant) structure, studying 3D-spatial pattern structures would be beneficial. The 3D-spatial structure might offer additional advantages compared to 2D structures only. For example, considering the biomechanical use, they might provide better sweat drainage and airflow, increasing the patients' comfort. Finally, a compliant structure could also be ensured by choosing another material (one with a lower Young's modulus than PA12), such as Thermoplastic Polyurethane (TPU).

It is necessary to mention here the cost-saving character of this virtual model. Should all configurations from Table 4 be evaluated experimentally, many more shells would have to be printed, which would immensely increase the costs. Using the presented approach, only a specific configuration is printed and evaluated experimentally, which would save roughly 90% of the costs—and this calculation is based on an experiment with a low number of specimens. To support a standard statistical evaluation, the experiment should use at least 5 to 10 specimens for each proposed structure, which would rocket the costs sky-high and the savings thanks to the use of our software solution could be over 98%.

It should be noted that the authors will continue their research on this topic and will expand on tested (2D and 3D-spatial) specimens in the following years.

Author Contributions: Conceptualization, P.M. and J.P.; methodology, P.M. and V.R.; software, V.R. and P.M.; validation, V.R., M.S., P.M. and D.R.; formal analysis, V.R.; investigation, P.M. and V.R., R.H. and M.F.; resources, V.R., P.M. and R.H.; data curation, P.M. and V.R.; writing—original draft preparation, V.R.; writing—review and editing, M.S. and P.M.; visualization, V.R. and M.S.; supervision, P.M. and M.F.; project administration, P.M.; funding acquisition, R.H. and P.M. All authors have read and agreed to the published version of the manuscript.

Funding: This work was supported by The Ministry of Education, Youth and Sports from the Specific Research Project SP2020/23, by The Technology Agency of the Czech Republic in the frame of the project TN01000024 National Competence Center-Cybernetics and Artificial Intelligence and by Structural Funds of the European Union within the project Innovative and additive manufacturing technology—new technological solutions for 3D printing of metals and composite materials, reg. no. CZ.02.1.01/0.0/0.0/17_049/0008407.

Institutional Review Board Statement: Not applicable.

Informed Consent Statement: Not applicable.

Data Availability Statement: Data sharing is not applicable to this article.

Conflicts of Interest: The authors declare no conflict of interest.

Abbreviations

The following abbreviations are used in this manuscript:

FS	Flexible Structures
2D	Two-dimensional
FEM	Finite Element Method
SLS	Selective Laser Sintering
MJF	Multi-Jet Fusion
IR	Infrared
AM	Additive manufacturing
PA12	Polyamide 12
CAD	Computer Aided Design
SW	Software
FFF	Fused Filament Fabrication
PLA	Polylactic Acid
RHN	Rhinoceros
SCDM	Ansys Spaceclaim (Design Modeler)
AWB	Ansys Workbench
DIC	Digital Image Correlation
TPU	Thermoplastic Polyurethane

References

1. Wang, S.; Wang, H.; Ding, Y.; Yu, F. Crushing behavior and deformation mechanism of randomly honeycomb cylindrical shell structure. *Thin-Walled Struct.* **2020**, *151*. [CrossRef]
2. Paik, J.K.; Thayamballi, A.K.; Kim, G.S. The strength characteristics of aluminum honeycomb sandwich panels. *Thin-Walled Struct.* **1999**, *35*, 205–231. [CrossRef]
3. Cheng, L.; Bai, J.; To, A.C. Functionally graded lattice structure topology optimization for the design of additive manufactured components with stress constraints. *Comput. Methods Appl. Mech. Eng.* **2019**, *344*, 334–359. [CrossRef]

4. Marsalek, P.; Grygar, A.; Karasek, T.; Brzobohaty, T. Virtual prototyping of 3D printed cranial orthoses by finite element analysis. In Proceedings of the Central European Symposium on Thermophysics 2019 (CEST), Banska Bystrica, Slovakia, 16–18 October 2019; p. 320010. [CrossRef]
5. Nagesha, B.; Dhinakaran, V.; Shree, M.V.; Kumar, K.M.; Chalawadi, D.; Sathish, T. Review on characterization and impacts of the lattice structure in additive manufacturing. *Mater. Today Proc.* **2020**, *21*, 916–919. [CrossRef]
6. Alberdi, R.; Dingreville, R.; Robbins, J.; Walsh, T.; White, B.C.; Jared, B.; Boyce, B.L. Multi-morphology lattices lead to improved plastic energy absorption. *Mater. Des.* **2020**, *194*, 1–10. [CrossRef]
7. Quoc, P.M.; Krzikalla, D.; Mesicek, J.; Petru, J.; Smiraus, J.; Sliva, A.; Poruba, Z. On Aluminum Honeycomb Impact Attenuator Designs for Formula Student Competitions. *Symmetry* **2020**, *12*, 1647. [CrossRef]
8. Sotola, M.; Stareczek, D.; Rybansky, D.; Prokop, J.; Marsalek, P. New Design Procedure of Transtibial Prosthesis Bed Stump Using Topological Optimization Method. *Symmetry* **2020**, *12*, 1837. [CrossRef]
9. Ligon, S.C.; Liska, R.; Stampfl, J.; Gurr, M.; Mülhaupt, R. Polymers for 3D Printing and Customized Additive Manufacturing. *Chem. Rev.* **2017**, *117*, 10212–10290. [CrossRef]
10. Lee, H.E.; Park, J.H.; Jang, D.; Shin, J.H.; Im, T.H.; Lee, J.H.; Hong, S.K.; Wang, H.S.; Kwak, M.S.; Peddigari, M.; et al. Optogenetic brain neuromodulation by stray magnetic field via flash-enhanced magneto-mechano-triboelectric nanogenerator. *Nano Energy* **2020**, *75*, 104951. [CrossRef]
11. Kim, D.H.; Shin, H.J.; Lee, H.; Jeong, C.K.; Park, H.; Hwang, G.T.; Lee, H.Y.; Joe, D.J.; Han, J.H.; Lee, S.H.; et al. In Vivo Self-Powered Wireless Transmission Using Biocompatible Flexible Energy Harvesters. *Adv. Funct. Mater.* **2017**, *27*, 1700341. [CrossRef]
12. Schumacher, C.; Marschner, S.; Gross, M.; Thomaszewski, B. Mechanical characterization of structured sheet materials. *ACM Trans. Graph.* **2018**, *37*, 1–15. [CrossRef]
13. Bickel, B.; Bächer, M.; Otaduy, M.A.; Lee, H.R.; Pfister, H.; Gross, M.; Matusik, W. Design and fabrication of materials with desired deformation behavior. *ACM Trans. Graph. (TOG)* **2010**, *29*, 63. [CrossRef]
14. Schumacher, C.; Thomaszewski, B.; Gross, M. Stenciling. *Comput. Graph. Forum* **2016**, *35*, 101–110. [CrossRef]
15. Hussain, G.; Khan, W.A.; Ashraf, H.A.; Ahmad, H.; Ahmed, H.; Imran, A.; Ahmad, I.; Rehman, K.; Abbas, G. Design and development of a lightweight SLS 3D printer with a controlled heating mechanism. *Int. J. Lightweight Mater. Manuf.* **2019**, *2*, 373–378. [CrossRef]
16. Cai, C.; Tey, W.S.; Chen, J.; Zhu, W.; Liu, X.; Liu, T.; Zhao, L.; Zhou, K. Comparative study on 3D printing of polyamide 12 by selective laser sintering and multi jet fusion. *J. Mater. Process. Technol.* **2021**, *288*, 116882. [CrossRef]
17. Řepa, V.; Maršálek, P.; Prokop, J.; Rybanský, D.; Halama, R. Modelling and Testing of 3D Printed Flexible Structures with Three-pointed Star Pattern Used in Biomedical Applications. In *Experimental Stress Analysis 2020*, 1st ed.; VSB—Technical University of Ostrava: Ostrava, Czech Republic, 2020; pp. 145–147.
18. Faustini, M.C.; Neptune, R.R.; Crawford, R.H.; Rogers, W.E.; Bosker, G. An Experimental and Theoretical Framework for Manufacturing Prosthetic Sockets for Transtibial Amputees. *IEEE Trans. Neural Syst. Rehabil. Eng.* **2006**, *14*, 304–310. [CrossRef]
19. Lammens, N.; De Baere, I.; Van Paepegem, W. On the orthotropic elasto-plastic material response of additively manufactured polyamide 12. In Proceedings of the 7th Bi-Annual International Conference of Polymers & Moulds Innovations, Ghent, Belgium, 21–23 September 2016; Ragaert, K., Delva, L., Cardon, L., Eds.; Academic Bibliography: Ghent, Belgium, 2016; p. 6.
20. Stoia, D.; Linul, E.; Marsavina, L. Influence of Manufacturing Parameters on Mechanical Properties of Porous Materials by Selective Laser Sintering. *Materials* **2019**, *12*, 871. [CrossRef]
21. Kminek, T.; Marsalek, P.; Karasek, T. Analysis of steel tanks for water storage using shell elements. In *AIP Conference Proceedings*; AIP Publishing LLC: Melville, NY, USA, 2019; Volume 2116, p. 320007. [CrossRef]
22. Lesnak, M.; Marsalek, P.; Horyl, P.; Pistora, J. Load-bearing capacity modelling and testing of single-stranded wire rope. *Acta Montan. Slovaca* **2020**, *25*, 192–200.
23. Horyl, P.; Snuparek, R.; Marsalek, P.; Poruba, Z.; Paczesniowski, K. Parametric studies of total load-bearing capacity of steel arch supports. *Acta Montanistica Slovaca* **2019**, *24*, 213–222.
24. Bai, J.; Song, J.; Wei, J. Tribological and mechanical properties of MoS2 enhanced polyamide 12 for selective laser sintering. *J. Mater. Process. Technol.* **2019**, *264*, 382–388. [CrossRef]
25. Dorčiak, F.; Vaško, M.; Handrik, M.; Bárnik, F.; Majko, J. Tensile test for specimen with different size and shape of inner structures created by 3D printing. *Transp. Res. Procedia* **2019**, *40*, 671–677. [CrossRef]
26. Dorčiak, F.; Vaško, M.; Bárnik, F.; Majko, J. Comparison of experimental flexural test with FE analysis for specimen with different size and shape of internal structure created by 3D printing's. *IOP Conf. Ser. Mater. Sci. Eng.* **2020**, *776*, 012078. [CrossRef]
27. Halama, R.; Pagáč, M.; Paška, Z.; Pavlíček, P.; Chen, X. Ratcheting Behaviour of 3D Printed and Conventionally Produced SS316L Material. In Proceedings of the ASME 2019 Pressure Vessels & Piping Conference, San Antonio, TX, USA, 14–19 July 2019. [CrossRef]

Article

On the Weldability of Thick P355NL1 Pressure Vessel Steel Plates Using Laser Welding

Jiří Čapek [1,*], Karel Trojan [1], Jan Kec [2], Ivo Černý [2], Nikolaj Ganev [1] and Stanislav Němeček [3]

1 Department of Solid State Engineering, Faculty of Nuclear Sciences and Physical Engineering, Czech Technical University in Prague, Trojanova 13, 120 00 Prague, Czech Republic; karel.trojan@fjfi.cvut.cz (K.T.); nikolaj.ganev@fjfi.cvut.cz (N.G.)
2 Laboratory of Material Properties, SVÚM a.s., Tovární 2053, 250 88 Čelákovice, Czech Republic; kec@svum.cz (J.K.); ivo.cerny@seznam.cz (I.Č.)
3 Department of Laser Material Processing, RAPTECH s.r.o., U Vodárny 473, 330 08 Zruč-Senec, Czech Republic; nemecek@raptech.cz
* Correspondence: jiri.capek@fjfi.cvut.cz; Tel.: +420-224-358-624

Abstract: Pipeline transport uses millions of kilometers of pipes worldwide to transport liquid or gas over long distances to the point of consumption. High demands are placed, especially on the transport of hazardous substances under high pressure (gas, oil, etc.). Mostly seamless steel pipes of various diameters are used, but their production is expensive. The use of laser-welded pipes could significantly reduce the cost of building new pipelines. However, sufficient mechanical properties need to be ensured for welded pipes to meet stringent requirements. Therefore, laser-welded 10 mm thick pressure vessel steel plates were subjected to various mechanical tests, including high-cycle fatigue tests. Furthermore, the microstructural parameters and the state of residual stresses were determined using X-ray and neutron diffraction, which could affect fatigue life, too. The critical areas for possible crack initialization, especially in and near the heat-affected zone, were found using different tests. The presented results outline the promising application potential of laser welding for the production of pipes for high-pressure pipelines.

Keywords: laser welding; pressure vessel steel; microstructure; X-ray and neutron diffraction; high-cycle fatigue tests

Citation: Čapek, J.; Trojan, K.; Kec, J.; Černý, I.; Ganev, N.; Němeček, S. On the Weldability of Thick P355NL1 Pressure Vessel Steel Plates Using Laser Welding. *Materials* **2021**, *14*, 131. https://doi.org/10.3390/ma14010131

Received: 30 November 2020
Accepted: 25 December 2020
Published: 30 December 2020

Publisher's Note: MDPI stays neutral with regard to jurisdictional claims in published maps and institutional affiliations.

Copyright: © 2020 by the authors. Licensee MDPI, Basel, Switzerland. This article is an open access article distributed under the terms and conditions of the Creative Commons Attribution (CC BY) license (https://creativecommons.org/licenses/by/4.0/).

1. Introduction

Welding has been widely used in fabrication industries producing ships, trains, steel bridges, pressure vessels, and more since the First World War. Increasing demands are being placed on the mechanical properties and durability of welds that are used to connect two or more components. There is also an increasing demand for productivity and cost-effectiveness of welding, which leads to the use of modern and progressive methods, including laser welding, which has an advantage at high welding speeds, a low thermal load of the surrounding material, precision and strength of the weld, and the possibility of joining components with a wide range of thickness (0.01 to 50 mm) [1,2].

Pipeline transport uses pipes to transport liquid or gas over long distances to the point of consumption. Millions of kilometers of pipes are used around the world for various applications. High demands are placed, especially on the transport of hazardous substances under high pressure (gas, oil, etc.). Therefore, the use of welded pipes could significantly reduce the cost of building new pipelines, if they replaced seamless pipes, which are expensive to manufacture. However, sufficient mechanical properties need to be ensured for welded pipes to meet stringent requirements to prevent accidents and natural disasters. Laser welding, with its advantages described above, could find an application in the welding of longitudinal welds on pipes in production. Another advantage of laser welding in comparison to conventional welding methods is the robotic automation of the welding process and to a defined wall thickness without additional material. Compared to electron welding, which also has excellent properties, there is no need to use a vacuum chamber, which would have to be up to 30 m long. However, welding of the segments during the construction of the pipeline itself will continue to be carried out mainly by manual welding, as a sufficiently precise fit cannot be guaranteed in the field.

With the application of new welding technology, it is necessary to describe the influence on the basic mechanical parameters. Tensile, Charpy impact, and hardness tests are usually an integral part of the set of performed tests. Furthermore, fatigue loading is a very important factor affecting service safety and reliability, mainly of dynamically loaded structures. Therefore, an evaluation of fatigue resistance together with the evaluation of damage mechanisms and affecting factors has to be considered in such cases. Note that an investigation of fatigue properties and phenomena is still missing in some works aimed at laser welding, e.g., high-pressure pipelines [3].

It is necessary to mention that the final mechanical properties are strongly dependent on the welding parameters, e.g., [4,5]. Zhang et al. [5] further found that some laser welding parameters could cause the mechanical properties of weld metal (WM) to not reach the properties of the base metal (BM). Mainly welding speed and laser power have a strong effect on the residual stresses (RS) among the studied parameters, e.g., increasing weld speed. The piping with laser power reduction decreases residual stresses with hardness and increases the toughness. However, it is important to monitor sufficient penetration and pressure fluctuations.

According to Guo et al. [6], the welding of low-alloy steel causes the generation of martensite in the WM and heat-affected zone (HAZ), which leads to an increase in hardness, but also a reduction of impact toughness in comparison with the BM. Moreover, laser welding causes microstructural changes of plastically deformed crystalline materials and decreases ductility and corrosion resistance due to the grain coarsening, carbide precipitation, and martensite formation [7]. These so-called microstructural notches are the critical areas for the potential crack initialization, e.g., at the interface between the WM and the HAZ.

Due to a heterogeneous application of energy and localized fusion that occurs during the welding process, high undesirable RS could be present in the region near the weld and in the weld itself. These RS are generated because of the superposition of thermal transformation processes [8] and could reach high values and subsequently cause fatigue or, in combination with crack-like defects, promote brittle fractures [9].

The influence of welding techniques on RS and the microstructural parameter FWHM (full width at half maximum) was investigated in Čapek et al. [10]. Comparable results for laser and electron welding were found; moreover, they were better than for conventional manual arc welding (MAG). The influence on the high-fatigue resistance was discussed, too, where a higher FWHM, i.e., higher microdeformation, higher dislocation density, or smaller crystallite size, could generally indicate a strengthening of the surface. The contribution Černý et al. [11] brought is the view of the high-cycle fatigue resistance of the laser welds. Partial attention was paid to the evaluation of crack initiation mechanisms and welding imperfections (internal lack of fusion, pores, etc.). These imperfections are a critical place for crack initialization, partly because of the reduction of the real effective cross-section. In the contribution of [12], the effect of re-calculation by considering only the real cross-section of welds without a lack of fusion or pores, as well as surface residual stresses, were discussed.

The research done by Moravec et al. [13] was focused on the evaluation of the laser welds under dynamic loading. It was shown that heat input has a place in the area of high-cycle fatigue. Higher heat input causes a longer thermal exposition; therefore, major changes could occur in the HAZ. The negative influence of these changes could have a very significant effect on the mechanical properties (hardness, toughness value, yield strength, etc.). Our team in [14] investigated the influence of a high-cycle fatigue test on the values of RS and microdeformation. The main conclusion was that the tensile RS could accelerate the fatigue initiation process and microdeformation could be a good indicator of future crack initialization and propagation.

The purpose of this investigation was to analyze the relationship between microstructure, mechanical properties, and RS not only of WM but also of surrounding areas. This article describes a comprehensive evaluation of mechanical and fatigue properties of laser-welded pressure vessel steel plates. Specifically, the 10 mm thick double-sided square butt welds performed on P355NL1 steel plates for high-pressure vessels were analyzed. This fine-grained low-alloy carbon steel is widely used for the fabrication of high-pressure vessels, steam boiler parts, pressure piping, compressors, etc. Based on the results of this research, it would be possible to continue the development of laser welding technology for pipes, which would find applications in the construction of pressure pipelines.

2. Materials and Methods

The P355NL1 hot-rolled steel plates of the dimensions 150 mm × 300 mm × 10 mm were studied. Flat plates were used for better processing of the results, especially fatigue tests. The chemical composition of this steel determined by glow-discharge optical emission spectroscopy is given in Table 1. The error at low concentrations was up to 10%. The determined chemical composition was in agreement with the European Standard EN 10028-3:2017 [15] with one exception, manganese, whose weight fraction was below the minimum value of 1.1 wt.% stated in the norm. This had a consequence in lowering the value of the carbon equivalent to the value of 0.21, according to IIW (International Institute of Welding).

Table 1. Chemical composition of the P355NL1 steel, determined using glow-discharge optical emission spectroscopy.

Element	Fe	C	Mn	Si	Ni	Cr	Cu	V	Al
Weight fraction (wt.%)	98.92	0.067	0.8	0.004	0.005	0.021	0.018	0.002	0.037

The double-sided square butt weld (the top side was welded first) was performed by the Laserline LDM3000-60 (Laserline GmbH, Koblenz, Germany) from the company RAPTECH s.r.o. The most important welding parameters, such as laser power P, welding speed v, welding mode, and laser wavelength λ, are given in Table 2.

Table 2. Welding parameters.

P (W)	v ($mm{\cdot}s^{-1}$)	Mode	λ (nm)
3000	5.5	Continuous	900–1080

For metallographic analysis, the sample was first ground and afterwards polished. The steel structure was induced by etching in 2% Nital (98 mL Ethanol + 2 mL HNO_3). The analysis was performed on the Zeiss Axio Observer light microscope (Carl Zeiss Microscopy GmbH, Berlin, Germany).

Lattice deformations of the diffraction lines $K\alpha_1$ of the planes {211} of the ferrite phase were analyzed using the Proto iXRD Combo diffractometer (Proto Manufacturing Inc., Taylor, MI, USA) in a ω-goniometer setup with chromium radiation. Diffraction angles $2\theta^{hkl}$ were determined by using the Gaussian function and Absolute peak method. The standard X-ray elastic constants $\frac{1}{2}s_2 = 5.75\ TPa^{-1}$, $s_1 = -1.25\ TPa^{-1}$, and the Winholtz and Cohen method [16] were used to calculate the surface RS. The FWHM parameter denoted the full width at half maximum of $K\alpha_1$ diffraction line.

The X'Pert PRO MPD diffractometer (Malvern Panalytical B.V., Almelo, The Netherlands) with cobalt radiation was used for the analyses by X-ray diffraction (XRD). The crystallite size (the size of coherently diffracted domains) and microdeformation were determined from the XRD patterns using the Rietveld refinement, performed in MStruct software(version from 2019) [17].

The sample was analyzed by XRD in both the perpendicular "T" and parallel "L" directions to the weld on both sides. The welded plate was analyzed in the middle area (see Figure 1). The irradiated volume was defined by experiment geometry, the effective penetration depth of the X-ray radiation (approximately 4–5 µm), and the pinhole size (4 mm × 0.25 mm).

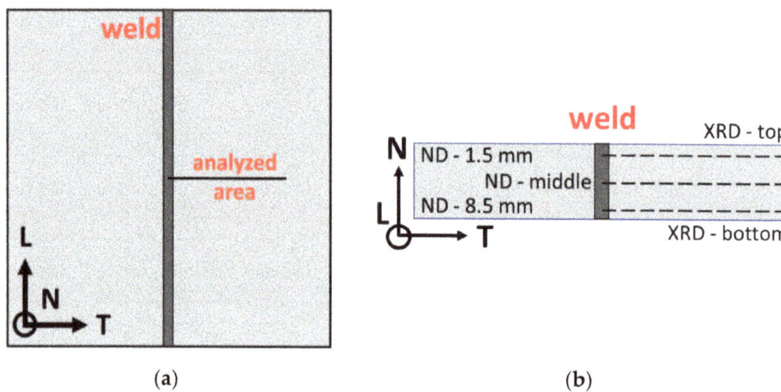

Figure 1. (**a**) Scheme of the sample and (**b**) scheme of marking the depths at which the residual stresses were determined by X-ray (XRD) and neutron (ND) diffraction.

Neutron diffraction was used for the determination of the RS profile across the weld in the different depths of the plate (see Figure 1). The measurements were performed at CANAM infrastructure in Nuclear Physics Institute of the Czech Academy of Sciences (Řež, Czech Republic) using the SPN-100 Strain scanner. For this purpose, the wavelength of the radiation used for the diffraction on *{110}* ferrite lattice planes was $\lambda = 0.213$ nm.

Samples were produced with dimensions corresponding to the ČSN EN ISO 6892-1 standard. Afterwards, tensile tests were performed according to the ČSN EN ISO 4136 standard on the EUS 40 testing machine (Werkstoffprüfmaschinen Leipzig GmbH, Leipzig, Germany). High-cycle fatigue tests were performed on the Schenck PHT resonance machine (Schenck PHT GmbH, Darmstadt, Germany) with a load capacity of 200 kN with Zwick computer-controlled electronics (Zwick Roell Group, Ulm, Germany) and performed at different stress ranges to obtain the whole Wöhler curve, including the endurance limit. The load asymmetry was R = 0.1 and the frequency was 33 Hz. The dog bone-shaped sample was used for fatigue and tensile tests. The weld was located across the center of the length of the analyzed samples where the surface of the sample was as welded.

A three-point bending impact test was performed according to the ČSN EN ISO 148-1 standard on the PSWO 30 impact hammer (Werkstoffprüfmaschinen Leipzig GmbH, Leipzig, Germany) at a temperature of 0 °C.

3. Results

The standard tests (metallographic and tensile) were supplemented with impact and fatigue tests. These findings were compared with non-destructive X-ray and neutron diffraction, and all gave an extensive and complex description of not only the weld itself but also the surrounding areas.

3.1. Metallographic Study

Figure 2 shows the macrostructure of a double-sided welded joint. The weld metal (WM), heat-affected zone (HAZ), and base material (BM) can be distinguished. The width of the WM and the HAZ were 2.2 mm and 1.2 mm, respectively. Apart from the pores, no other defects of the macrostructure, such as cracks, lack of fusion, etc., were observed. Pores arose in the WM due to the instability of the keyhole [18].

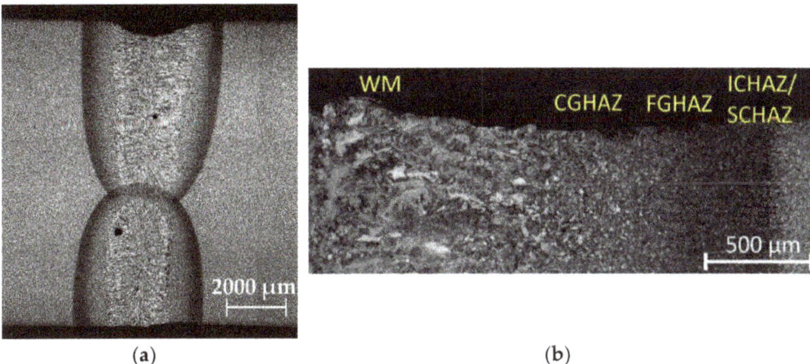

Figure 2. Macrostructure of (**a**) the weld cross-section and (**b**) particular zones after etching.

The microstructure of the BM consisted of mainly polygonal ferrite and perlite. The ferrite grains were very fine, with a mean grain size of 5.7 μm (G12). No significant changes in ferritic grain size at the surface and or the center of the plate were found. The volume fraction of perlite in the ferrite matrix was estimated at 1%.

The HAZ could be divided into four different zones (see Figure 2b): coarse-grained (CGHAZ), fine-grained (or recrystallized zone—FGHAZ), intercritical (ICHAZ), and subcritical (or over-tempering zone—SCHAZ) [19]. The closest to the WM was the CGHAZ. This zone was relatively narrow, and the microstructure was heated nearly to the solidification temperatures, resulting in the coarsening of the primary austenitic grains and the decomposition of the carbides and carbonitrides of the microalloying elements, which led to the formation of coarse bainite and small volume fraction of low-carbon martensite (Figure 3a). On the other hand, the FGHAZ was wide in comparison to other zones in the HAZ and consisted of fine ferrite and bainite. The last two zones (ICHAZ and SCHAZ) were characterized by heating from the laser source to temperatures below A_3, which led to partial recrystallization of the ferrite and transformation of the pearlite.

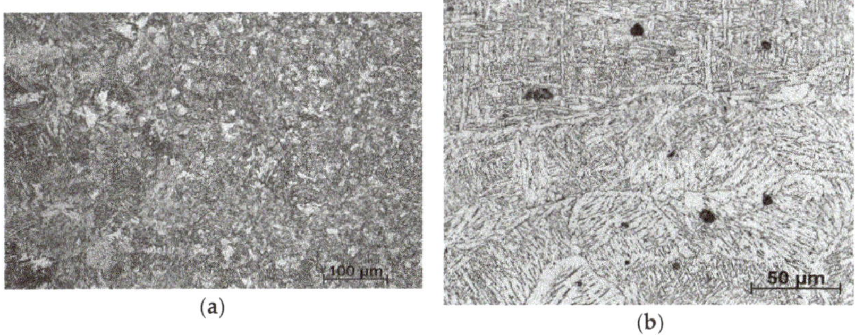

Figure 3. Microstructure in (**a**) the heat-affected zone and (**b**) the weld metal.

The WM microstructure was oriented preferentially in the direction of the solidification gradient and consisted of acicular ferrite and grain boundary ferrite (see Figure 3b).

3.2. X-ray and Neutron Diffraction

In our previous work [20], the stability of the laser welding process along the entire length of the weld was fulfilled. Therefore, only one area, the middle part of the plate (see Figure 1), was analyzed.

3.2.1. Microstructure Parameters

Resulting from a previous study by Čapek et al. [14], the analysis of the microstructural parameters (e.g. microdeformation e, crystallite size D, or FWHM parameters) by non-destructive X-ray diffraction in the surface layers indicated that this method could be a good indicator of future microcrack initialization and propagation.

The comparison of microstructural parameters depending on the distance from the weld axis is depicted in Figure 4. Especially on the bottom side, the typical zones were possible to identify:

(1) WM (up to 1 mm from weld axis): The values of e and D were the highest, approximately 18×10^{-4} and above 500 nm (due to the maximum value of the D given by the MStruct program [17]), respectively. This zone was very coarse-grained; moreover, each grain was distorted due to the quenching of the weld. Paying attention to the FWHM parameter, it could be stated that the microstructure of the WM was finer on the top side and was more coarse-grained than the CGHAZ.

(2) CGHAZ (up to 2 mm from weld axis): The values of D were still very high (around 500 nm) and e decreased about 30%. The microstructure was still coarse-grained but, due to the smaller heat-load, the grain distortion, i.e., e, was smaller.

(3) FGHAZ (up to 3.5 mm from weld axis): The steep decreasing of D values was observed and the values of e started gently decreasing. Parallel with D decreasing, the FWHM increased.

(4) BM with high e (up to 8 mm from weld axis): The values of D reached the minimum (bulk) value and e was still gently decreasing. In this region, there was a fine-grained microstructure but with relatively high grain distortion. The temperature was not high enough to enlarge the grains; on the other hand, the heat influence and cooling rate were enough to retain sufficient energy for grain distortion.

(5) BM with high FWHM (up to 15 mm from weld axis): e reached the bulk values. On the other side, the FWHM parameter still had a high value, which indicated the relatively high dislocation density. This zone was still influenced by welding, but the microstructure itself was not.

(6) BM (above 15 mm from weld axis): All microstructure parameters reached the constant values. The microstructure was not influenced by welding.

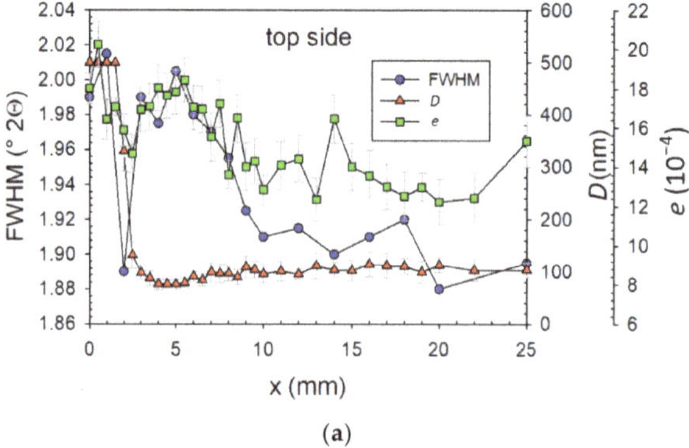

(a)

Figure 4. *Cont.*

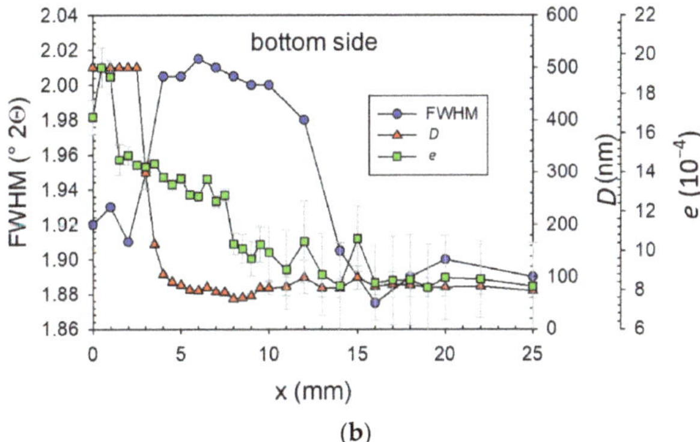

(**b**)

Figure 4. (**a**,**b**) Dependence of FWHM parameter, crystallite size *D*, and microdeformation *e* on the distance from the weld axis.

It is necessary to mention that the analyzed widths of the zones were strongly affected by the real irradiated area. Considering the divergence of X-ray radiation and experiment geometry, it was approx. two times larger than the pinhole size. Therefore, very narrow ICHAZ and SCHAZ were not observed and the mentioned zone widths were narrower. The differences between the top and bottom side were caused by double-side welding, where the first weld was annealed during the second pass.

The boundary between the HAZ and BM created a microstructural notch because of the rapid decrease in *D* values. Another microstructural notch was on the boundary of the WM and HAZ (i.e., fusion zone, FZ). The FZ identification was limited by the high *D* value, so therefore the decreasing of *e* could be appropriate for identification. These areas were the most critical for the potential initialization of surface fatigue crack.

3.2.2. Residual Stresses

From the point of view of residual stress (RS) generation, the thickness of welded plates is a very important parameter because thicker material means more volume of materials for heat conducting (see previous research [20]). Moreover, according to [20], the absolute value of RS also increases with the increasing thickness, which is caused by the mentioned heat conducting. Therefore, in the case of the analyzed sample, the highest thermal influence consequent of the higher shrinkage was closer to the weld in the L direction (see Figure 5).

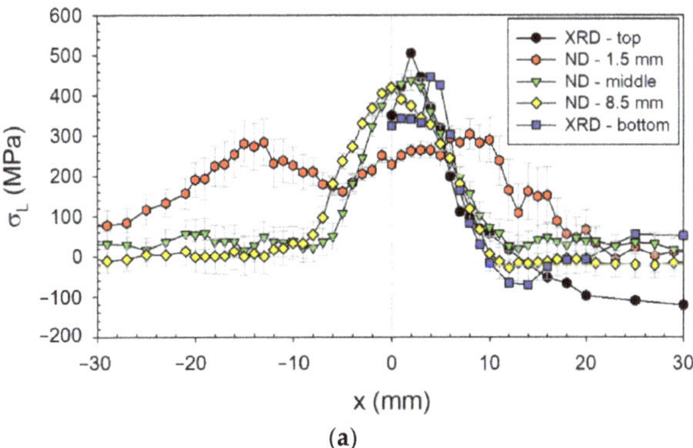

(**a**)

Figure 5. *Cont.*

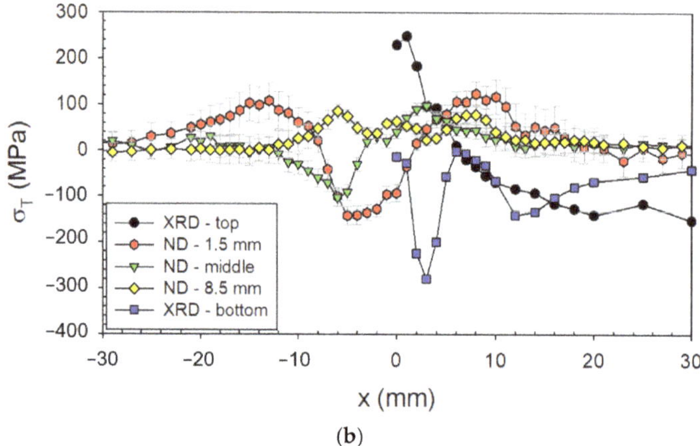

(**b**)

Figure 5. (**a**,**b**) Residual stresses RS (analyzed by XRD and ND) depending on the distance from the weld axis.

From the obtained RS profiles, it could be said that the typical trends were generated, i.e., tensile RS in the L direction, and both compressive and tensile RS in the T direction (see Figure 5). Contrary to the expectation, the high tensile RS were found on the top side in the T direction. These stresses were probably generated during the second pass when the deformation occurred (warping).

The comparison of RS trends in different depths (surface measured by XRD and bulk material by neutron diffraction) showed that longitudinal components have typical tensile character. On the other hand, in the case of transversal components, substantial differences across the thickness were observed. These differences could be explained by the fact that the length of the heated material and the subsequent shrinkage in the T direction was much shorter than in the L direction, and therefore, transformation (compressive) RS could occur in the T direction. These volume changes had a different impact on the surface layers and bulk. Another significant influence was double-side welding and the associated warping in the T direction. Considering these influences, the non-symmetric trends of bulk RS around the weld were observed by neutron diffraction.

The RS trends also showed the size of the so-called stress affected (tempering) zone (SAZ). This zone did not differ from the BM by microstructure but only by RS; it was much wider for the bulk material than for the surface (see FWHM parameter) and extended approx. 25 mm from the weld axis.

3.3. Mechanical Tests

3.3.1. Tensile Test

All tested samples reached a satisfying tensile strength of $R_m = 556.5 \pm 0.5$ MPa; thus they showed at least the minimum strength of the BM (490 MPa). Despite the porosity of the weld (see Figure 2), the failure occurred outside the WM. This finding could be explained by the combination of the real effective cross-section and the mechanical properties of the weld, which were (in this case) more dominant than the porosity.

3.3.2. Hardness

Not exceeding the limit values of the hardness was one of the fundamental conditions for approving the welding procedure. Figure 6 depicts the HV5 (Vickers hardness test, load 5 kg) through the laser weld. In general, there are two different standard methods of evaluation, namely, (1) performing indentations at exact regular distances, and (2) a simplified method of performing indentations at each weld zone—the BM, HAZ, WM, second HAZ, and BM again. The latter method was used. According to the standards, three indents were made in each zone. The diagram validates the excellent quality, stability, and reproducibility of the laser welding in the WM and HAZ—the three dependencies, namely in the bottom surface, middle, and top layer, were almost perfectly identical.

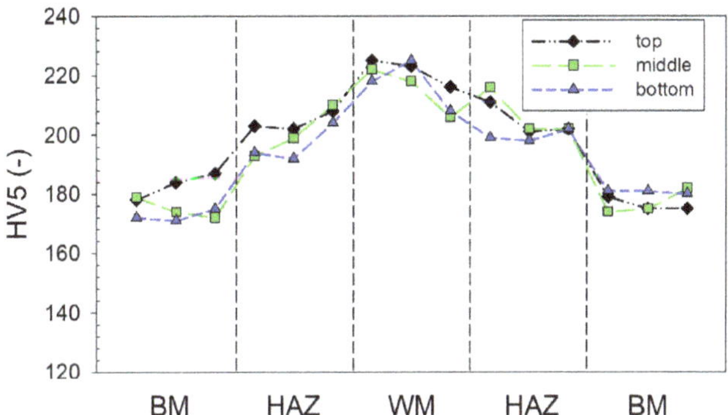

Figure 6. Course of the hardness of the HV through laser welding in the specimen.

The diagram also shows a very uniform and smooth course of hardness from the BM through the HAZ and the WM. When the hardness was recalculated to the strength according to EN standards, the differences of the strength of the weld (around 710 MPa) and base material (550 MPa) were unusually low, which can be considered a very positive factor and significantly smaller than usual for conventional welds. Compared to [6], the difference in hardness between the WM and HAZ was negligible. The reason was in the low heat input during the welding and consequent quenching and the lack of hard martensite formation. It is necessary to note that the increasing hardness could have been caused by a fine-grained microstructure or higher dislocation density.

3.3.3. Impact Test

The average values with the standard deviation of the absorbed energy were high for all tested areas (WM, HAZ, and BM) and exceeded 80 J (see Figure 7). However, for the WM samples, it must be noted that the individual values had a significant scatter; the minimum measured value was 42 J and the highest 125 J. This was caused by a different amount of porosity in the weld and mainly by the deviation of the main crack from the WM to the HAZ and BM. The deviation of the crack from the WM caused the values of the absorbed energy to be artificially increased. This issue was also widely discussed in all the technologies that created narrow weld beads such as laser and electron beam welding. Many of the articles, e.g., [21,22], discussed this problem in detail and it was generally called fracture path deviation (FPD). FPD arose not only in impact tests but also in static three-point bend tests and fracture toughness tests. Different fracture morphology was also observed depending on the FPD (see Figure 8). In the case of the sample with a minimum measured value of 42 J, it can be seen that the fracture had a limited plasticity represented by lateral expansion (see Figure 8a). Dimple morphology also occurred on the samples with higher absorbed energy and cleavage facets on the samples with lower absorbed energy.

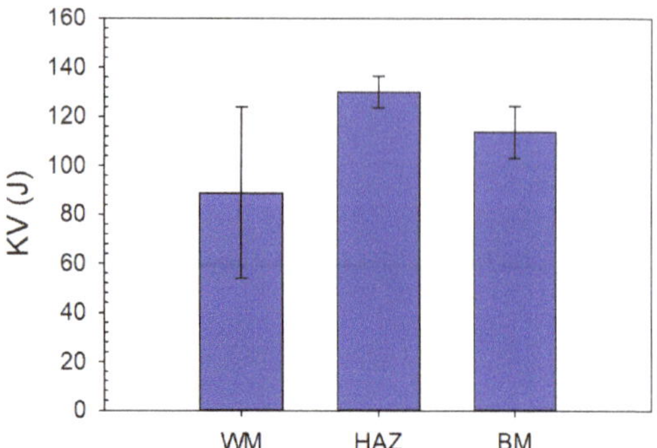

Figure 7. Absorbed energy of the WM, HAZ, and BM during the impact test.

(**a**) (**b**)

Figure 8. Differences in the fracture morphology of the WM samples: (**a**) 42 J and (**b**) 125 J.

3.3.4. High-Cycle Fatigue Test

The results of the high-cycle fatigue (HCF) tests are shown in Figure 9, where the fatigue resistance of the laser welds is compared with conventional arc-welding of the same material and same thickness. The arc welds were investigated in our previous study [10,12].

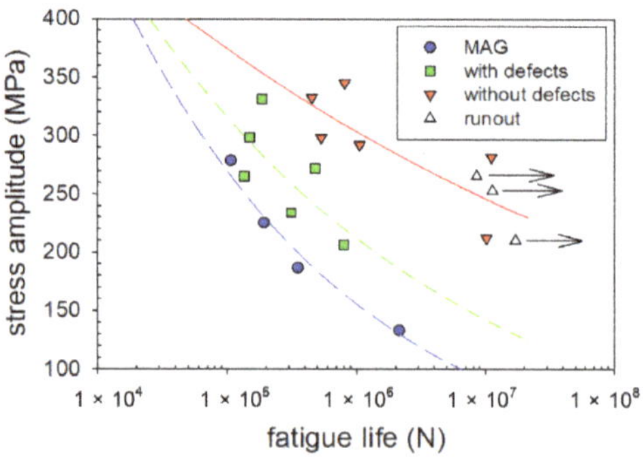

Figure 9. Wöhler curves of laser welded samples with and without internal defects compared with manual arc welding (MAG).

The HCF results of the laser-welded samples were burdened with a very big scatter. However, the analysis of the fracture surfaces enabled the specimens to be divided into two groups. The first one contained significant weld defects—a lack of fusion in the center (see

Figure 10a). These defects were areas of quick fatigue crack initiation and growth to failure. The crack initiation in specimens without welding defects occurred on the surface in the FZ (the boundary of the WM and HAZ) (see Figure 10b). Therefore, the crack initiation and growth process were significantly slower, resulting in better fatigue resistance.

(**a**) (**b**)

Figure 10. (**a**) Fatigue crack initiation on the weld defect in the center and (**b**) the fracture surface without weld defects—initiation on the top side of the FZ.

4. Discussion

From the fatigue life point of view, the critical areas for a potential crack initialization were essential to investigate. Comparing destructive (Figures 2 and 3) and non-destructive (Figures 4 and 5) methods, the typical microstructure zones were observed: the weld metal (WM), the coarse-grained (CGHAZ), fine-grained (FGHAZ), intercritical (ICHAZ), sub-critical (SCHAZ) heat-affected zones, the stress affected zone (SAZ), and the base material (BM). Each zone was recognized according to the microstructure, crystallite size D, microdeformation e, and residual stresses (RS): WM (coarse-grained, the highest D and e), CGHAZ (coarse-grained, high D, medium e), FGHAZ (fine-grained, small D, medium e), SAZ (fine-grained BM microstructure, small e, high FWHM parameter, or non-constant RS), and BM (unaffected by welding).

Considering the microstructure from Figure 3, it could be stated that there was harder and more brittle acicular ferrite in the WM compared to the softer and tougher bainite in the CGHAZ.

Regarding the RS, the T direction was more important from the fatigue point of view. In this direction, the main stresses occurred on the longitudinal weld of the pressurized pipe. The higher tensile RS were determined on the surface of the top side, reaching 300 MPa. In combination with the stress amplitude applied during the fatigue test (Figure 9), the real stress could exceed the yield strength and the fatigue life could be reduced.

Therefore, this boundary between the WM and CGHAZ, the so-called FZ, which contained surface notches, high surface tensile RS (on the top side), and grain-size change, was the most critical area for a potential surface crack initialization.

Another critical area was located on the boundary of the CGHAZ and FGHAZ, mainly due to the microstructural changes where the coarse-grained bainite and fine-grained ferrite with bainite were in the CGHAZ and FGHAZ, respectively. This area could be weakened by the presence of surface tensile RS on the top side. Contrarily, the presence of compressive RS strengthened this area on the bottom side.

The last and least critical area was located between the HAZ (more precisely the FGHAZ) and the BM (more precisely the SAZ). These zones had different microstructure (fine-grained ferrite-perlite in the BM) and hardness, too, but the toughness was similar within the error. These changes were moderated by the presence of the very narrow transitional zones ICHAZ and SCHAZ. However, a detailed study of the hardness [6] showed the soft area in the SCHAZ. The measurement in this study was not so detailed, but the soft area should be taken into account.

The crack initialization across the weld was not impossible, but the occurrence was not frequent; therefore, the mechanical tests in the L direction were not performed. This type of fracture could be caused by a combination of large external forces and a small effective cross-section, i.e., the presence of weld imperfections, different notches, or very high tensile RS. The analyzed RS near the weld in the L direction (see Figure 5) was higher than the yield strength (approx. 355 MPa for the BM), and approached the tensile strength R_m. This effect was caused by the higher hardness of the weld (see Figure 6), which indicates the occurrence of hard phases with higher yield strength.

From the fatigue life point of view, the imperfections were usually considered to be the most qualitative deficiency of the weld. In case of the lack of fusion or big pores (i.e., smaller effective cross-section), the energy accumulated along these imperfections and afterwards, the internal crack was initialized. For this case, the final rupture could occur across the WM and reduce the fatigue life (see Figure 9). Despite the defects in the weld, the fatigue of the welds with defects was better than the fatigue of the arc-welded samples. The most apparent reasons were a huge heat input in combination with microstructural and surface notches.

Detailed analyses showed that in all cases without defects, cracks were initiated in the FZ on the top side (see Figure 10b). The previous text explained this phenomenon. High surface tensile RS in the transversal direction in combination with surface and microstructure notches, and changes in hardness, toughness, and microstructural parameters *D*, *e*, and FWHM, led to the weakening this zone against fatigue. Only the presence of the tensile character of RS was unusual. These undesirable stresses were probably generated during the second pass when the deformation occurred (warping). This effect indicates that the role of RS on fatigue crack initiation can be considered significant.

5. Conclusions

The purpose of this investigation was to explain the relationship of the microstructure, mechanical properties, and residual stresses of laser-welded pressure vessel steel plates. Specifically, 10 mm thick double-sided square butt welds performed on P355NL1 steel plates for high-pressure vessels were investigated. Various experimental techniques were used for the analysis of the weld properties and the critical areas for a potential surface crack initialization.

The major experimentally obtained knowledge can be summarized in the following points:

(1) Despite the internal pores, the failure occurred outside the weld metal (WM) during the tensile test.
(2) A uniform and smooth course of hardness from the base material through the heat-affected zone (HAZ) and the WM was found.
(3) Using metallography and X-ray diffraction, the microstructure notches were found near the weld. The main notch, the fusion zone (FZ), was located in the boundary of the WM and HAZ.
(4) The state of residual stresses (RS) was satisfactory in the bulk and on the bottom surface of the welded plate in the T direction. The unusual high tensile RS were found on the top side in the T direction.
(5) The toughness of the WM was smaller due to the presence of hard bainite.
(6) The lack of fusion, which was found on the fracture surface, was the primary reason for the fatigue crack in the WM. Nevertheless, the fatigue of the welds with defects was better than the fatigue of the arc-welded samples.
(7) The fatigue cracks of the samples without a lack of fusion were always initialized on the top side in the FZ.

The FZ was marked as the most critical area for the potential surface crack initialization because of the higher hardness, different microstructure, presence of notches, various toughness, and unusually high surface tensile RS. All fatigue cracks were initialized on this marked area. Some of the notches could be removed by a smaller heat input during welding. In the case of the RS, the warping of the weld was necessary to prevent.

Based on these and our previous results [10], laser welding is comparable to electron welding with less manufacturing limitations and, furthermore, has many advantages against conventional metal arc welding. Therefore, the laser is a suitable and promising tool for welding pipelines.

Author Contributions: Conceptualization, J.Č. and K.T.; resources, I.Č., N.G., and S.N.; methodology, investigation, and writing—original draft preparation, J.Č., K.T., J.K., and I.Č.; writing—review, editing, and visualization, J.Č., K.T., J.K., I.Č., N.G., and S.N.; supervision, J.Č.; project administration and data curation, I.Č.; funding acquisition, J.Č. and K.T. All authors have read and agreed to the published version of the manuscript.

Funding: This research was funded by the Center for Advanced Applied Science, grant number CZ.02.1.01/0.0/0.0/16_019/0000778. "Center for Advanced Applied Science" within the Operational Program Research, Development and Education supervised by the Ministry of Education, Youth and Sports of the Czech Republic and the project TH02010664 of the Technology Agency of the Czech Republic. K.T.'s work was supported by the Grant Agency of the Czech Technical University in Prague, grant number SGS19/190/OHK4/3T/14. Neutron diffraction measurements were carried out at the CANAM infrastructure of the NPI ASCR Rez supported through MEYS project No. LM2015056. Presented neutron diffraction results were obtained with the use of infrastructure Reactors LVR-15 and LR-0, which is financially supported by the Ministry of Education, Youth and Sports—project LM2018120.

Institutional Review Board Statement: Not applicable.

Informed Consent Statement: Not applicable.

Data Availability Statement: Data sharing is not applicable to this article.

Conflicts of Interest: The authors declare no conflict of interest.

References

1. Oyyaravelu, R.; Kuppan, P.; Arivazhagan, N. Metallurgical and mechanical properties of laser welded high strength low alloy steel. *J. Adv. Res.* **2016**, *7*, 463–472. [CrossRef] [PubMed]
2. Karayama, S. Introduction: Fundamentals of Laser Welding. In *Handbook of Laser Welding Technologies*; Woodhead Publishing: Sawston, UK, 2013.
3. Grünenwald, S.; Seefeld, T.; Vollertsen, F.; Kocak, M. Solutions for joining pipe steels using laser-GMA-hybrid welding processes. *Phys. Procedia* **2010**, *5*, 77–87. [CrossRef]
4. Anawa, E.; Olabi, A.-G. Control of welding residual stress for dissimilar laser welded materials. *J. Mater. Process. Technol.* **2008**, *204*, 22–33. [CrossRef]
5. Zhang, L.; Lu, J.; Luo, K.; Feng, A.; Dai, F.; Zhong, J.; Luo, M.; Zhang, Y. Residual stress, micro-hardness and tensile properties of ANSI 304 stainless steel thick sheet by fiber laser welding. *Mater. Sci. Eng. A* **2013**, *561*, 136–144. [CrossRef]
6. Guo, W.; Crowther, D.; Francis, J.A.; Thompson, A.; Liu, Z.; Li, L. Microstructure and mechanical properties of laser welded S960 high strength steel. *Mater. Des.* **2015**, *85*, 534–548. [CrossRef]
7. Khorrami, M.S.; Mostafaei, M.A.; Pouraliakbar, H.; Kokabi, A.H. Study on microstructure and mechanical characteristics of low-carbon steel and ferritic stainless steel joints. *Mater. Sci. Eng. A* **2014**, *608*, 35–45. [CrossRef]
8. Nitschke-Pagel, T.; Digler, K. Sources and consequences of residual stresses due to welding. *Mater. Sci. Forum.* **2014**, *783*, 2777–2785. [CrossRef]
9. Radaj, D. *Heat Effects of Welding: Temperature Field, Residual Stress, Distortion*; Springer: Berlin/Heidelberg, Germany, 2012.
10. Čapek, J.; Ganev, N.; Trojan, K.; Němeček, S.; Kolařík, K. Investigation of the real structure using X-ray diffraction as a toll for laser welding optimization. In Proceedings of the Advances in X-ray Analysis: The Proceeding of Denver X-ray Conference 2019, Lombard, IL, USA, 5–9 August 2019; Volume 63.
11. Černý, I.; Kec, J. Evaluation of Fatigue Resistance of Laser Welded High Pressure Vessels Steel P355 Considering Fracture Mechanics Approach. *Key Eng. Mater.* **2019**, *827*, 428–433. [CrossRef]
12. Čapek, J.; Trojan, K.; Kec, J.; Černý, I.; Ganev, N.; Kolarik, K.; Němeček, S. Comparison of residual stresses and mechanical properties of unconventionally welded steel plates. In Proceedings of the Experimental Stress Analysis 2020, Online Conference, 19–22 October 2020.
13. Moravec, J.; Sobotka, J.; Solfronk, P.; Thakral, R. Heat Input Influence on the Fatigue Life of Welds from Steel S460MC. *Metals* **2020**, *10*, 1288. [CrossRef]

14. Čapek, J.; Černý, I.; Trojan, K.; Ganev, N.; Kec, J.; Němeček, S. Investigation of residual stresses in high cycle loaded laser steel welds. In Proceedings of the Experimental Stress Analysis 2019, Luhačovice, Czech Republic, 3–6 June 2019.
15. European Committee for Standardization. *EN 10028-3:2017. Flat Products Made of Steels for Pressure Purposes—Part 3: Weldable Fine Grain Steels, Normalized*; European Committee for Standardization: Brussels, Belgium, 2017; p. 22.
16. Winholtz, R.A.; Cohen, J.B. Generalised least-squares determination of triaxial stress states by X-ray diffraction and the associated errors. *Aust. J. Phys.* **1988**, *41*, 189–200. [CrossRef]
17. MTEX Toolbox. Available online: https://mtex-toolbox.github.io (accessed on 20 November 2020).
18. Šebestová, H.; Horník, P.; Mrňa, L.; Doležal, P.; Mikmeková, E. The Effect of Arc Current on Microstructure and Mechanical Properties of Hybrid LasTIG Welds of High-Strength Low-Alloy Steels. *Met. Mater. Trans. A* **2018**, *49*, 3559–3569. [CrossRef]
19. Layus, P.; Kah, P.; Khlusova, E.; Orlov, V. Study of the sensitivity of high-strength cold-resistant shipbuilding steels to thermal cycle of arc welding. *Int. J. Mech. Mater. Eng.* **2018**, *13*, 3. [CrossRef]
20. Čapek, J.; Ganev, N. Evaluation of Residual Stresses in Laser Welded High-Pressure Vessels Steels by X-ray Diffraction. *Key Eng. Mater.* **2019**, *827*, 165–170. [CrossRef]
21. Ohata, M.; Morimoto, G.; Fukuda, Y.; Minami, F.; Inose, K.; Handa, T. Prediction of ductile fracture path in Charpy V-notch sample for laser beam welds. *Weld. World* **2015**, *59*, 667–674. [CrossRef]
22. Üstündağ, Ö.; Gook, S.; Gumenyuk, A.; Rethmeier, M. Hybrid laser arc welding of thick high-strength pipeline steels of grade X120 with adapted heat input. *J. Mater. Process. Technol.* **2020**, *275*, 116358. [CrossRef]

 materials

Article

Validation of Multiaxial Fatigue Strength Criteria on Specimens from Structural Steel in the High-Cycle Fatigue Region

František Fojtík [1,*], Jan Papuga [2,3], Martin Fusek [1] and Radim Halama [1]

1 Faculty of Mechanical Engineering, Department of Applied Mechanics, VŠB—Technical University of Ostrava, 17. listopadu 2172/15, 70800 Ostrava, Czech Republic; martin.fusek@vsb.cz (M.F.); radim.halama@vsb.cz (R.H.)
2 Department of Mechanics, Biomechanics and Mechatronics, Faculty of Mechanical Engineering, Czech Technical University in Prague, Technická 4, 16607 Prague, Czech Republic; jan.papuga@fs.cvut.cz
3 Center of Advanced Aerospace Technology, Faculty of Mechanical Engineering, Department of Instrumentation and Control Engineering, Czech Technical University in Prague, Technická 4, 16607 Prague, Czech Republic
* Correspondence: frantisek.fojtik@vsb.cz

Abstract: The paper describes results of fatigue strength estimates by selected multiaxial fatigue strength criteria in the region of high-cycle fatigue, and compares them with own experimental results obtained on hollow specimens made from ČSN 41 1523 structural steel. The specimens were loaded by various combinations of load channels comprising push–pull, torsion, bending and inner and outer pressures. The prediction methods were validated on fatigue strengths at seven different numbers of cycles spanning from 100,000 to 10,000,000 cycles. No substantial deviation of results based on the selected lifetime was observed. The PCRN method and the QCP method provide best results compared with other assessed methods. The results of the MMP criterion that allows users to evaluate the multiaxial fatigue loading quickly are also of interest because the method provides results only slightly worse than the two best performing solutions.

Keywords: multiaxial fatigue; high-cycle fatigue; multiaxial fatigue experiments; S-N curve approximation

Citation: Fojtík, F.; Papuga, J.; Fusek, M.; Halama, R. Validation of Multiaxial Fatigue Strength Criteria on Specimens from Structural Steel in the High-Cycle Fatigue Region. *Materials* **2021**, *14*, 116. https://doi.org/10.3390/ma14010116

Received: 27 November 2020
Accepted: 23 December 2020
Published: 29 December 2020

Publisher's Note: MDPI stays neutral with regard to jurisdictional claims in published maps and institutional affiliations.

Copyright: © 2020 by the authors. Licensee MDPI, Basel, Switzerland. This article is an open access article distributed under the terms and conditions of the Creative Commons Attribution (CC BY) license (https://creativecommons.org/licenses/by/4.0/).

1. Introduction

To validate the multiaxial fatigue strength criteria, the prediction results should be compared with experimental results. Any experimental campaign that would cover the mean normal stress effect, the mean shear stress effect, phase shift effect, etc., in various lifetimes for a single material (or more materials) presents a lengthy and costly process. The validation is thus often realized on experimental data retrieved from other sources: conference papers, papers in journals, books, PhD theses or technical reports. If such sources are used, they must be carefully verified, as to whether the data to be adopted are credible and usable for such validation.

The prediction quality of multiaxial fatigue strength criteria is often assessed on experiments with various load combinations of push–pull and torsion [1–4]. By some experiments, axial loading of specimens can be induced by bending, which causes non-constant stress distribution over the cross-section [5–9]. Relatively rarely, the inner pressure is applied on hollow specimens [10–12]. The stress gradient differs in this load mode—it is higher at the inner surface and lower at the outer surface. It is, therefore, important to know, which of those two surfaces is critical due to the other co-acting load channels, and both surfaces should be evaluated in some cases. The pressure is usually acting as a constant load. Such setup is simpler, because it allows the experimenter to run the desired load history quicker. If more load channels are superposed, the multiaxial stress state is likely to be induced. Such more complicated combinations are very interesting for validation purposes. Experimental sets comprising multiple various load cases on different specimen designs and sizes, that also provide a detailed information on material properties to tune the multiaxial fatigue strength criteria, are published only rarely. The experiments described in this paper cover various combinations of push–pull, bending, torsion, inner and outer pressures. The broad spectrum of various multiaxial load combinations with varying stress ratios on individual channels are an important condition for any adequate validation of calculation methods used by different multiaxial fatigue strength criteria.

One of the few papers, which looks for a general solution covering more different load modes in one campaign is the paper by Morel and Palin-Luc [13], in which they propose the use of the non-local model averaging the stress quantities over the critical volume. The paper by Papuga et al. [14] does not treat the same problem more generally and simply uses the axial load modes in the analysis according to assumed stress distribution. If multiaxial loading includes bending, the plane bending fatigue strengths are used in analyses. If it involves any other load case than bending, the fatigue strengths relevant to push–pull are used. This paper presents an extensive experimental campaign that was realized on hollow specimens made from ČSN 41 1523 structural steel. There are results of 24 uniaxial and multiaxial load cases of very diverse setups. The paper validates several multiaxial fatigue strength criteria on fatigue strengths obtained at 750,000 cycles from the Kohout-Věchet approximations [15] for each load case. The new MMP method usable for multiaxial fatigue strength analysis is published in [14] for the first time. As MMP is an extension of the Manson–McKnight criterion (MMK) [16,17], its biggest advantage is the simplicity of the computational analysis. It can be easily run using a common spreadsheet program such as MS Excel. The newly introduced MMP criterion significantly improves the prediction quality found for the MMK solution to a point that the MMP criterion could reach the quality of the output comparable with much more complex multiaxial fatigue strength criteria. Papuga et al. used the same data set in [18] to describe the validation results by selected multiaxial fatigue strength criteria at three additional lifetimes (the analyses were run between 100,000 and 750,000 cycles).

The current paper increases the scope of tested load cases from 24 to 34. This enlargement brings along some important load cases missing previously for some sizes of specimens—above all, the reversed torsion on smaller specimens or the repeated bending load case. These previously non-existent load cases had to be in some way substituted in the previous papers. Their inclusion into the computational scheme should result in a more consistent validation process.

The recent paper by Karolczuk et al. [19] opened a question, whether the same material parameters weighting the effect of stress parameters in the multiaxial criteria could be used over bigger ranges of lifetimes. They proved that the actual material parameters valid for the given final lifetime should result in a superior output than some fixed constant parameters could provide. To confirm or to deny that finding, the validation campaign on all 34 load cases is performed in this paper on fatigue strengths derived at seven different lifetimes spanning from 100,000 cycles to 10,000,000 cycles. To also cover the high lifetime levels, a special approximation FF formula is adopted in this paper. With all these changes involved, the paper focuses on validating 11 different multiaxial fatigue strength criteria of various types of solutions in order to assess their credibility for an accurate fatigue strength evaluation.

To reach this goal, the experimental campaign is first described in Section 2, including also the regression models used to derive the fatigue strengths from the S-N curves. The various multiaxial models processed in the validation are described in Section 3, together with the way the quality of the regression is assessed. Section 4 discusses the obtained results and Section 5 concludes the outcome of the presented paper.

2. Experiments and Processing of Their Results

2.1. Material and Specimens

The specimens were manufactured from ČSN 41 1523 structural steel (equivalent to S355JR or St52-3) delivered in bars retrieved from the single T31052 melt. The static material properties are provided in Table 1, and the chemical composition can be found in Table 2.

Table 1. Static material parameters and chemical composition of the studied material.

Designation	Tensile Strength [MPa]	Tensile Yield Stress [MPa]	Elongation at Fracture [%]	Reduction of Area at Fracture [%]	True Fracture Strength in Torsion [MPa]
ČSN 41 1523	560	400	31.1	74.0	516.6

Table 2. Chemical composition of the studied material.

Chemical Composition:					
C [%]	Mn [%]	Si [%]	P [%]	S [%]	Cu [%]
0.18	1.38	0.4	0.018	0.006	0.05

Three different specimen types: S1, S2 and S3 have already been used in [14]. The diameters (D—outer diameter, d—inner diameter) in the critical cross-sections were: (S1) D = 11 mm and d = 8 mm, (S2 and S3) D = 20 mm and d = 18 mm—see Figure 1. The newly introduced specimen type S4 is also shown in the same figure with its cross-sectional parameters slightly modifying the S1 configuration: D = 12 mm and d = 8 mm. All specimens were polished on the outer surface, while the inner surface was reamed.

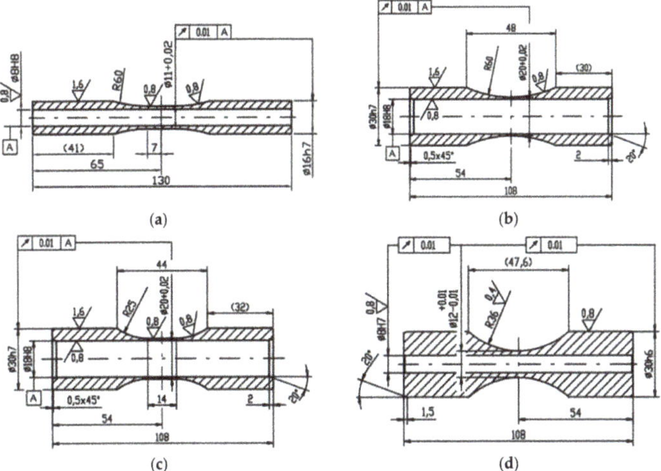

Figure 1. Drawings of specimens used in this campaign: (**a**) S1—top left, (**b**) S2—top right, (**c**) S3—bottom left, (**d**) S4—bottom right.

2.2. Load Cases

The original experimental campaign from [14] is extended in this paper by 10 further new load cases. The summary of all load cases imposed in the total of 34 configurations is provided in Figures 2 and 3. The detailed description of individual tests sets, of their types, of applied test frequencies and of geometric parameters of used specimens can be found in Table 3.

The new experimental load cases concern:

- Load case FF041—repeated plane bending.
- Load cases FF048-FF051—repeated bending with constant inner pressure imposed in the mode of a pressure vessel.

- Load cases FF062-FF063—fully reversed push–pull combined with inner pressure.
- Load cases FF064-FF065—repeated push–pull combined with inner pressure.
- Load case FF092—fully reversed torsion on S4 specimen (see Figure 1).

The S4 specimens were not manufactured on purpose after publishing [14]—they were prepared before the whole test campaign in the moment, when the search for an optimum geometry of specimens was targeted. As other specimen types were later selected for testing, the results of S4 specimens tested in fully reversed torsion were put aside. They were again uncovered, when the lack of the S-N curve for fully reversed torsion on smaller specimen type, S1, became obvious.

Figure 2 provides schematic drawings explaining the way the individual specimens were loaded for specific load cases marked FFXXX, where XXX is replaced by a unique ID number for each load case. All load cases were run under the load control. The ranges of forces, moments and pressures applied to individual load cases are provided in Table 4, and the information on the phase shift between the axial load channel and the torsion load channels accompanies them there.

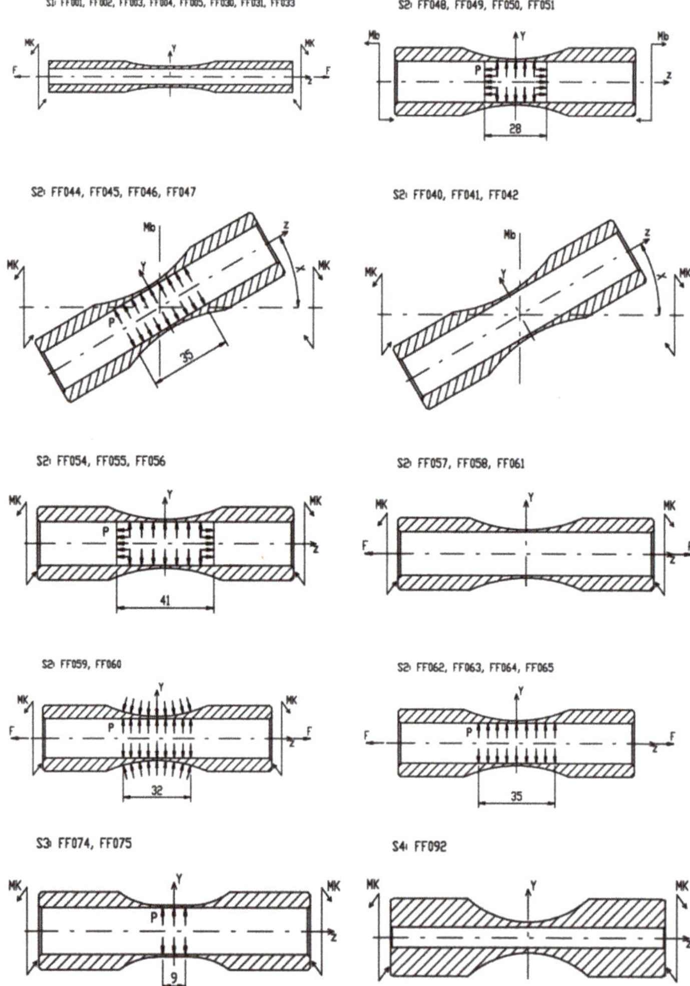

Figure 2. Overview of various setups of experiments. *F*—push/pull, *Mb*—bending moment, *Mk*—torque and *P*—pressure.

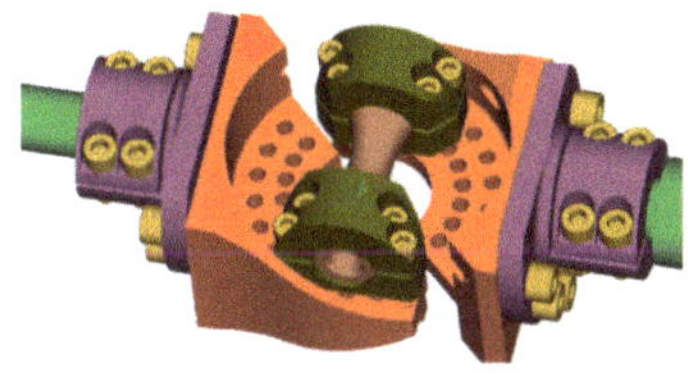

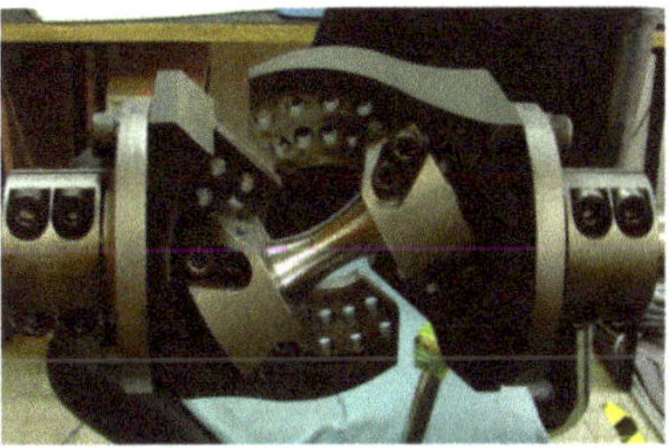

Figure 3. Setup of the experiments for load cases of torsion and bending on the torque controlled FF044 (right) and bending load case FF040 (left), both using S2 specimens.

Table 3. Summary of all load cases. Abbreviations used: Ten—tension, To—torsion, RP—pressurized, PV—pressure vessel mode, PB—plane bending, A—amplitude, M—mean value, *D*—outer diameter, *d*—inner diameter.

Mark	Specimen Type	Machine No. (Frequency [Hz])	Load Combination	Measured Diameters		Input Load Channels		
				D [mm]	*d* [mm]	Ten PB	RP PV	To
FF001	S1	1 (10)	Ten + To	10.95	8.02	A		A
FF002	S1	1 (10)	Ten + To	10.95	8.02	A		A
FF003	S1	1 (10)	Ten + To	10.95	8.02	A		A
FF004	S1	1 (10)	Ten + To	10.95	8.02	A		A
FF005	S1	1 (10)	Ten + To	10.95	8.02	A, M		A, M
FF030	S1	1 (10)	Ten	10.95	8.02	A		
FF031	S1	1 (10)	Ten	10.95	8.02	A, M		
FF033	S1	1 (10)	To	10.95	8.02			A, M
FF040	S2	2 (25)	PB	19.99	18.05	A		
FF041	S2	3 (10)	PB	19.96	18.06	A, M		
FF042	S2	2 (25)	To	19.99	18.05			A
FF044	S2	2 (25)	PB + To	19.99	18.05	A		A
FF045	S2	2 (25)	PB + To	19.99	18.05	A		A
FF046	S2	2 (25)	PB + RP	20.00	18.03	A	M	
FF047	S2	2 (25)	PB + RP	20.02	18.02	A	M	
FF048	S2	4 (20)	PB + PV	19.93	18.05	A, M	M	
FF049	S2	4 (20)	PB + PV	19.94	18.05	A, M	M	
FF050	S2	4 (20)	PB + PV	19.94	18.05	A, M	M	
FF051	S2	4 (20)	PB + PV	19.92	18.05	A, M	M	
FF054	S2	2 (25)	To + PV	19.97	18.05		M	A
FF055	S2	2 (25)	To + PV	19.94	18.09		M	A
FF056	S2	2 (25)	To + PV	19.94	18.07		M	A
FF057	S2	2 (25)	To + Ten	19.99	18.05	M		A
FF058	S2	2 (25)	To + Ten	19.99	18.05	M		A
FF059	S2	2 (25)	To + Ten + RP	20.04	18.25	M	M	A
FF060	S2	2 (25)	To + Ten + RP	19.96	18.29		M	A
FF061	S2	3 (4)	To + Ten	19.99	18.05	A		A
FF062	S2	4 (20)	Ten + RP	20.00	18.02	A	M	
FF063	S2	4 (20)	Ten + RP	20.01	18.02	A	M	
FF064	S2	4 (20)	Ten + RP	20.01	18.02	A, M	M	
FF065	S2	4 (20)	Ten + RP	20.03	18.02	A, M	M	
FF074	S3	2 (25)	To + RP	20.00	18.05		M	A
FF075	S3	2 (25)	To + RP	20.00	18.05		M	A
FF092	S4	2 (25)	To	12.00	8.00			A

Table 4. Summary of all load cases as regards to the range of applied forces, moments and pressures; also, the phase shift φ_{at} between the axial and torsion load signals is stated.

Load Case	Axial Force [kN]		Torque [Nm]		Bending Moment [Nm]		Phase Shift [°] φ_{at}	Pressure [MPa]
	F_a	F_m	Mk_a	Mk_m	Mb_a	Mb_m		
	From–To	From–To	From–To	From–To	From–To	From–To		P_m
FF001	4.34	0	34–25	0	0	0	0	0
FF002	8.5	0	24.5–16	0	0	0	0	0
FF003	4.34	0	34–26.15	0	0	0	90	0
FF004	8.5	0	33–23	0	0	0	90	0
FF005	5	5	24.1–17.5	24.1–17.5	0	0	0	0
FF030	12.85–10.4	0	0	0	0	0	0	0
FF031	9.5–8.2	9.5–8.2	0	0	0	0	0	0
FF033	0	0	37.5–27.7	37.5–27.7	0	0	0	0
FF040	0	0	0	0	99.6–81.6	0	0	0
FF041	0	0	0	0	73.3–64.7	73.3–64.7	0	0
FF042	0	0	94.5–83.9	0	0	0	0	0
FF044	0	0	81.1–63.7	0	60.5–47.5	0	0	0
FF045	0	0	42.2–33	0	94.4–73.7	0	0	0
FF046	0	0	0	0	90.1–76.5	0	0	23.3
FF047	0	0	0	0	88.0–74.0	0	0	36.0
FF048	0	0	0	0	65.6–59.6	65.6–59.6	0	10.5
FF049	0	0	0	0	67.7–57.6	67.7–57.6	0	20.0
FF050	0	0	0	0	68.5–56.5	68.5–56.5	0	30.0
FF051	0	0	0	0	66.7–54.1	66.7–54.1	0	40.0
FF054	0	0	93.0–76.2	0	0	0	0	15.0
FF055	0	0	95.0–78.0	0	0	0	0	10.0
FF056	0	0	75.5–62.8	0	0	0	0	20.2
FF057	0	14	98.8–71.6	0	0	0	0	0
FF058	0	10.2	100.6–76.6	0	0	0	0	0
FF059	0	0	91.0–77.5	0	0	0	0	40.0
FF060	0	5.9	84.7–77.0	0	0	0	0	40.0
FF061	16.6–12.3	0	27.1–20.1	0	0	0	0	0
FF062	17.0–14.6	0	0	0	0	0	0	20.0
FF063	16.5–13.0	0	0	0	0	0	0	40.0
FF064	7.2–6.3	7.2–6.3	0	0	0	0	0	20.0
FF065	7.3–6.4	7.3–6.4	0	0	0	0	0	40.0
FF074	0	0	87.6–74.9	0	0	0	0	13.0
FF075	0	0	85.7–48.2	0	0	0	0	27.0
FF092	0	0	55.7–45.2	0	0	0	0	0

To obtain the described load cases, various experimental machines had to be used as noted in Table 3, and different special fixtures to impose the desired load configurations had to be applied. Figure 3, left, depicts the example of the fixture model with the test specimen in grips as used for the load case of reversed bending marked as FF040. The combined bending and torsion loading as applied in FF044 test series is depicted in Figure 3, right. The FF046 load case combining the reversed bending with constant internal pressure can be found in Figure 4, left. The test specimens loaded by reversed torsion with inner and outer pressure in the FF059 test case can be seen in Figure 4, right. These tests were performed on the reconstructed and modernized biaxial testing machine Schenck type PWXN, which is originally equipped by the control of torque. It is extended by the possibility to apply additional axial constant force. Table 3 refers to this machine type by number 2. Number 1 in Table 3 corresponds to the biaxial servohydraulic pulsator INSTRON 8802. Another used testing machine is the biaxial servohydraulic pulsator LABCONTROL 100 kN/1000 Nm, which is equipped by a combined hydraulic actuator able to impose push–pull and torque. This machine was derived during the reconstruction of the original INOVA ZUZ 200 machine. It is marked by number 3 in Table 3. Number 4 in Table 3 concerns the uniaxial hydraulic pulsator INOVA FU-63-930-V1.

Figure 4. The load case of bending and pressurizing with the pressure chamber and the pressure sensor—FF046 load case (left). The setup of the test case with the pressure chamber inducing inner and outer pressure—FF059 load case (right).

The necessary input into all multiaxial fatigue strength criteria are local stresses. The purely elastic material response is assumed to derive them. To locate the hot-spot on more complicated testing specimens for various superposed load channels' acting, the finite element (FE) solution is necessary in order to deliver stress tensor components induced by individual load channels. All test cases and specimens were modeled and computed within the Ansys FE-solver. The stress tensors were obtained for unit loads acting on individual load channels, and the obtained stress components were then multiplied by the factor related to the ratio between the actual load and the unit load. Experiments FF062-FF065 are special, because the pressure causes higher (tangential) stress on the inner surface, while the stress response to axial loading induces more or less uniform axial stress distribution over the cross-section. For these specimens, thus, inner and outer surfaces were evaluated in the stress analysis and also in the subsequent fatigue analysis.

During the fatigue tests, responses to individual load channels were measured by certified sensors, which are regularly checked by the Czech Metrology Institute. To set up the load parameters by individual tests, strain gages were installed on chosen tested specimens; see the examples in Figure 5. The strain gages provided the information on strains and stresses attained on the surface of test specimens for individual load cases. These data items then could be compared with results of the finite element analyses to verify the applied boundary conditions.

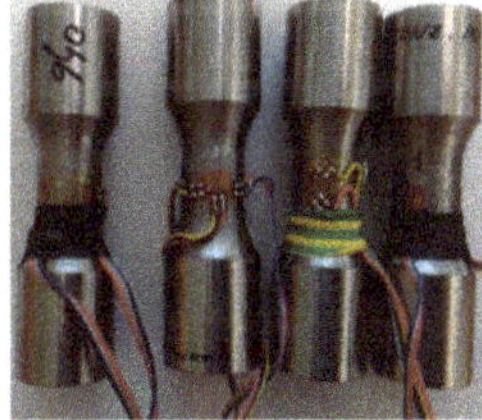

Figure 5. Examples of the load setup validations taken to ensure the applied stress are conforming to the expectations.

2.3. Regression Analyses

In the study by [14], experimental results were processed to obtain the regression S-N curves either by the linear Basquin model or by the Kohout-Věchet [15] non-linear model. Each load case was covered by at least 5 finished experiments on different load levels, and by one run-out test, which was left unfinished at 10 million cycles. Some experimental test cases were not described well above 750,000 cycles by any of the two mentioned models—see Figure 12 in [14] as regards to the FF033 test case (or see Figure 6 hereafter). This was the main reason why the fatigue strength analysis was covered only at 750,000 cycles in [14]. The newer paper by [16] documents a similar analysis of chosen multiaxial fatigue strength criteria in the limited lifetime region; more precisely at 100,000, 200,000, 500,000 and 750,000 cycles. The tests were performed in compliance with the valid ČSN 42 0362 standard [20]. Every experimental load case is completed by at least one run-out test, for which the specimen did not break even at 10 million cycles. The difference between amplitudes of the dominant stress channel for the run-out specimen and for the last broken specimen with the longest lifetime does not exceed 10 MPa for most cases. The ČSN 42 0363 standard [20] in its paragraph No. 49 recommends choosing this difference in applied stress levels in dependency on the expected fatigue limit of the evaluated tests' case. The basic number of cycles to determine the fatigue limit is set to 10 million by ČSN 42 0363 for steels.

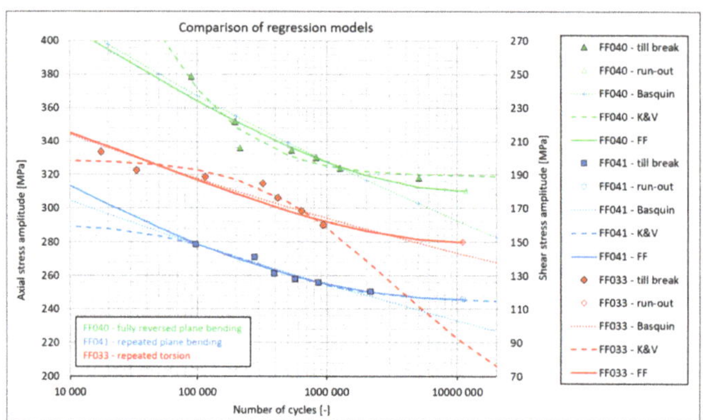

Figure 6. Comparison of the three approximation models for three different load cases.

The value of 10 million cycles is therefore set as the limit of the possible approximation domain. This paper compares results of three approximation methods for the S-N curve data. The first, and the most commonly used method is the Basquin approximation, which should be optimally used only within the experiments with limited lifetime. It is formulated by the following [21]:

$$\sigma = \sigma_f' \cdot (2N)^b \tag{1}$$

where σ_f' is the coefficient of fatigue strength, and b is the exponent of fatigue strength. Though this model is more than 100 years old, it presents the part of standards for steel structures (e.g., Eurocode 3, ISO12107).

If the experimental points are selected to be included in the approximation, two tendencies cause some bias. The first one concerns the requirement to include as many experimental points into the regression analysis, as only possible. This requirement ensures that the curve will really correspond to the behavior of material and will not describe the response of only several experimental points. The second tendency to bear in mind are the attempts to omit those data points, which do not conform to the expected material model in the limited lifetime region. For the Basquin model, such data points then can be a part of the S-N curve transition into the quasi-static domain or to the fatigue limit domain, which cannot be approximated by it reasonably well.

A suitable approximation, which can integrate into the model most of experimental data from all those domains, is the Kohout-Věchet regression model ([15], also, K&V hereafter):

$$\sigma = a \cdot \left(C \cdot \frac{N+B}{N+C} \right)^{\beta} \tag{2}$$

The non-linear regression analysis demands a certain setup for initial estimates of material parameters a, β, B, C. They can be set based on the previously obtained Basquin regression curve:

$$a = 2^b \cdot \sigma_f', \ \beta = b, \ B = 10^{[max_i(log\sigma_i) - loga]/b}, C = 10^{[min_i(log\sigma_i) - loga]/b} \tag{3}$$

Thanks to the additional two parameters available in this model in comparison with the Basquin formula, the curve can follow the trend of the S-N data with two bends—one in the transition to the horizontal line at the quasi-static domain, and the other in the transition to the horizontal line at the fatigue limit region. To define better the quasi-static response, the tensile strength at $\frac{1}{4}$ or $\frac{1}{2}$ cycles can be taken as an additional regression input to other experimental data points. Such inclusion affects, above all, the shape of the curve in the domain of low-cycle fatigue and in the quasi-static domain. However, the Kohout-Věchet model is not suitable for regression of multiple experiments per the outer load levels (load level tests), because the outermost points substantially affect the trend of the curve and its transition to horizontal lines.

To approximate the experimental data, the FF approximation first published in [22] was used. The proposed approximation function is:

$$\sigma = \sigma_0 - (\sigma_0 - \sigma_C)\cdot \sin\left\{\frac{\pi}{2}\cdot[log(4\cdot N_0)/log(4\cdot N_C)]^{a_2}\right\}. \quad (4)$$

This approximation is here used as a one-parametric, where a_2 is the only fitted parameter and σ_0 is tensile strength. The σ_C parameter corresponds to the highest stress level, at which the specimen did not break until the lifetime $N_C = 10^7$ cycles. This stress level is in accordance with [19], assumed to correspond to the fatigue limit. Thanks to the use of the sinus function, the approximations are twice bent, which enables to follow the S-N curve trends in both transitions to the horizontal lines. Materials and specimens leading to S-N curve data items that show such S-like trend can thus be suitably modeled by the FF function. On the other hand, the same function limits the use of this formula for lifetimes longer than $N_C = 10^7$ cycles, where the function would start to increase again to higher stresses, which is unlikely for any material. The FF approximation is also not suitable for load level tests. The examples of approximations by the three mentioned formulas can be compared in Figure 6. The functions of the first two formulas (Basquin—Equation (1), Kohout-Věchet—Equation (2)) are shown also outside the interpolation domain of analyzed data, so that their general trends were clearer. Due to the mentioned character of the sinus function, the FF approximation is shown only until $N_C = 10^7$ cycles, and not at higher lifetimes.

The shape of the Kohout-Věchet curve, e.g., for FF033 experiment in repeated torsion (Figure 6), is caused by the model properties, where the transition to the quasi-static region best follows the experimental data items, and it leads to the smallest coefficient of determination R^2. As only data items related to broken specimens are used for this regression model, the obtained regression curves for most evaluated load cases are limited in their interpolation region only to the lifetimes up to 2 million cycles. Some test cases show the limitation even more stringently, as e.g., the FF033 test case that is usable only up to approx. 900,000 cycles. To get the reasonably set fatigue strengths for each load case, another approximation rather than the Kohout-Věchet model should be applied. To increase the multiaxial fatigue strength analyses reported hereafter, also to the lifetimes of 1, 2, 5 and 10 million cycles, the FF approximation was chosen.

The quality of each of the approximations described in Equations (1), (2) and (4) is compared in Figure 7. The chosen characteristics shown for each load case is the coefficient of determination R^2. In this comparison, the Kohout-Věchet model clearly attains the best results, and the Basquin model and FF model are comparable one to another, but are weaker than the Kohout-Věchet approximation. It should be anyhow reminded, that R^2 parameters are computed on different sets of experimental points—the Basquin curve is regressed only on points in the inclined part of the S-N curve, the Kohout-Věchet curve excludes all run-outs, and only the FF model covers all data points. Logically, its results can be comparably worse to the Kohout-Věchet model in R^2 parameters due to the largest scope of regression inputs.

The parameters of the FF model for each tested load case are summarized in Table 5. To also show the limitation on the scope of usable lifetimes that should be imposed when dealing with the regression curves, the shortest fatigue life obtained experimentally for each load case is documented in Table 5 as N_{min} parameter.

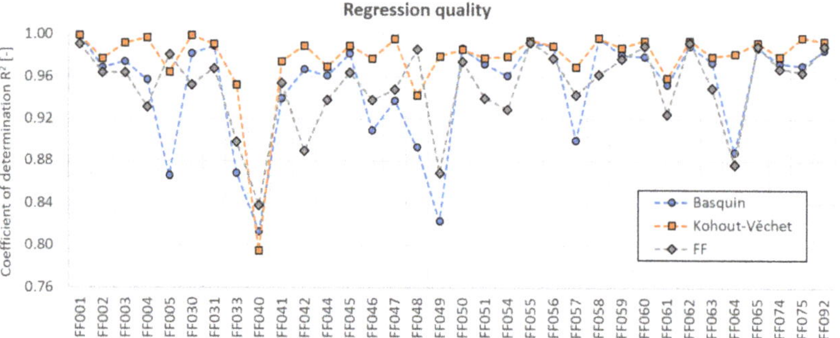

Figure 7. Comparison of the coefficient of determination R^2 for 3 evaluated regression formulas and individual load cases.

Table 5. Parameters of the FF approximations (Equation (4)) for each load case, including also the lowest measured experimental lifetime N_{min}, which together with the run-out level at 10,000,000 cycles define the complete interpolation region of each regression curve.

Mark	a_2 [-]	σ_0 [MPa]	σ_C [MPa]	N_0 [-]	N_{min} [-]
FF001	1.366	405.5	135.6	0.25	59,663
FF002	1.6989	319.9	87.0	0.25	101,102
FF003	3.108	215.2	141.8	0.25	38,206
FF004	3.061	206.1	124.7	0.25	19,604
FF005	3.551	153.6	95.3	0.25	31,769
FF030	0.92	734.1	238.9	0.25	28,562
FF031	1.388	327.0	187.3	0.25	29,150
FF033	2.153	280.9	150.2	0.25	17,553
FF040	2.001	520.5	310.3	0.25	89,700
FF041	1.35	480.4	245.8	0.25	97,324
FF042	2.349	230.1	159.5	0.25	92,200
FF044	2.369	217.8	121.1	0.25	56,102
FF045	2.09	107.8	62.7	0.25	33,100
FF046	0.911	837.6	286.9	0.25	81,900
FF047	0.873	805.4	273.4	0.25	66,537
FF048	1.163	464.3	234.5	0.25	123,147
FF049	0.512	1362.4	225.2	0.25	79,393
FF050	1.304	540.5	220.8	0.25	72,732
FF051	1.207	610.6	214.1	0.25	75,425
FF054	2.509	247.4	146.6	0.25	96,540
FF055	2.729	231.7	155.3	0.25	51,330
FF056	8.14	153.2	123.9	0.25	175,000
FF057	2.135	233.4	136.1	0.25	8886
FF058	2.404	240.7	145.7	0.25	18,210
FF059	5.815	191.3	157.1	0.25	82,310
FF060	5.774	189.2	167.2	0.25	97,500
FF061	6.216	294.7	212.7	0.25	45,723
FF062	1.818	402.2	244.9	0.25	69,061
FF063	0.828	661.9	217.0	0.25	15,656
FF064	1.656	367.4	208.6	0.25	135,447
FF065	6.263	252.3	210.6	0.25	220,139
FF074	2.996	204.4	142.5	0.25	120,600
FF075	2.588	304.0	91.6	0.25	80,310
FF092	2.962	234.7	166.2	0.25	30,800

3. Multiaxial Fatigue Strength Criteria

Though the validation program of multiaxial fatigue strength criteria on the presented experimental data concerned 24 calculation methods, this paper presents only 11 of them due to space limits. Mostly, the methods frequently appearing in papers on multiaxial fatigue strength analyses or those implemented in commercial fatigue solvers were selected in addition to some others performing well in other such comparisons [14,23–25]. The chosen methods are summarized in Table 6, where each is described by the appropriate reference, by its abbreviation used hereafter and, above all, by its formula.

The multiaxial fatigue strength criteria summarized in Table 6 cover a different approach to multiaxial prediction. The most common solution presently is the critical plane model. If a given hot-spot in which the fatigue crack initiates is evaluated, the critical plane model expects that there is some unique plane on which the stress parameters, when evaluated, result in the highest equivalent stress amplitude, and that this critical plane thus manifests the fatigue response of the whole specimen. From the models selected for the documented validation, the Dan Van criterion [26], the Findley criterion [27], the McDiarmid criterion [28], the Papuga QCP criterion [24,29] and the Papuga PCRN criterion [23,29] belong to this family of criteria. The same concept of one decisive critical plane can use another option to set the critical plane—this is the orientation for which the maximum shear stress range is found on the plane during the load cycle. This solution is here represented by the Matake criterion [30].

Another approach does not look for a critical plane but evaluates the composition of stress parameters on all planes defined by two Euler angles: φ and θ resulting in its integral mean value to be input into the final equivalent stress amplitude $\sigma_{eq,a}$. These criteria are usually called integral criteria. Two representatives were chosen in this validation—the Liu and Zenner criterion [31] and the Papadopoulos criterion [32]. The latter one processes the projection of the shear stress path on the evaluated plane into a specific direction given by the third angle χ. This projection is called resolved shear stress T.

In most of these criteria, the processed stress parameters relate to the examined plane—these are normal stress N (normal to the plane), and shear stress C (lying in the examined plane). All analyses, results of which are described hereafter, were done with the analysis of the stress path parameters via the minimum circumscribed circle concept, as described by Papadopoulos et al. in [25]. It is important to note that there are alternatives to this solution, as discussed by Meggiolaro et al. [33] or by Papuga et al. [34]—e.g., the minimum circumscribed ellipse method or the maximum prismatic hull. The difference in the method of processing the shear stress path would, however, concern only the two load cases of FF003 and FF004 run with the non-zero phase shift (see Table 4).

In addition to critical plane criteria and integral criteria, two more methods remain. There are two criteria—by Crossland [35] and by Sines [36] —that process the history of stress tensor components separated to the description of the stress deviator J_2 and the hydrostatic stress σ_H. To minimize the computation costs, the six stress tensor components can be reduced to five, thanks to the dependency of stress deviator components on its trace. This reduction to five parameters allows to project the load history into 5D Ilyushin's deviatoric space and the concept of the minimum circumscribed hyperball (or hyperellipsoid) which can be applied to it to provide the amplitude value as the radius of the hyperball.

The last criterion is the recently proposed MMP criterion, which offers the unique simplicity of processing the load history (but only then, when the load cycle is clearly defined). Each of the stress components is treated separately to define its amplitude and mean values, which are then completed into the formula in Equation (12). The validation of this method in [14] showed that despite the simple approach, the results are highly competitive with other commonly used criteria.

The material parameters necessary for each criterion are derived from fatigue strengths in axial loading and in torsion loading. They are reported in Table 7. The stress components that correspond to the fatigue strengths at given N_x cycles are derived from the FF approximation and from the outputs of the FE-analyses. All fatigue strength analyses were run in PragTic fatigue solver [37].

For the validation purposes, equivalent stress amplitude $\sigma_{eq,a}(N_x)$ in Equations (6)–(16) was compared with fatigue strengths in fully reversed push–pull loading $p_{-1}(N_x)$ or in bending $b_{-1}(N_x)$, which replaced, in the formulas in Equations (6)–(16), the more general fatigue strength in fully reversed axial loading $s_{-1}(N_x)$. The fatigue strength in fully reversed bending was applied in all multiaxial load cases including the non-zero bending load channel, while all other test cases were analyzed while using the fatigue strength $p_{-1}(N_x)$.

In the validation process, the prediction quality was not assessed directly for each experimental data point, but on fatigue strengths derived from the FF regression curves within whole interpolation domain of each test case. Seven levels of lifetimes (numbers of cycles) were chosen for the analysis in this paper: 0.1, 0.2, 0.5, 1.0, 2.0, 5.0 and 10.0 million cycles.

The relative error between the computed equivalent stress amplitude $\sigma_{eq,a}(N_x)$ and the given material response in fully reversed axial fatigue strength $s_{-1}(N_x)$ corresponds to the fatigue index error $\Delta FI(N_x)$:

$$\Delta FI(N_x) = \left(\frac{\sigma_{eq,a}(N_x) - s_{-1}(N_x)}{s_{-1}(N_x)} \right) \cdot 100\% \tag{5}$$

Table 6. Formulas of methods used within the validation program. Parameters used: s_{-1}—fatigue strength in fully reversed axial loading, S_u—tensile strength, J_2—second invariant of the deviatoric stress tensor, σ_H—hydrostatic stress, C—shear stress on the examined plane, N—normal stress on the examined plane, θ, φ—Euler angles defining the orientation of the examined plane, T—resolved shear stress (shear stress projection into a direction described by χ angle), w—Walker's mean stress effect parameter. Indexes a and m designate the amplitude and mean values, respectively. Formulas for computing a, b, c, d parameters are provided in Table 7.

Criterion	Abbrev.	Formulation of the Equivalent Stress Amplitude	Equation
Crossland	CROSS	$\sigma_{eq,a} = a_C \cdot \left(\sqrt{J_2}\right)_a + b_C \cdot \sigma_{H,max}$	(6)
Dang Van	DV	$\sigma_{eq,a} = \max\limits_{\varphi,\theta}(a_{DV} \cdot C_a + b_{DV} \cdot \sigma_{H,max})$	(7)
Findley	FIN	$\sigma_{eq,a} = \max\limits_{\varphi,\theta}(a_F \cdot C_a + b_F \cdot N_{max})$	(8)
Liu-Zenner	LZ	$\sigma_{eq,a} = \sqrt{\int_{\varphi=0}^{2\pi} \int_{\theta=0}^{\pi} \left[a_L C_a^2 (1 + c_L C_m^2) + b_L N_a^2 (1 + d_L N_m)\right] sin\theta \, d\theta \, d\varphi}$	(9)
Matake	MATA	$\sigma_{eq,a} = a_M \cdot C_a + b_M \cdot N_{max}$	(10)
McDiarmid	MCD	$\sigma_{eq,a} = \max\limits_{\varphi,\theta}\left(\frac{s_{-1}}{t_{AB}} \cdot C_a + \frac{s_{-1}}{2S_u} \cdot N_{max}\right)$	(11)
Manson-McKnight-Papuga	MMP	$\sigma_{eq,a} = \sigma_{aP}^{w} \cdot (\sigma_{aP} + \beta_P \cdot \sigma_{mP})^{1-w} \le s_{-1}$ $\sigma_{aP} = \sqrt{\frac{1}{2}\left[\begin{array}{c}(\sigma_{x,a} - \sigma_{y,a})^2 + (\sigma_{y,a} - \sigma_{z,a})^2 + (\sigma_{z,a} - \sigma_{x,a})^2 \\ +2 \cdot \kappa^2 \left(\sigma_{xy,a}^2 + \sigma_{yz,a}^2 + \sigma_{zx,a}^2\right)\end{array}\right]}$ $\sigma_{mP} = \sqrt{\frac{1}{2}\left[\begin{array}{c}(\sigma_{x,m} - \sigma_{y,m})^2 + (\sigma_{y,m} - \sigma_{z,m})^2 + (\sigma_{z,m} - \sigma_{x,m})^2 \\ +2 \cdot X_m^2 \left(\sigma_{xy,m}^2 + \sigma_{yz,m}^2 + \sigma_{zx,m}^2\right)\end{array}\right]}$	(12)
Papadopoulos	PAPADO	$\sigma_{eq,a} = \sqrt{a_P \cdot T_a^2} + b_P \cdot \sigma_{H,max}$ $\sqrt{T_a^2} = \sqrt{\frac{5}{8\pi^2} \int_{\varphi=0}^{2\pi} \int_{\theta=0}^{\pi} \int_{\chi=0}^{2\pi} (T_a(\varphi,\theta,\chi))^2 d\chi \, sin\theta \, d\theta \, d\varphi}$	(13)
Papuga QCP	QCP	$\sigma_{eq,a} = \max\limits_{\varphi,\theta}\sqrt{a_Q \cdot C_a (C_a + c_Q \cdot C_m) + b_Q \cdot N_a (N_a + d_Q \cdot N_m)}$	(14)
Papuga PCRN	PCRN	$\sigma_{eq,a} = \max\limits_{\varphi,\theta}\sqrt{a_I \cdot C_a (C_a + c_I \cdot C_m) + b_I \cdot \sqrt{N_a (N_a + d_I \cdot N_m)}}$	(15)
Sines	SINES	$\sigma_{eq,a} = a_S \cdot \left(\sqrt{J_2}\right)_a + b_S \cdot \sigma_{H,m}$	(16)

The multiaxial criteria often include the effect of the mean normal stress N_m in their formula, but the effect of the mean shear stress C_m is usually neglected. The exact formulation of the mean stress effect depends on each criterion. Crossland [35] covers the mean stress effect only by $\sigma_{H,max}$, which corresponds to the maximum hydrostatic stress during the loading cycle. Focusing on the mean normal stress only is probably historically caused by Sines [36], who postulated on the basis of gathered experimental data, that the mean shear stress has only limited effect on the fatigue limit, unless it exceeds the value of yield stress in shear. Papadopoulos et al. [25] applied this assumption as the requirement for assessing the suitability of various fatigue strength methods, and their work affected many researchers who did not object to it.

Table 7. Material parameters of individual criteria and their limitations. s_{-1}—fatigue strength in fully reversed axial loading, t_{-1}—fatigue strength in fully reversed torsion, t_0—fatigue strength in repeated torsion (maximum stress of the cycle), s_0—fatigue strength in repeated axial loading (maximum stress of the cycle), σ_1 is maximum principal stress, σ_3 is minimum principal stress, κ is ratio of fatigue strengths in fully reversed loadings (s_{-1}/t_{-1}), κ_0 is ratio of fatigue strengths in repeated loadings (s_0/t_0).

Criterion	Formula	Equation
Crossland	$a_C = \frac{s_{-1}}{t_{-1}},\ b_C = 3 - \sqrt{3}\cdot\frac{s_{-1}}{t_{-1}}$	(17)
Dang Van	$a_{DV} = \frac{s_{-1}}{t_{-1}},\ b_{DV} = 3 - \frac{3}{2}\cdot\frac{s_{-1}}{t_{-1}}$	(18)
Findley	$a_F = 2\cdot\sqrt{\frac{s_{-1}}{t_{-1}} - 1},\ b_F = 2 - \frac{s_{-1}}{t_{-1}}$	(19)
Liu-Zenner	$a_L = \frac{3}{2}\left[3\left(\frac{s_{-1}}{t_{-1}}\right)^2 - 4\right],\ c_L = \frac{28}{3a_L\cdot t_0^4}\left[s_{-1}^2 - \left(\frac{\frac{s_{-1}}{t_{-1}}\cdot t_0}{2}\right)^2\right]$ $b_L = 3\left[3 - \left(\frac{s_{-1}}{t_{-1}}\right)^2\right],\ d_L = \frac{28}{15b_L\cdot s_0}\left[\left(\frac{2s_{-1}}{s_0}\right)^2 - \frac{4}{21}c_L\cdot a_L\cdot\left(\frac{s_0}{2}\right)^2 - 1\right]$	(20)
Matake	$a_M = \frac{s_{-1}}{t_{-1}},\ b_M = 2 - \frac{s_{-1}}{t_{-1}}$	(21)
Manson-McKnight-Papuga	$w = \frac{log\frac{s_0}{s_{-1}}}{log2},\ X_m = 2\cdot\kappa\cdot\left[\left(\frac{2\cdot t_{-1}}{t_0}\right)^{\frac{1}{1-w}} - 1\right]$ $\lvert\sigma_{1,max}\rvert \geq \lvert\sigma_{3,min}\rvert : \beta_P = \frac{\sigma_{1,max}}{\sigma_{1,max}-\sigma_{3,min}}$ $\lvert\sigma_{1,max}\rvert < \lvert\sigma_{3,min}\rvert : \beta_P = \frac{\sigma_{1,min}}{\sigma_{1,max}-\sigma_{3,min}}$	(22)
Papadopoulos	$a_P = \left(\frac{s_{-1}}{t_{-1}}\right)^2,\ b_P = 3 - \sqrt{3}\cdot\frac{s_{-1}}{t_{-1}}$	(23)
Papuga QCP	$a_{QCP} = \kappa^2,\quad c_{QCP} = \frac{4\cdot s_{-1}^2}{b\cdot t_0^2} - 1$ $\kappa < \sqrt{2}:\quad b_{QCP} = 1,\ d_{QCP} = \frac{4\cdot s_{-1}^2}{b_{QCP}\cdot s_0^2} - 1$ $\kappa \geq \sqrt{2}:\quad b_{QCP} = \kappa^2 - \frac{\kappa^4}{4},\ d_{QCP} = \frac{4\cdot s_{-1}^2}{b_{QCP}\cdot t_0^2}\cdot\left(1 - \frac{\kappa^2}{4}\right) - 1$	(24)
Papuga PCRN	$1 \leq \kappa\ (\kappa_0) < \sqrt{\frac{4}{3}} : a_I = \frac{\kappa^2}{2} + \frac{\sqrt{\kappa^4-\kappa^2}}{2},\ b_I = s_{-1}$ $c_I = \frac{2s_{-1}^2}{a_I\cdot t_0^2}\cdot\left(1 + \sqrt{1 - \frac{1}{\kappa_0^2}}\right) - 1,\ d_I = \left(\frac{2s_{-1}^2}{b_I\cdot s_0}\right)^2 - 1$ $\kappa\ (\kappa_0) \geq \sqrt{\frac{4}{3}} : a_I = \left(\frac{4\cdot\kappa^2}{4+\kappa^2}\right)^2,\ b_I = 8\cdot s_{-1}\cdot\kappa^2\cdot\frac{4-\kappa^2}{(4+\kappa^2)^2},\ c_I = \frac{z}{a_I} - 1,$ $d_I = \frac{z}{b_I^2}\cdot(4\cdot s_{-1}^2 - z\cdot t_0^2) - 1,\ z = \left[\frac{8\cdot\kappa_0\cdot s_{-1}}{t_0\cdot(4+\kappa_0^2)}\right]^2$	(25)
Sines	$a_S = \frac{s_{-1}}{t_{-1}},\ b_S = 6\cdot\frac{s_{-1}}{s_0} - \sqrt{3}\cdot\frac{s_{-1}}{t_{-1}}$	(26)

Experiments and references cited by Papuga and Halama in [35] document that including the mean shear stress into the criterion can improve the prediction quality, which is proven on PCRN and QCP criteria. Acceptance of this effect and its inclusion into the formulas imposes another requirement on material parameters to be used—the S-N curve in repeated torsion is necessary. In the test set presented here, the FF033 test case performed on S1 configuration of specimens is available for these smaller specimens. To get the appropriate fatigue strengths in repeated torsion, also for bigger S2 and S3 specimens, the formula proposed by Zenner et al. [28] is used:

$$\frac{4 \cdot t_{-1}}{t_0} - \frac{2 \cdot s_{-1}}{s_0} = 1 \tag{27}$$

The same formula was used by Papuga in [35]. Papuga described that the validation of the formula in Equation (27) for test sets for which all four fatigue strengths were available showed that the relative error of the estimated fatigue strength in repeated torsion did not exceed 5%.

4. Discussion of Results

Results of chosen multiaxial fatigue strength estimation methods from Equations (6)–(16) are statistically processed and provided to the reader in Tables 8–10. Table 8 describes the mean ΔFI fatigue index errors. It also summarizes the sum of ΔFI squares over all evaluated lifetimes, which nicely documents the overall prediction quality. To support a quicker evaluation of this output, the conditional formatting of this parameter over all validated methods is shown, with the red color marking the worst results and the green color highlighting the best results. The same system of conditional formatting is also used in Tables 9 and 10. The mean ΔFI errors are provided at each evaluated lifetime and also for all of them together to document the overall trend and the potential deviations from it for individual lifetimes. The results gathered in Table 8 highlight the good prediction quality of PCRN, MMP and QCP formulas, and it also clearly shows the much worse results of SINES, CROSS, PAPADO and FINDLEY criteria.

Table 8. Results of the ΔFI statistics for 11 methods at different N_x—sum of squares and mean values.

Comp. Method	Sum of ΔFI Squares	Mean Value of ΔFI for N_x							
		All	100,000	200,000	500,000	1×10^6	2×10^6	5×10^6	1×10^7
CROSS	1334%	−9.1%	−9.2%	−8.9%	−8.5%	−8.2%	−9.1%	−9.8%	−10.1%
DV	841%	2.4%	1.9%	2.2%	2.8%	3.2%	2.6%	2.1%	1.8%
FINDLEY	1134%	7.8%	7.0%	7.7%	8.5%	8.7%	8.1%	7.5%	7.1%
LZ	295%	2.0%	3.6%	3.3%	2.8%	1.7%	1.5%	0.7%	0.2%
MATA	864%	4.5%	4.0%	4.5%	5.1%	5.2%	4.7%	4.1%	3.8%
MCDMD	279%	−0.2%	2.7%	1.7%	0.5%	−0.9%	−1.1%	−1.9%	−2.2%
MMP	148%	0.3%	1.6%	1.4%	1.0%	0.0%	−0.1%	−0.8%	−1.2%
PAPADO	1341%	−8.3%	−8.6%	−8.1%	−7.7%	−7.4%	−8.2%	−8.8%	−9.1%
QCP	165%	0.7%	2.3%	2.0%	1.5%	0.4%	0.3%	−0.4%	−0.8%
PCRN	98%	2.8%	4.0%	3.9%	3.4%	2.6%	2.5%	1.9%	1.6%
SINES	1561%	11.7%	13.1%	13.0%	12.5%	11.6%	11.3%	10.5%	10.0%

Table 9. Results of the ΔFI statistics for 11 methods at different Nx—sample standard deviations.

Comp. Method	Sample Standard Deviation of ΔFI for Nx							
	All	100,000	200,000	500,000	1×10^6	2×10^6	5×10^6	1×10^7
CROSS	21.9%	22.8%	22.8%	22.8%	22.0%	22.1%	21.5%	21.2%
DV	18.7%	18.2%	18.9%	19.5%	19.4%	19.2%	18.7%	18.4%
FINDLEY	20.4%	20.1%	20.7%	21.3%	21.3%	21.0%	20.4%	20.0%
LZ	11.0%	10.9%	11.0%	11.1%	10.8%	11.3%	11.3%	11.1%
MATA	18.6%	18.9%	19.2%	19.5%	19.1%	18.8%	18.2%	17.8%
MCDMD	11.1%	11.1%	10.9%	10.8%	10.5%	10.8%	10.9%	10.9%
MMP	7.9%	8.1%	8.0%	8.0%	7.6%	8.0%	8.0%	7.9%
PAPADO	22.3%	23.2%	23.2%	23.2%	22.4%	22.5%	22.0%	21.6%
QCP	8.3%	7.1%	7.4%	7.9%	8.3%	8.8%	9.2%	9.3%
PCRN	5.8%	5.4%	5.4%	5.6%	5.6%	6.0%	6.2%	6.3%
SINES	22.8%	22.7%	23.1%	23.5%	23.2%	23.4%	23.1%	22.6%

Table 10. Results of the ΔFI statistics for 11 methods at different N_x—minimum and maximum values and the variation range.

Comp. Method	Minimum	Maximum	Variation Range ΔFI
CROSS	−70.6%	30.5%	101.1%
DV	−30.1%	65.2%	95.3%
FINDLEY	−30.4%	58.4%	88.8%
LZ	−30.2%	42.8%	73.0%
MATA	−30.4%	49.4%	79.8%
MCDMD	−28.3%	22.7%	51.1%
MMP	−18.5%	25.2%	43.6%
PAPADO	−70.6%	30.5%	101.1%
QCP	−31.8%	27.5%	59.3%
PCRN	−16.5%	17.0%	33.5%
SINES	−25.0%	92.7%	117.6%

A quite important question is whether the individual criteria will show some systematic change of the ΔFI output at different lifetimes. Tables 8 and 9 show that there are changes for different lifetime levels, but the values remain quite stable—the typical range of values of 2% can be found for most cases. This confirms the expectation by Karolczuk et al. [36] that the fatigue strengths at the evaluated lifetime (and not for some hypothetical fatigue limit) should be used for computing the material parameters of individual multiaxial fatigue criteria. The relative insensitiveness of the ΔFI level to N_x fatigue life, at which it is computed, demonstrates also that the regression by the FF function did not affect the output in a negative way.

Table 9 states the sample standard deviations of ΔFI, both at individual lifetimes and over all of them, where the conditional formatting is again used. The evaluation of individual methods do not differ substantially from the conclusion made for Table 8.

The smallest sample standard deviation can be detected by PCRN, MMP and QCP methods. Its highest value is the result of applying SINES, PAPADO and CROSS methods. If the overall prediction quality is compared over all evaluated lifetimes, the PCRN, QCP and MMP methods provide better output than other validated methods. If the sample standard deviation of PCRN is compared with the Dang Van method (DV), which is the most often used solution in the engineering practice, it can be noted that the sample standard deviation of ΔFI for the PCRN criterion is three times smaller than the sample standard deviation of the DV criterion.

Table 10 shows minimum and maximum ΔFI values and its variation range. It can be concluded that even this parameter results in a very similar ranking of individual fatigue strength criteria as found in Tables 8 and 9. Histograms of ΔFI occurrence in all experiments and at all evaluated lifetimes are depicted in Figure 8 for all validated multiaxial fatigue strength criteria.

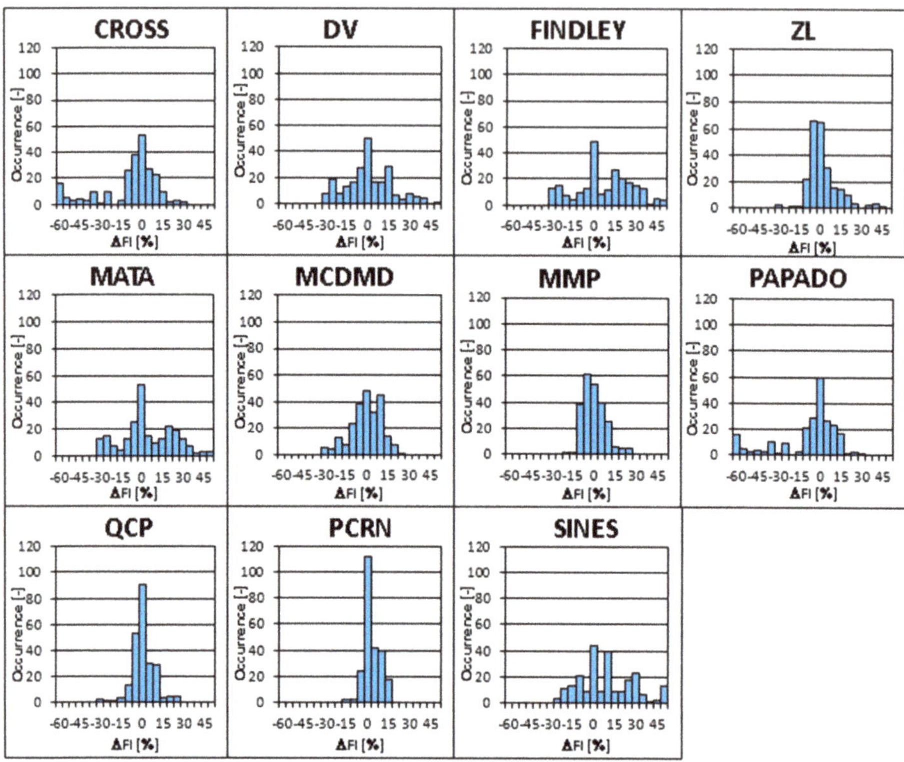

Figure 8. Histograms of ΔFI errors for all 34 test cases, 7 evaluated lifetimes and all 11 multiaxial fatigue strength criteria validated according to Table 6.

5. Conclusions

The analysis presented in this paper focuses on validating 11 different multiaxial fatigue strength criteria on the own test set composed of 34 S-N curves describing the fatigue response for different multiaxial load cases. The S-N curves were approximated on the experimental data based on the FF regression function. All test cases relate to hollow specimens manufactured from a single melt of the ČSN 41 1523 structural steel equivalent to S355JR.

Eleven chosen fatigue strength criteria comprise commonly used criteria and some new criteria, which were recently proven to provide good prediction results. For each load case, the validation is realized on fatigue strengths retrieved from the FF approximation at seven different lifetimes between 0.1 and 10 million cycles. The results are assessed based on the defined fatigue index error ΔFI and its statistical processing over all evaluated lifetimes and checked load cases. The mean value of ΔFI, the sum of its squares, sample standard deviation or its variation range are assessed. Based on the comparison for various evaluated multiaxial criteria, for the tested material and for tested load cases, it can be concluded:

- PCRN, QCP and MMP methods result in the best prediction quality, and their sample standard deviation is multiple times lower than the sample standard deviation of the Dang Van method, though the Dang Van method is commonly used in the engineering practice.
- In the range between 0.1 and 10 million cycles, the prediction quality is stable, and the variability of ΔFI over this interval is negligible if compared with the overall prediction scatter. This holds true only then, when the complete calculation of the equivalent

stress amplitude and of the material parameters specific to each multiaxial fatigue strength criterion, as described in Tables 6 and 7, is done at the same fatigue life.

Author Contributions: Conceptualization, J.P. and F.F.; methodology, J.P.; software, J.P. and F.F.; validation, J.P. and F.F.; formal analysis, J.P. and F.F.; investigation, J.P. and F.F.; data curation, F.F.; writing—original draft preparation, J.P. and F.F.; writing—review and editing, J.P. and F.F visualization, J.P., F.F, M.F. and R.H; supervision, M.F. and R.H.; project administration, F.F., M.F. and J.P. All authors have read and agreed to the published version of the manuscript.

Funding: The research by František Fojtík, Martin Fusek and Radim Halama was supported by 19-03282S project of the Czech Science Foundation, and by the specific research SP2020/23 project, supported by the Ministry of Education, Youth and Sports of the Czech Republic. Jan Papuga acknowledges support from the ESIF, EU Operational Programme Research, Development and Education, from the Center of Advanced Aerospace Technology (CZ.02.1.01/0.0/0.0/16_019/0000826); Faculty of Mechanical Engineering, Czech Technical University in Prague, and from the Grant Agency of the Czech Technical University in Prague, grant number SGS20/158/OHK2/3T/12.

Institutional Review Board Statement: Not applicable.

Informed Consent Statement: Not applicable.

Data Availability Statement: The data presented in this study are available on request from the corresponding author. The data are not publicly available due to their complexity.

Conflicts of Interest: The authors declare no conflict of interest. The funders had no role in the design of the study; in the collection, analyses, or interpretation of data; in the writing of the manuscript, or in the decision to publish the results.

Nomenclature

a_2	[-]	parameter for FF approximation
a, b, c, d	[-]	material parameters for various multiaxial fatigue limit estimation methods
a, β, B, C	[-]	material parameters of Kohout-Věchet regression
b_{-1}	[MPa]	fatigue strength in fully reversed bending loading
b	[MPa]	exponent of fatigue strength
β	[-]	mean stress coefficient
C	[MPa]	shear stress on an examined plane
d	[mm]	inner diameter
D	[mm]	outer diameter
ΔFI	[%]	fatigue index error
F	[N]	axial force
φ, θ	[°]	Euler angles defining the orientation of the examined plane
φ_{at}	[°]	phase shift between signals on axial and torsion load channels
J_2	[MPa]	second invariant of the deviatoric stress tensor
κ	[-]	ratio of fatigue strengths in fully reversed loadings $\kappa = s_{-1}/t_{-1}$
κ_0	[-]	ratio of fatigue strengths in repeated loadings $\kappa_0 = s_0/t_0$
Mb	[Nmm]	bending moment
Mk	[Nmm]	torque
N	[MPa]	normal stress on an examined plane
N_x	[-]	number of cycles
N_C	[-]	number of cycles at σ_c
p_{-1}	[MPa]	fatigue strength in fully reversed push–pull
P	[MPa]	pressure
R^2	[-]	coefficient of determination
s_{-1}	[MPa]	fatigue limit in fully reversed axial loading (stress amplitude of the cycle)
s_0	[MPa]	fatigue limit in repeated axial loading (maximum stress of the cycle)
S_u	[MPa]	tensile strength
σ_1	[MPa]	maximum principal stress
σ_3	[MPa]	minimum principal stress
σ_c	[MPa]	the highest stress level, at which the specimen did not break until $N_c = 10^7$ cycles

$\sigma_{eq,a}$	[MPa]	equivalent stress amplitude
σ'_f	[MPa]	coefficient of fatigue strength
σ_H	[MPa]	hydrostatic stress
t_0	[MPa]	fatigue limit in repeated torsion (maximum stress of the cycle)
t_{-1}	[MPa]	fatigue limit in fully reversed torsion (stress amplitude of the cycle)
T	[MPa]	resolved shear stress (shear stress projection into a given direction)
X_m	[-]	shear stress weight coefficient in the MMP method
w	[-]	Walker exponent

Abbreviations

K&V	Kohout-Věchet regression curve
PB	plane bending
PV	pressure vessel mode
RP	pressure loading
Ten	tension-compression (push–pull) loading
To	torsion loading

Indexes

a, A	amplitude
m, M	mean value
max	maximum value
0	referring to repeated loading (from zero to maximum value)
−1	referring to fully reversed loading (from −x to +x)

References

1. Baier, F. Zeit- und Dauerfestigkeit Bei Uberlagerter Statischer und Schwingender Zug-Druck und Torsionbeanspruchung. Ph.D. Thesis, Universitat Stuttgart, Stuttgart, Germany, 1970.
2. Davoli, P.; Bernasconi, A.; Filippini, M.; Foletti, S.; Papadopoulos, I.V. Independence of the torsional fatigue limit upon a mean shear stress. *Int. J. Fatigue* **2003**, *25*, 471–480. [CrossRef]
3. Bomas, H.; Bacher-Hochst, M.; Kienzler, R.; Kunow, S.; Lowisch, G.; Muehleder, F.; Schroder, R. Crack initiation and endurance limit of a hard steel under multiaxial cyclic loads. *Fatigue Fract. Eng. Mater. Struct.* **2009**, *33*, 126–139. [CrossRef]
4. Heidenreich, R.; Zenner, H. Schubspannungsintensitatshypothese–Erweiterung und experimentelle Abschatzung einer neuen Festigkeitshypothese fur schwingende Beanspruchung. *Forschungshefte FKM* **1979**, *77*. (In German)
5. Gough, H.J. Engineering steels under combined cyclic and static stresses. *J. Appl. Mech.* **1950**, *17*, 113–125. [CrossRef]
6. Froustey, C.; Lasserre, S. Multiaxial fatigue endurance of 30NCD16 steel. *Int. J. Fatigue* **1989**, *11*, 169–175. [CrossRef]
7. Kluger, K.; Karolczuk, A.; Robak, G. Validation of multiaxial fatigue criteria application to lifetime calculation of S355 steel under cyclic bending-torsion loading. In Proceedings of the 9th International Conference on Materials Structure & Micromechanics of Fracture MSMF9, Brno, Czech Republic, 26–28 June 2019; Volume 23, pp. 89–94.
8. Findley, W.N. *Combined-Stress Fatigue Strength of 76S-T61 Aluminum Alloy with Superimposed Mean Stresses and Corrections for Yielding*; National Advisory Committee for Aeronautics: Washington, DC, USA, 1953.
9. Lüpfert, H.P.; Spies, H.J. Fatigue Strength of Heat-treatable Steel Under Static Multiaxial Compression Stresses. *Adv. Eng. Mater.* **2004**, *6*, 544–550. [CrossRef]
10. Bennebach, M. Fatigue Multiaxiale D'une Fonte Gs. Influence De L'entaille Et D'un Traitement De Surface. Ph.D. Thesis, ENSAM, Paris, France, 1993. (In French).
11. Heidenreich, R. *Schubspannungsintensitätshypothese—Dauerschwingfestigkeit bei mehrachsiger Beanspruchung*; FKM: Frankfurt, Germany, 1983. (In German)
12. Troost, A.; Akin, O.; Klubberg, F. Dauerfestigkeitsverhalten metallischer Werkstoffe bei zweiachsiger Beanspruchung durch drei phasenverschoben schwingende Lastspannungen. *Konstruktion* **1987**, *39*, 479–488.
13. Morel, F.; Palin-Luc, T. A non-local theory applied to high cycle multiaxial fatigue. *Fatigue Fract. Eng. Mater. Struct.* **2002**, *25*, 649–665. [CrossRef]
14. Papuga, J.; Fojtík, F. Multiaxial fatigue strength of common structural steel and the response of some estimation methods. *Int. J. Fatigue* **2017**, *104*, 27–42. [CrossRef]
15. Kohout, J.; Věchet, S. A new function for fatigue curves characterization and its multiple merits. *Int. J. Fatigue* **2001**, *23*, 175–183. [CrossRef]
16. Krgo, A.; Kallmeyer, A.R.; Kurath, P. Evaluation of HCF multiaxial fatigue life prediction methodologies for Ti-6Al-4V. In Proceedings of the 5th National Turbine Engine High Cycle Fatigue Conference, Chandler, AZ, USA, 7–9 March 2000.
17. Kallmeyer, A.R.; Krgo, A.; Kurath, P. Multiaxial fatigue life prediction methods for notched bars of Ti-6Al-4V. In Proceedings of the 6th National Turbine Engine High Cycle Fatigue Conference, Jacksonville, FL, USA, 6–10 June 2001.

18. Papuga, J.; Fojtík, F.; Fusek, M. Efficient Lifetime Estimation Techniques for General Multiaxial Loading. In Proceedings of the 56th International Scientific Conference on Experimental Stress Analysis, Harrachov, Czech Republic, 5–7 June 2018; Volume 2018, pp. 96–101.
19. Karolczuk, A.; Papuga, J.; Palin-Luc, T. Progress in fatigue life calculation by implementing life-dependent material parameters in multiaxial fatigue criteria. *Int. J. Fatigue* **2020**, *134*, 105509. [CrossRef]
20. Methods of Fatigue Testing of Metals. Available online: http://www.technicke-normy-csn.cz/inc/nahled_normy.php?norma=420363-csn-42-0363&kat=27461 (accessed on 27 December 2020).
21. Basquin, O.H. The exponential law of endurance test. *Am. Soc. Test. Mater.* **1910**, *10*, 625–630.
22. Fojtík, F.; Fuxa, J. New modification of conjugated strength criterion. *Trans. VŠB Tech. Univ. Ostrav. Mech. Ser.* **2010**, *LVI*, 53–60.
23. Papuga, J. A survey on evaluating the fatigue limit under multiaxial loading. *Int. J. Fatigue* **2011**, *33*, 153–165. [CrossRef]
24. Papuga, J.; Halama, R. Mean stress effect in multiaxial fatigue limit criteria. *Arch. Appl. Mech.* **2018**, *89*, 1–12. [CrossRef]
25. Papadoupolos, I.V.; Davoli, P.; Gorla, C.; Filippini, M.; Bernasconi, A. A comparative study of multiaxial high-cycle fatigue criteria for metals. *Int. J. Fatigue* **1997**, *19*, 219–235. [CrossRef]
26. Dang Van, K. Sur la résistance a la fatigue des métaux. Ph.D. Thesis, Université de Paris, Paris, France, 1973; p. 647.
27. Findley, W.N.; Coleman, J.J.; Hanley, B.C. Theory for combined bending and torsion fatigue data for SAE 4340 steel. In Proceedings of the International Conference on Fatigue of Metals, London, UK, 10–14 September 1956.
28. McDiarmid, D.L. A general criterion for high cycle multiaxial fatigue failure. *Fatigue Fract. Eng. Mater. Struct.* **1991**, *14*, 429–453. [CrossRef]
29. Papuga, J. Improvements of two criteria for multiaxial fatigue limit evaluation. *Bull. Appl. Mech.* **2009**, *5*, 80–86.
30. Matake, T. An explanation on fatigue limit under combined stress. *Bull. Jpn. Soc. Mech. Eng.* **1977**, *20*, 257–263. [CrossRef]
31. Zenner, H.; Simburger, A.; Liu, J. On the fatigue limit of ductile metals under complex multiaxial loading. *Int. J. Fatigue* **2000**, *22*, 137–145. [CrossRef]
32. Papadopoulos, I.V. A new criterion of fatigue strength for out-of-phase bending and torsion of hard metals. *Int. J. Fatigue* **1994**, *16*, 377–384. [CrossRef]
33. Meggiolaro, M.A.; de Castro, J.T.P. An improved multiaxial rainflow algorithm for non-proportional stress or strain histories—Part I: Enclosing surface methods. *Int. J. Fatigue* **2012**, *42*, 217–226. [CrossRef]
34. Papuga, J.; Nesládek, M.; Jurenka, J. Differences in the response to in-phase and out-of-phase multiaxial high-cycle fatigue loading. *Frat. Integeità Strutt.* **2019**, *13*, 163–183. [CrossRef]
35. Crossland, B. Effect of large hydrostatic pressure on the torsional fatigue strength of an alloy steel. In Proceedings of the International Conference on Fatigue of Metals, Institution of Mechanical Engineers, London, UK, 10–14 September 1956; pp. 138–149.
36. Sines, G. Behavior of metals under complex static and alternating stresses. In *Metal Fatigue*; McGraw Hill: New York, NY, USA, 1959; pp. 145–469.
37. Papuga, J. Manual Program PragTic. Available online: http://www.pragtic.com/program.php#help (accessed on 27 December 2020).

Article

Billet Straightening by Three-Point Bending and Its Automation

Radim Halama [1,*], Jan Sikora [2,*], Martin Fusek [1,*], Jaromír Mec [3], Jana Bartecká [1] and Renata Wagnerová [2]

1 Department of Applied Mechanics, Faculty of Mechanical Engineering, VŠB—Technical University of Ostrava, 17. listopadu 2172/15, 70800 Ostrava, Czech Republic; jana.bartecka@vsb.cz

2 Department of Control Systems and Instrumentation, Faculty of Mechanical Engineering, VŠB—Technical University of Ostrava, 17. listopadu 2172/15, 70800 Ostrava, Czech Republic; renata.wagnerova@vsb.cz

3 ELCOM, a. s., Lomnického 1705/9, 14000 Praha 4-Nusle, Czech Republic; Jaromir.Mec@elcom.cz

* Correspondence: radim.halama@vsb.cz (R.H.); jan.sikora@vsb.cz (J.S.); martin.fusek@vsb.cz (M.F.)

Abstract: This paper presents the current results of cooperation focused on automatic billet straightening machine development. First, an experimental study of three-point bending realized on small specimens is presented to explain the basic ideas of the straightening. Then, the main regimes of straightening and the algorithm itself are described together. Subsequent finite element simulations of operational experiments show the applicability of the developed theory. The significance of material parameters estimation is depicted in this work. At least four parameters have to be properly determined for a new material in the straightening process.

Keywords: straightening process; three-point bending; FEM; control strategy; billet straightening

Citation: Halama, R.; Sikora, J.; Fusek, M.; Mec, J.; Bartecká, J.; Wagnerová, R. Billet Straightening by Three-Point Bending and Its Automation. *Materials* **2021**, *14*, 90. https://dx.doi.org/10.3390/ma14010090

Received: 30 November 2020
Accepted: 22 December 2020
Published: 28 December 2020

Publisher's Note: MDPI stays neutral with regard to jurisdictional claims in published maps and institutional affiliations.

Copyright: © 2020 by the authors. Licensee MDPI, Basel, Switzerland. This article is an open access article distributed under the terms and conditions of the Creative Commons Attribution (CC BY) license (https://creativecommons.org/licenses/by/4.0/).

1. Introduction

Modern steel factories and enterprises of heavy industry, whose field of activity includes the production of long metallic articles, meet the issue of effective straightening of such products. Typical products that involve straightening during their production technology are billets [1,2], strips [3], railway rails [4], elevator guide rails [5], or more general long linear guideways that enable precise linear motion of machines [6]. For the mentioned commodities, there are only two straightening principles that mostly used in technical practice. The first option is continuous straightening [7,8], where the bar is straightened between two cross-rolling straighteners [9,10] or inside a multi-roller straightening machine [11]. This option, however, is very problematic for straightening bars with large cross-sections [12,13], mainly owing to the requirements of employing mighty bearings.

This article is devoted to the issue of billet straightening, where the second type of straightening is commonly used. The principle of this straightening type relies on three-point bending [14,15], which is more accurate and admits higher dimension variability of straightened billets cross-sections [16]. In ironworks, three-point bending is a necessary operation performed before grinding billets. The straightening of billets is usually done manually by operators in a manual regime based on human vision and joystick control [1].

The straightening process can be automated in accordance with the Industry 4.0 strategy, but this is a challenging task [17,18]. An automatic straightening machine can achieve optimal effectivity only if the straightening algorithm is adopted to the various profile curvatures of the billet (e.g., single-arc shape, "S" shape, or shape with multiple vertices [19]). Each type of billet shape requires a unique approach to the straightening, which minimalizes the time of the process. This is the so-called multi-step straightening mechanism [6,19], for which functionality is necessary to correctly determine the velocity of the straightening force/stroke, the distance of supports, the number of straightening steps, and so on. Different parameter settings return differently straightened billets [5].

The crucial thing is to achieve accurate prediction of spring back [20,21] after releasing the straightening force. To achieve fast calculations, analytical and semianalytical approaches are currently used. The finite element method (FEM) is time-consuming and the solution is dependent on many parameters such as element type, and thus shape functions, geometry, and time discretization (according to the material model implementation), among others. For the purposes of the development of the straightening algorithm, the analytical approach could be inspired by other research works using an analytical solution for spring back prediction. During the last two decades, the strategy of multistep straightening was enhanced for deflected shafts with the circular cross-section by the fuzzy self-learning method [22], for steel wires using genetic programming [23] and for T-section beams using neural networks [24]. The latter approach required finite element simulations to develop the artificial neural network approach. The straightening history should be considered for the prediction of residual stresses, which play an important role in the service of the final products. Ling et al. [25] published an interesting study in this field including the prediction of residual stresses after grinding.

A significant benefit of analytical methods is also the accuracy of the solution, especially when a robust material model is considered in the analysis. Eggertsen and Mattiasson evaluated six cyclic plasticity models for spring back prediction [21]. They showed that the Yoshida–Uemori model [26,27] and its modification can correctly describe the Bauschinger effect, a transient behavior, a permanent softening, and a workhardening stagnation. Hajbarati and Zajkani [28] used the modified Yoshida–Uemori two-surface hardening model [21] to predict the spring back of an advanced high-strength steel. High-strength steels reveal significant spring back. FE analyses of three-point bending experiments were presented, for instance, by Zhao and Lee [29].

The following chapters of this article present the current results in the frame of a long-term project devoted to the development of an automatic billet straightening machine. The machine was designed, constructed, and manufactured by KOMA—Industry s.r.o. for TŘINECKÉ ŽELEZÁRNY a.s. The camera system and visualisation of the straightening was developed by experts from ELCOM, a.s. The focus of the article is to show the basic ideas of the newly proposed algorithm and to explain the necessary optimisation procedure needed to obtain some process parameters. This is very important for achieving reliable and robust straightening.

2. Three-Point Bending

First of all, the terminology for three-point bending straightening should be introduced. The simplified situation of the three-point bending case is shown in Figure 1, where the initial shape of the billet is depicted by a dotted line. The maximal deflection w caused by the applied force F can be visualised by the deformed shape of the billet drawn with a dashed line. In the ideal case, the billet shape is straight after spring back, as displayed by the solid line.

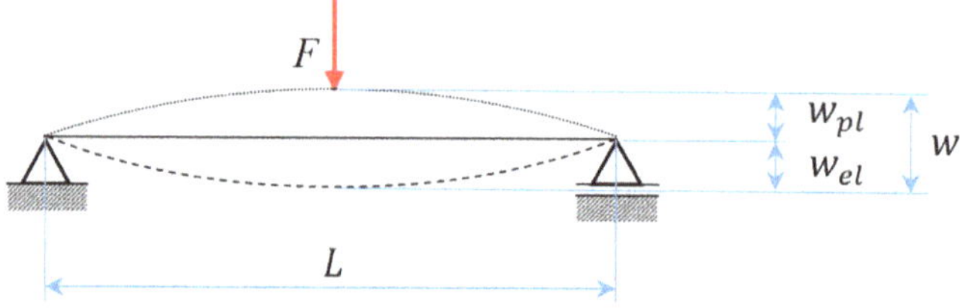

Figure 1. A scheme of a three-point bending case.

In accordance with the additive rule, the total deflection w is composed of plastic deflection w_{pl} and elastic deflection w_{el}, thus

$$w = w_{el} + w_{pl} \quad (1)$$

The irreversible deflection w_{pl} is also important input for the algorithm, and it is supposed that it can be accurately measured by a sensory system of an automatic straightening machine for the given distance of supports L.

The plastic deflection w_{pl} can be calculated considering an elastic stiffness k_{el} and applied bending force F according to the analogy to Hooke's law.

$$w_{pl} = w - w_{el} = w - \frac{F}{k_{el}} \tag{2}$$

The elastic stiffness k_{el} is a function of the Young modulus E, moment of inertia I_z, and support distance L. We will consider just a square cross-section of the billet in this work, i.e., $I_z = D^4/12$, where D is the dimension of the square cross-section.

A prediction of required total deflection (output quantity of the algorithm) is proposed to be determined from the linear relationship.

$$w = k_w w_{pl} + w_y, \tag{3}$$

where w_y and k_w are material parameters. Substituting (3) into (2), one can obtain the linear relation between the bending force and total deflection.

$$F = k_{el}\frac{w_y}{k_w} + k_{el}\left(1 - \frac{1}{k_w}\right)w = A + B \times w. \tag{4}$$

It can be noted that the parameter w_y expresses the total deflection of the billet corresponding to the maximal bending stress in the cross-section for the elastic region of loading, i.e., yield stress σ_y.

3. Laboratory Experiments and Their Numerical Simulations

In order to show the idea of the approximation of material response during straightening by three-point bending, an experimental study on three-point bending performed on 51CrV4 material at room temperature will be presented. First, the basic mechanical properties were determined by tensile test; see Table 1. The bending tests were realised on specimens with the square cross-section of variety of dimensions D and distances of supports L. The proper ratio of D/L for each bending test had to be determined analytically or numerically.

Table 1. Mechanical properties of 51CrV4 material obtained from the tensile tests.

Quantity	Averaged Values
Yield strength $R_{p0,2}$ (MPa)	523
Ultimate strength R_m (MPa)	1005
Young modulus E (MPa)	207,000
Ductility (%)	15.6

In this study, finite element method (FEM) was used. The material model introduces the nonlinear kinematic hardening rule of Chaboche [30]. According to Chaboche's superposition, two back-stress parts are considered to express the back-stress.

$$\alpha = \sum_{i=1}^{2} \alpha_i = \alpha_1 + \alpha_2 \tag{5}$$

and the evolution equation of Armstrong and Frederick [31] for uniaxial loading is

$$d\alpha_i = C_i d\varepsilon_p - \gamma_i \alpha_i dp \tag{6}$$

where C_i and γ_i are material parameters, $d\varepsilon_p$ is the increment of longitudinal plastic strain, and dp is the increment of accumulated plastic strain.

The constitutive equation of the Chaboche model for uniaxial tension is

$$\sigma = \sigma_y + \alpha_1 + \alpha_2 = \sigma_y + \frac{C_1}{\gamma_1}\left(1 - e^{-\gamma_1 \varepsilon_p}\right) + \frac{C_2}{\gamma_2}\left(1 - e^{-\gamma_2 \varepsilon_p}\right) \tag{7}$$

The tensile curve of the investigated material is used to calibrate the Chaboche model [30] for preliminary simulations by FEM; see Figure 2. All material parameters resulting from a non-linear least-square method application are stated in Table 2. Poisson's ratio ν = 0.3 was considered in the simulations too.

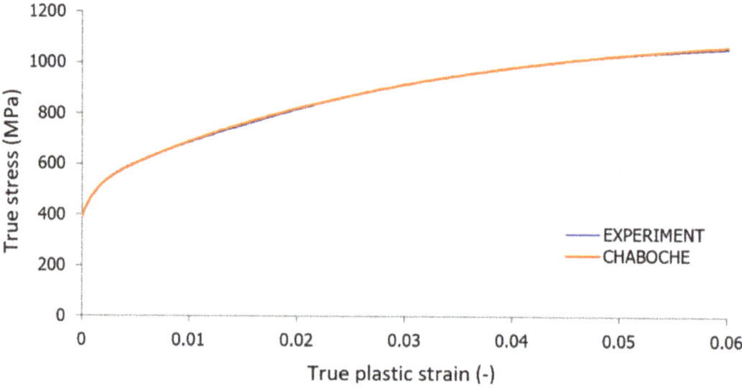

Figure 2. Deformation curve of 51CrV4 material and its approximation by Equation (7).

Table 2. Material parameters of the Chaboche model for 51CrV4 material.

E	σ_y (MPa)	C_1 (MPa)	γ_1 (-)	C_2 (MPa)	γ_2 (-)
207,000	391	97,000	877	22,000	34

All FE simulations within this paper were done in ANSYS 2020R1. The goal of the numerical study was to find a proof of the relationship between the total deflection and the plastic deflection described by Equation (3). The square cross-sections of 6 × 6, 8 × 8, 10 × 10, 12 × 12, and 14 × 14 were considered.

For the discretisation of geometry, the BEAM188 element was used. Boundary conditions applied to the FE model are shown in Figure 3. All nodes of the model are fixed in rotations around the x-axis. Ramped displacement with time is applied in the middle of the model in the y-direction, leading to maximal displacement of U_y = 4 mm at the end of the computation.

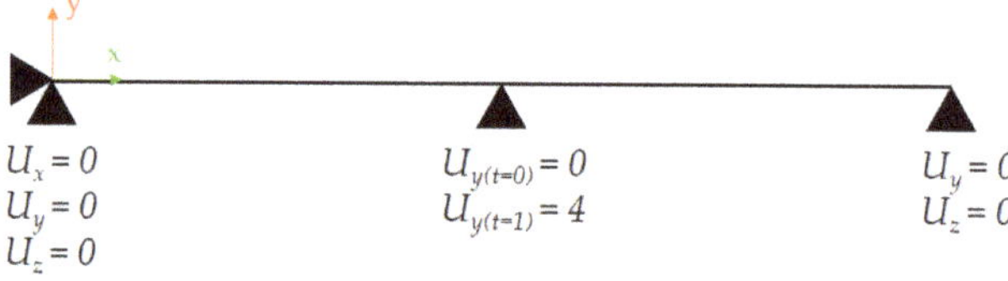

Figure 3. Finite element (FE) model with boundary conditions.

An optimization task was done (parametric study) to get proper distance of the supports for each cross-section dimension D. Initially, the distance of supports of 80 mm was chosen for the 6 × 6 specimen. After performing the FE analysis for this case, the dependency of the total deflection on the plastic deflection was evaluated using Equation (2) and approximated by the linear function (3). Then, the largest cross-section of 14 × 14 was considered for simulations by trial and error to gain acceptable correlation with the approximated curve of the first case (total deflection vs. plastic deflection). Other cases, 8 × 8, 10 × 10, and 12 × 12, were solved by repeated FE simulation with an initial guess of the support length supposing the linear relationship between the support distance and cross-section dimension from previous two limit cases.

The resulting curves, which describe the relation between the total deflection and the plastic deflection, are shown in Figure 4. Good overall correlation is achieved for particular cases of cross-sectional dimensions. The dependency is pretty linear in the interval between 0.5 and 2.5 mm of plastic deflection, which confirms the validity of Equation (3). The optimal distances of supports are as follows: 80, 90, 100, 110, and 120 mm (for cross-sectional dimensions of 6 × 6, 8 × 8, 10 × 10, 12 × 12, and 14 × 14).

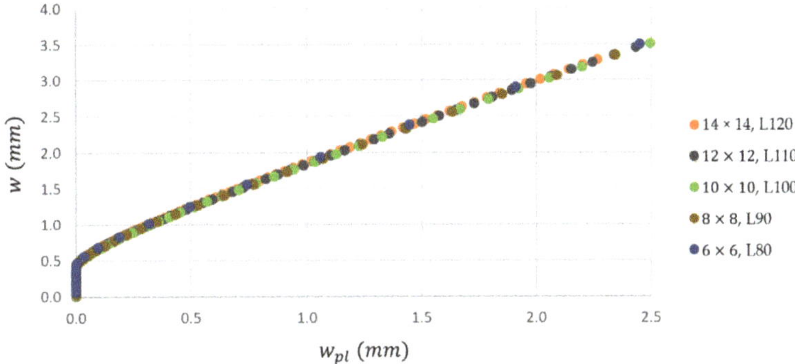

Figure 4. The dependency of total deflection on plastic deflection from the numerical study concerning optimal distances of supports (*L* in the legend) for each cross-section size *D* × *D*.

Based on the numerical study, the curve describing the dependency of the optimal support distance *L* on the cross-sectional dimension *D* of specimens is constructed; see Figure 5. It is clear that the idea of linear dependency of the optimal support distance on the cross-sectional dimension is true. Concerning the available material of billet, the following appropriate dimensions of specimens were selected: 2.9, 4.75, 7.45, 9.55, and 14 mm. The corresponding support distances are as follows: 65, 73, 89, 98, and 120 mm. The specimens for experiments were made by electric discharge machining (EDM) using a portion of the material chosen from the same position of the billet cross-section as for tensile tests.

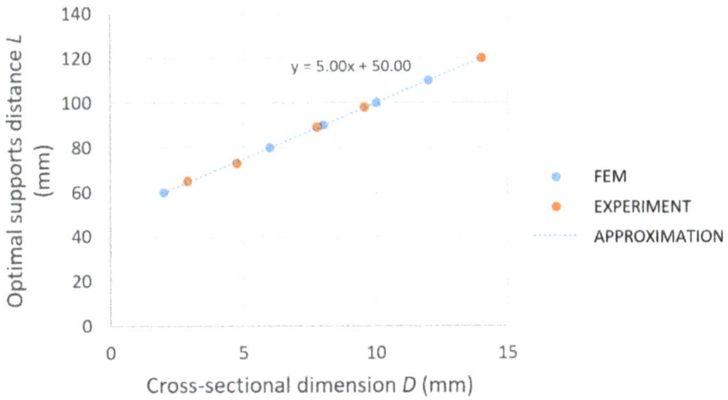

Figure 5. The dependency of the optimal distance of supports *L* on cross-section size *D*. FEM, finite element method.

All experiments were realized using a TESTOMETRIC M500-50CT universal testing machine. The position rate was 5 mm per minute. Deflection was measured as the position of the crossbar. A photo from a three-point bending test realization is shown in Figure 6, where the deformed shapes of specimens are also presented.

(a) (b)

Figure 6. Photos from the three-point bending test: the whole setup (**a**) and selected deformed specimens (**b**).

Obtained bending force versus total deflection diagrams are shown in Figure 7. The target total deflection (position of crossbar) was 3 mm for D = 2.9, 5 mm for D = 14, and 4 mm for all others.

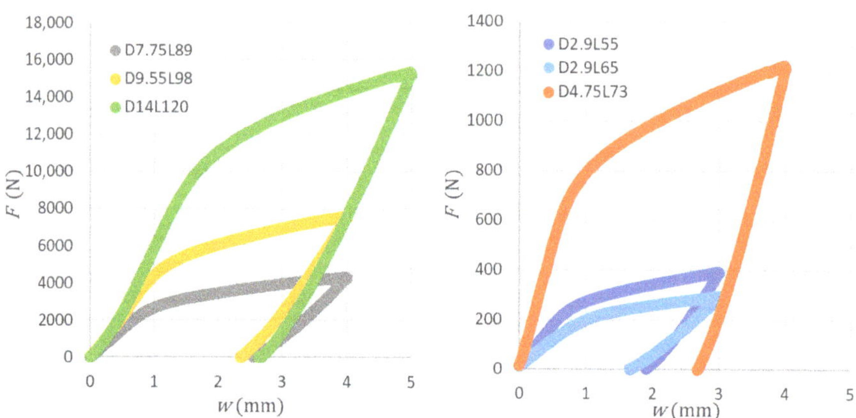

Figure 7. Force response to total deflection for all considered cases.

For eventual straightening of billets with different cross-sectional dimensions, it is important to investigate how the dependences of the total deflection on the plastic deflection differ for individual cross-sections, as presented in Figure 8.

An important finding from the performed experimental study is the fact that the slope k_w remains approximately the same even though the cross-sections are significantly different in their dimensions. The curves on the graph shown in Figure 8 differ only in the vertical offset. It should be noted that a slight nonlinearity is present in the initial part of the curve of total deflection versus plastic deflection. However, the straightening of billets will be done only in positions where it makes sense. The interventions will be proposed only for significant deviation from a straight line created between supports based on billet shape captured by the camera system.

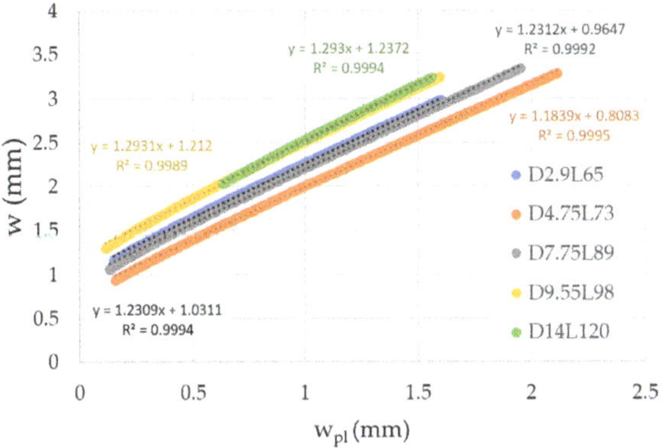

Figure 8. Dependences of total deflection on plastic deflection evaluated from three-point bending tests.

4. Camera System

As mentioned above, accurate measurement of the initial billet shape is important to achieve reliable results in the straightening process. At the beginning of the straightening process, a profile of the billet is scanned by the camera system for a given side of the billet. The sensory system is composed of eight 2D monochrome cameras. Each camera is paired with a projector that projects a strip pattern on the scanned billet (Figure 9). The projectors are involved into the scanning process to eliminate poor contrast between the billet and the straightening machine and improve the overall quality of received data that directly influence the quality of the curve representing the billet shape.

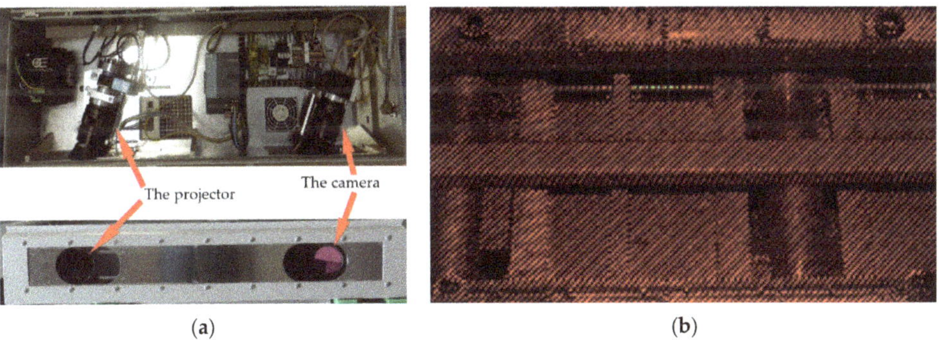

Figure 9. The case for a camera-projector subsystem (**a**), and scanned sector of the straightening machine with the billet (**b**).

The cameras are equally distributed above the straightening machine to capture the whole area of the press technology (14 m × 1 m). Each camera captures a sector of technology with a length of 2.2 m. Pictures from neighbour cameras are overlapping, so we can get a picture of the whole billet by continuous junction of pictures from individual sections. By processing the picture of the whole billet, we can detect one of the upper edges of the billet. This is crucial for obtaining the curve representing the profile of the billet. An example of a screen visible for operators with subsequently proposed two strokes is shown in the Figure 10.

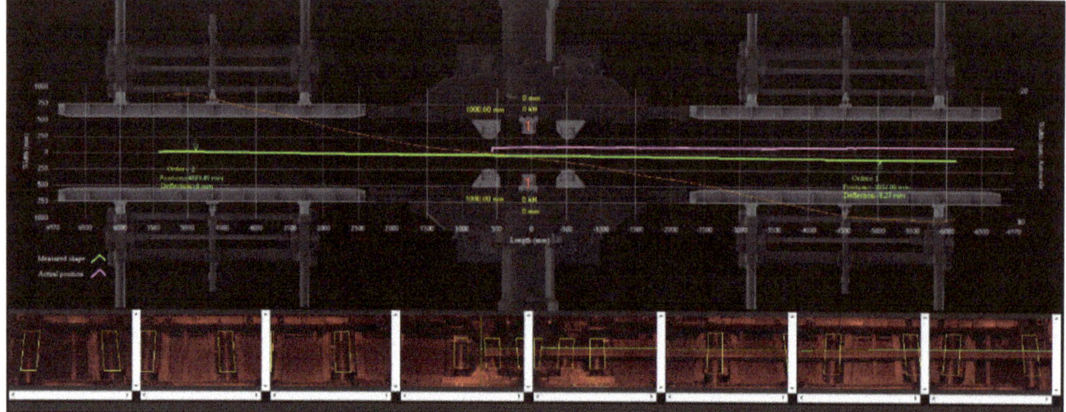

Figure 10. Visualization of initial billet shape and proposed strokes in the technology.

5. Straightening Algorithm

The straightness of the billet is defined by two criteria that determine the type of the billet based on its shape. Both parameters direct the straightening regime subsequently applied in the algorithm.

The first parameter is the sum of the maximum and minimum deviation from the linear regression line (Figure 11) considering the whole curve of the billet. It is marked as p_1 in the algorithm. The critical value of parameter p_1 is marked as p_{1crit} and should be appropriately chosen according to the current billet length.

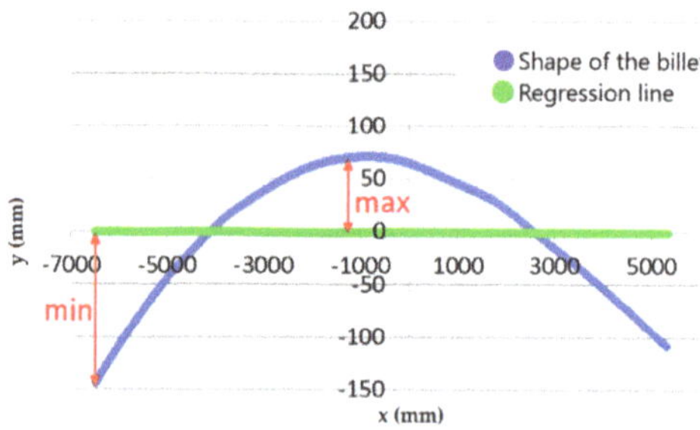

Figure 11. Scheme of gathering parameter p_1.

The second parameter called p_2 contains the value of maximum deviation on a 1 m segment. This is obtained when the 1 m segment is virtually moved along the whole length of the curve. The deviation on the 1 m segment is determined by the maximum deviation of the curve point from the line connecting the two ending points of the segment. The critical value of the parameter p_2 will be marked as p_{2crit} and influences the output accuracy of the straightening process.

The objective of the straightening algorithm is to straighten the billet i.e., to reduce both billet parameters below their critical values. The definition of a straight billet depends on subsequent technological processes and customer requirements. The most commonly applied technological process is grinding. Currently acceptable values by customers are p_{1crit} = 15 mm (12 m billet length) and p_{2crit} = 2 mm.

Four different straightening regimes of the algorithm are currently applied. The regime of the straightening algorithm is chosen based on the parameters mentioned above, supplemented by p_0 and p_{max}, which help to distinguish slightly curved billets and strongly crooked ones, respectively. The values of p_0 and p_{max} are constant for a given material. The regime of the algorithm is chosen based on the billet shape according to schema of the algorithm; see Figure 12.

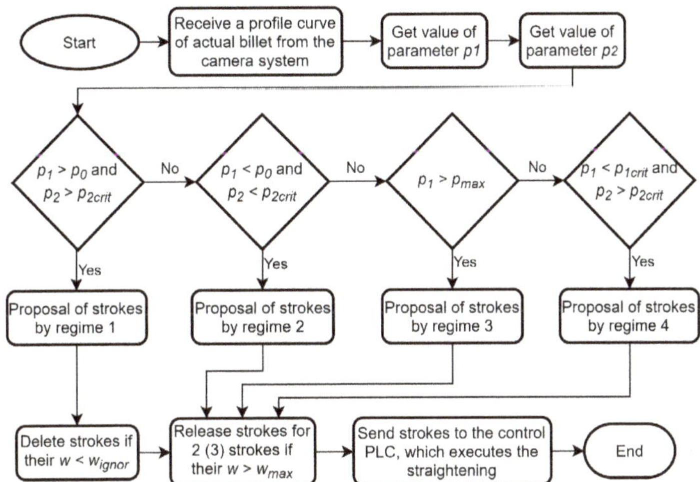

Figure 12. Flowchart of the straightening algorithm part proposing strokes (PLC—Programmable Logic Controller).

The first regime is applied in the case of valid conditions $p_1 > p_0$ and $p_2 > p_{2crit}$. This variant is usually the most effective one for "snake-like" billets. The billet is divided into particular sections with a length of 1 m. In each section, a regression line is determined and the value of w is calculated by Equation (3) based on the value of w_{pl}, which is given from the measured shape within the 1 m segment. If $w > w_{ignor}$ then an intervention is performed in the given position. This variant of straightening is usually quite time-consuming for a large number of interventions, thus the value of w_{ignor} should be optimized to achieve an acceptable speed of straightening without compromising accuracy. The parameter w_{ignor} has the meaning of the minimal applied stroke in the first regime.

The second regime of the algorithm is chosen for $p_1 < p_0$ and $p_2 < p_{2crit}$. The billet can be categorized as slightly curved "S-shaped" billet or "single-arc" type billet. Therefore, it is straightened either by two strokes or just one.

The third regime of straightening is used in the interval $p_1 > p_{max}$. The condition corresponds to a strongly crooked billet. The straightening is boosted according to the given material. An empirically determined multiplier is used for all strokes calculated by Equation (3).

The last regime is when $p_1 < p_{1crit}$ and $p_2 > p_{2crit}$. It is evident from the condition that it usually corresponds to the case where the billet is curved in just one place. The largest deviation on the 1 m segment is found and the stroke is proposed using Equation (3).

The detailed flowchart of the complete straightening algorithm is shown in Figure 12. The part of the algorithm determining the positions and stroke proposals was written in NI LabView 2014 interface.

6. Operational Experiments and Their Numerical Simulations

To show the efficiency of the straightening algorithm in the second regime of the algorithm, two exemplar billets made from 100Cr6 material with a cross-section of 150 mm × 150 mm corresponding to "single-arc" and "S-shaped" type were selected for reporting as operational experiments.

Input parameters of the first billet shape were p_1 = 29.8 mm and p_2 = 2.4 mm. After one stroke application (w = 7.76 mm, in the position x = −3889 mm), the output parameters evaluated by the sensory system were 11.9 and 1.3 mm. The initial and final shape of the first exemplar billet is shown in Figure 13a. The input parameters of the second billet shape were p_1 = 20.8 mm and p_2 = 2.4 mm. After two strokes application (w = 8.37 mm, in the position x = −4350 mm and w = −8.1 mm, in the position x = −1580 mm), the output parameters evaluated by the sensory system were 14.4 and 1.8 mm. The initial and final shape of the second exemplar billet is shown in Figure 13b.

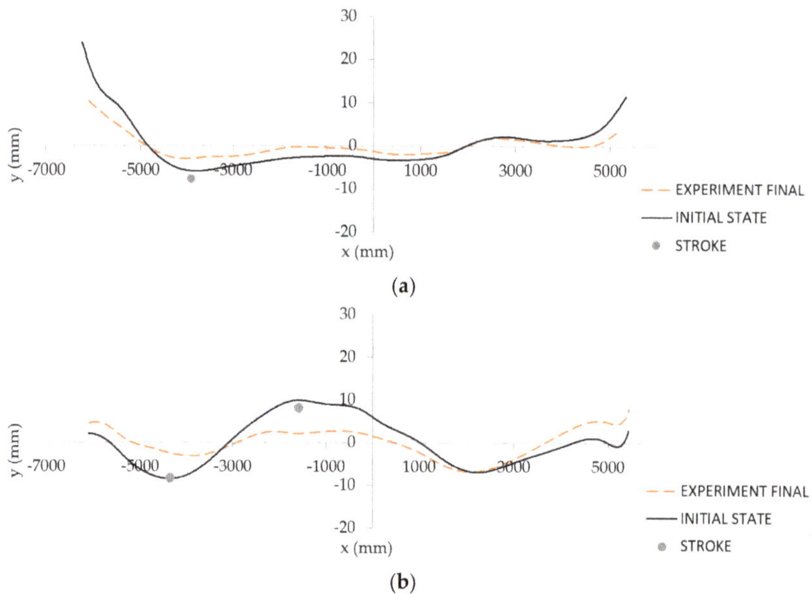

Figure 13. Initial and final shapes of two exemplary billets from operational experiments: "single-arc" type (**a**) and "S-shaped" type (**b**).

Finite element simulations were performed using the same strategy as in Section 3. Each billet was modelled using a spline curve created from points with an increment of 10 mm in the x-axis based on data obtained from the camera system. All nodes of the FE model are fixed in rotations around the x-axis. In the simulation of the "S-shaped" billet, two load steps were used. First, the boundary conditions of load step one will be described. Displacement boundary conditions were applied according to Figure 14. The force applied in the middle of the support distance (L = 1 m) was applied as a linear function of time. The maximal size of force is reached for 1 s with the corresponding value calculated from Equation (4). Then, a linear decrease of force to 10 N (because of convergency) is applied during the unloading phase, which ends after 2 s. In the second load step of the simulation, the maximal force is applied for 3 s considering Equation (4), and displacements were fixed similarly as shown in Figure 14 (supports moved to the new positions). The unloading phase is finished at 4 s with 10 N of force in the computation. The boundary conditions for the "single-arc" billet straightening simulation were analogous to those described for load step one of the "S-shaped" billet straightening simulation.

The Chaboche material model was calibrated to give an acceptable response of force for a given total deflection and to give a similar curve of total deflection versus plastic deflection; see Figure 15. Poisson's ratio ν = 0.3 was considered in both simulations. All other material parameters of the Chaboche model are stated in Table 3.

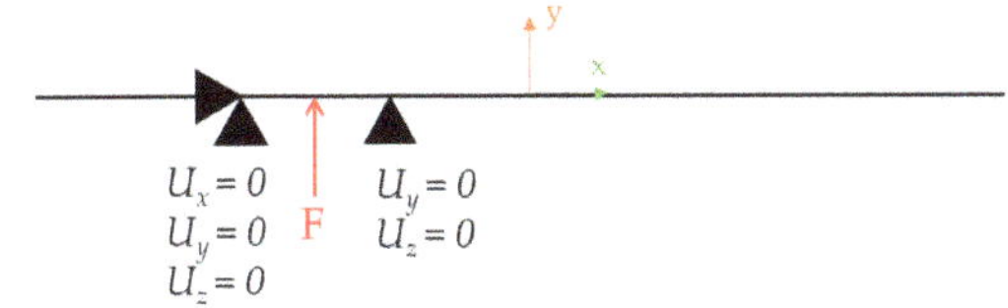

Figure 14. Boundary conditions for the first step of "S-shaped" billet simulation.

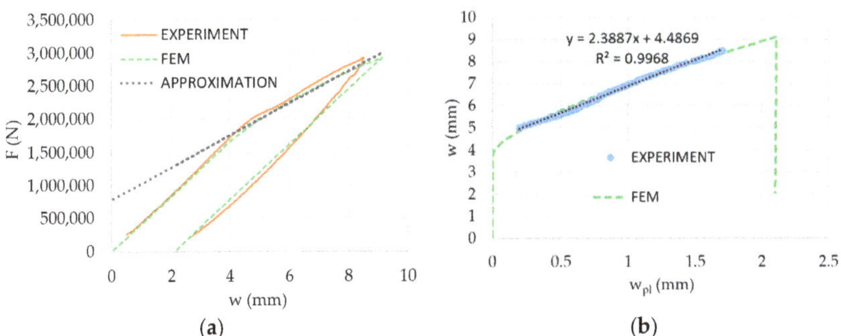

Figure 15. Prediction by the Chaboche model: force vs. total deflection including approximation by Equation (4) (**a**) and total deflection vs. plastic deflection (**b**).

Table 3. Material parameters of the Chaboche model for 100Cr6 material.

E	σ_y (MPa)	C_1 (MPa)	γ_1 (-)	C_2 (MPa)	γ_2 (-)
220,000	550	202,000	802	61,600	44

A comparison of experimental and predicted final billet shapes is provided in Figure 16. It is clearly shown that the strategy for numerical prediction gives acceptable results.

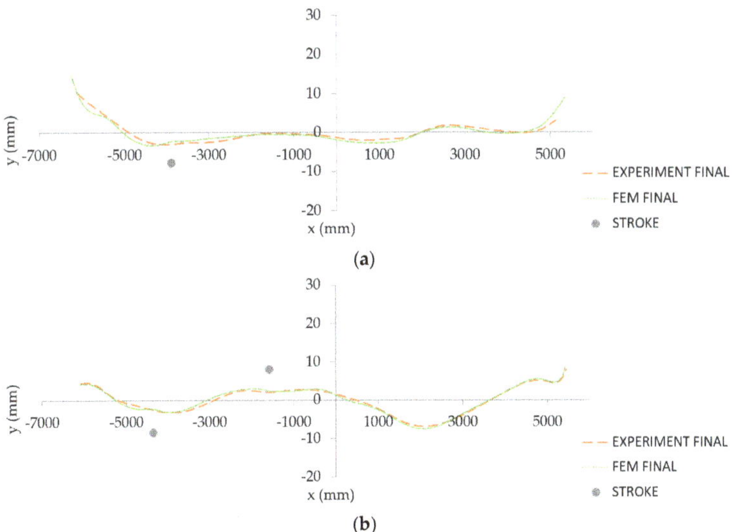

Figure 16. Comparison of experimental and predicted final shapes of two exemplary billets: "single-arc" type (**a**) and "S-shaped" type (**b**).

An exemplary result of regime 3 application on a very curved billet is shown in Figure 17. The input parameters of the third considered billet were p_1 = 95.4 mm and p_2 = 2.7 mm. After straightening, the output parameters evaluated by the sensory system were 17.8 mm and 1.8 mm. Thus, the straightening in regime 2 followed.

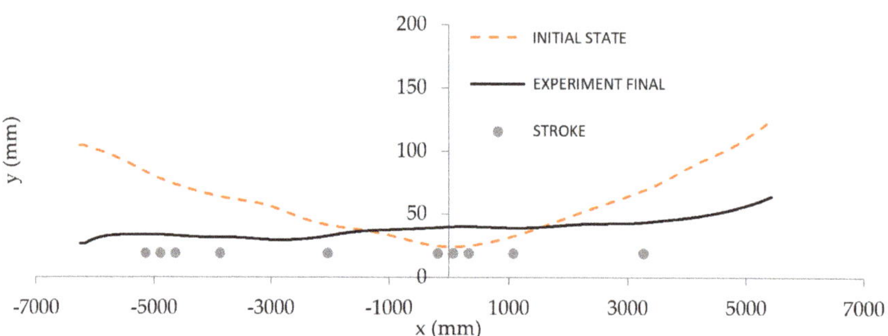

Figure 17. Shape of the third exemplary billet before straightening (solid curve) and after straightening (dashed curve) in regime 3.

The results of the fourth regime, which treats the situation of significant curvature in one place, will be presented on the exemplary billet with input parameters p_1 = 10.3 mm and p_2 = 2.4 mm. The stroke of $w = 7.9$ mm was realized in the position $x = -4575$ mm . The output values observed after straightening were p_1 = 10.5 mm and p_2 = 1 mm. The initial and final shapes are displayed together with the symbol of applied stroke in Figure 18.

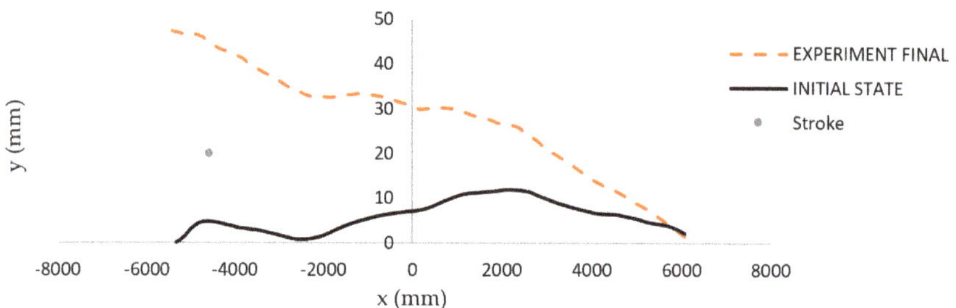

Figure 18. Shape of the fourth exemplary billet before straightening (solid curve) and after straightening (dashed curve) in regime 4.

The efficiency of the first regime in the straightening algorithm will be shown on the exemplar "snake-like" billet with input parameters p_1 = 66.2 mm and p_2 = 3.9 mm. After the first application of regime 1 (strokes $w = 9$ mm for $x = -5731$ mm, $w = 11$ mm for $x = -5211$ mm, $w = 10.3$ mm for $x = -3951$ mm, $w = 9.3$ mm for $x = -2981$ mm, $w = 8.9$ mm for $x = -1801$ mm, $w = 9$ *mm* for $x = -401$ mm, and $w = 8.4$ mm for $x = 839$ mm), the output parameters evaluated by the sensory system were 45.8 mm and 3 mm, which means that the first regime was applied again. The second application of regime 1 (strokes $w = 9$ mm for $x = -5731$ mm, $w = 11$ mm for $x = -5211$ mm, $w = 10.3$ mm for $x = -3951$ mm, $w = 9.3$ mm for $x = -2981$ mm, $w = 8.9$ mm for $x = -1801$ mm, $w = 9$ mm for $x = -401$ mm, and $w = 8.4$ mm for $x = 839$ mm) gave acceptable output parameters of p_1 = 8.6 mm and p_2 =1 mm.

The initial and straightened experimental shapes of the "snake-like" exemplar billet are shown in Figure 19. The results of corresponding FE simulations are presented in Figure 20. Strokes proposed by the algorithm in reality were applied in particular load-steps in a stroke by stroke manner. The boundary conditions used in each load-step of simulations were analogous to those presented in Figure 14.

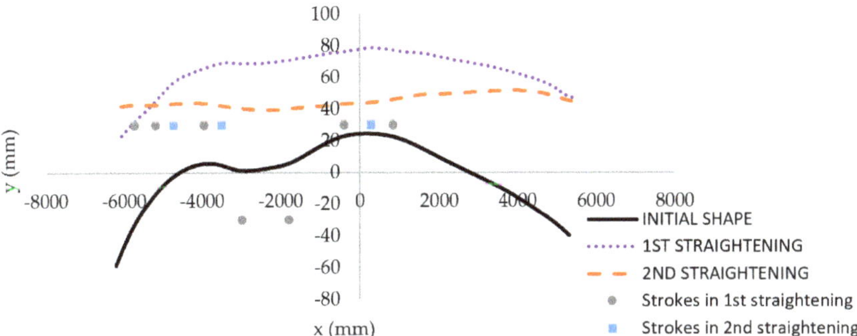

Figure 19. Shape of the "snake-like" exemplary billet before straightening (solid curve), after first straightening (dotted curve), and after second straightening (dashed curve) in regime 1.

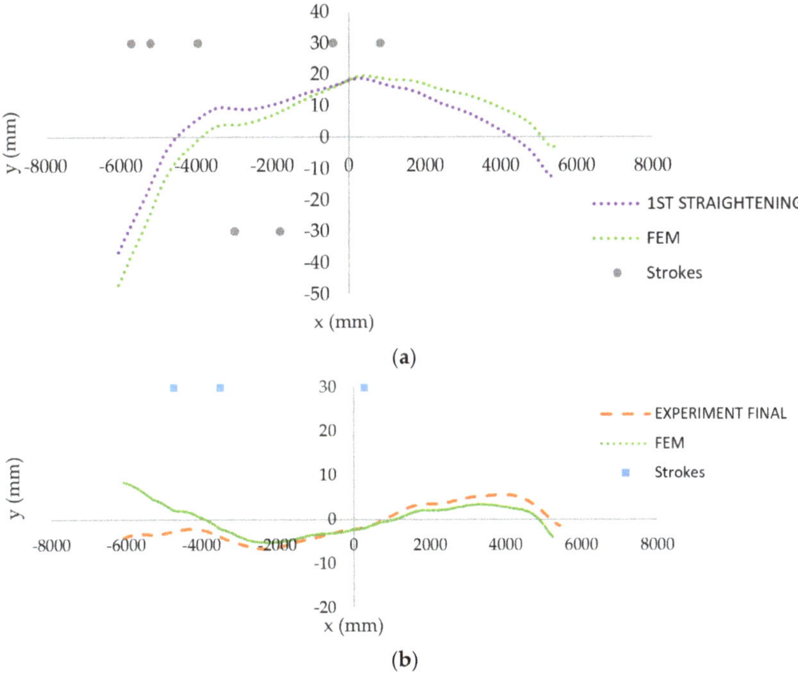

Figure 20. Comparison of experimental and predicted shapes of the exemplary "snake-like" billet: after first straightening (**a**) and after second straightening (**b**).

7. Conclusions

An automatic billet straightening machine was developed in cooperation between the university and industrial companies. The nature of the algorithm proposing interventions during the straightening process is described in this scientific work.

While the algorithm currently works for a 150 × 150 mm^2 cross-section, it can be expanded into a more general form based on findings shown in the laboratory experiments

(Section 3) to be applicable for the straightening of billets with various cross-section sizes. In that case, the most important outcome of the laboratory study is the possibility of constant value consideration for the plastic hardening parameter k_w. Then, it is necessary to increase the support distance for a larger cross-section according to the linear approximation shown in the Figure 5. The material parameter w_y depends on the yield strength of the material, Young modulus E, and the cross-section dimension of the billet. Both material parameters, k_w and w_y, must be properly identified for the considered material of billet from the force versus total deflection curve obtained for the chosen cross-section size. The algorithm itself is based on the assumption of a linear relationship between the total deflection and the plastic deflection. In fact, there is a slight nonlinearity for very small values of plastic deflection and this interval corresponds to the nonlinear part of the force versus total deflection diagram (for example in Figure 7). However, this interval is rarely used in the straightening algorithm. There is the parameter w_{ignor}, which corresponds to the minimal applied stroke for regime 1. In other regimes, it was experimentally proven that even a small intervention can help to straighten the billet ("single-arc" or "S-shaped" billets, usually).

Numerical simulations of operational experiments were done based on the Chaboche material model with two backstress parts to show the relevance of the algorithm. The basic regimes of the straightening algorithm considering different shapes of the billet were described. The algorithm was adopted on chosen steels in The New Long Billet Treatment Plant of Třinecké železárny a.s. The process and material parameters are optimised using a Python code. The billet straightening strategy currently works properly for ten materials under consideration.

The next step of research is the application of rigid body movement calculations (a simplified approach to predict the impact of performed stroke on the billet shape change) to speed up the straightening process by minimizing the necessity of scanning.

Author Contributions: Conceptualization, R.H., J.S., and R.W.; methodology, R.H., M.F., and J.M.; software, J.M., M.F., J.S., and J.B.; validation, J.B., R.H., and M.F.; formal analysis, M.F.; investigation, R.H., M.F., J.S., and J.B.; resources, J.M. and M.F.; data curation, J.M., M.F., and R.H.; writing—original draft preparation, J.S.; writing—review and editing, R.H.; visualization, J.M. and J.B.; supervision, R.H. and R.W. All authors have read and agreed to the published version of the manuscript.

Funding: This work was supported by The TECHNOLOGY AGENCY OF THE CZECH REPUBLIC in the frame of the project TN01000024 National Competence Center-Cybernetics and Artificial Intelligence; by the EUROPEAN REGIONAL DEVELOPMENT FUND in the Research Centre of Advanced Mechatronic Systems project, CZ.02.1.01/0.0/0.0/16_019/0000867; by the project SP2020/57 Research and Development of Advanced Methods in the Area of Machines and Process Control; and by the project SP2020/23 Application of numerical and experimental modeling in industrial practice financed by the MINISTRY OF EDUCATION, YOUTH, AND SPORTS OF THE CZECH REPUBLIC.

Acknowledgments: The authors appreciate the assistance of TŘINECKÉ ŽELEZÁRNY a.s. and KOMA—Industry s.r.o. companies during the preparation and realization of experiments and documentation support throughout the article preparation.

Conflicts of Interest: The authors declare no conflict of interest.

References

1. Fusek, M.; Halama, R.; Poruba, Z. Calibration of material parameters during billet straightening. In Proceedings of the EAN 2017-55th Conference on Experimental Stress Analysis 2017, Novy Smokovec, Slovakia, 30 May–1 June 2017; pp. 611–615.
2. Halama, R.; Sikora, J.; Fusek, M.; Marek, M.; Bartecká, J.; Guráš, R.; Wagnerová, R.; Mahdal, M. Algorithms for Automatic Billet Straightening Machine. In Proceedings of the EAN 2020—58th Conference on Experimental Stress Analysis 2020, Online, Czech Republic, 19–22 October 2020; pp. 107–112.
3. Yi, G.; Wang, Z.; Hu, Z. A novel modeling method in metal strip leveling based on a roll-strip unit. *Math. Probl. Eng.* **2020**, *2020*, 1–16. [CrossRef]
4. Kaiser, R.; Stefenelli, M.; Hatzenbichler, T.; Antretter, T.; Hofmann, M.; Keckes, J.; Buchmayr, B. Experimental characterization and modelling of triaxial residual stresses in straightened railway rails. *J. Strain Anal. Eng. Des.* **2015**, *50*, 190–198. [CrossRef]
5. Wang, K.; Wang, B.; Yang, C. Research on the Multi-Step Straightening for the Elevator Guide Rail. *Procedia Eng.* **2011**, *16*, 459–466. [CrossRef]

6. Zhang, Y.; Lu, H.; Zhang, X.; Ling, H.; Fan, W.; Wei, Q.; Lian, Y. A novel control strategy for the multi-step straightening process of long/extra-long linear guideways. *Proc. Inst. Mech. Eng. Part. C J. Mech. Eng. Sci.* **2019**, *233*, 2959–2975. [CrossRef]
7. Petruška, J.; Návrat, T.; Šebek, F. Novel approach to computational simulation of cross roll straightening of bars. *J. Mater. Process. Technol.* **2016**, *233*, 53–67. [CrossRef]
8. Wu, B.J.; Chan, L.C.; Lee, T.C.; Ao, L.W. A study on the precision modeling of the bars produced in two cross-roll straightening. *J. Mater. Process. Technol.* **2000**, *99*, 202–206. [CrossRef]
9. Fan, Q.-H.; Ma, Z.-Y.; Ma, L.-D.; Lei, J.-Y. Study on roller shape of two-roll straightening of titanium alloy based on rolling and reverse bending theory. *Suxing Gongcheng Xuebao J. Plast. Eng.* **2020**, *27*, 212–220. [CrossRef]
10. Ma, L.; Du, Y.; Liu, Z.; Ma, L. Design of continuous variable curvature roll shape and straightening process research for two-roll straightener of bar. *Int. J. Adv. Manuf. Technol.* **2019**, *105*, 4345–4358. [CrossRef]
11. Petruška, J.; Návrat, T.; Šebek, F.; Benešovský, M. Optimal intermeshing of multi roller cross roll straightening machine. In Proceedings of the 19th International ESAFORM Conference on Material Forming, Nantes, France, 27–29 April 2016; Volume 1769, p. 120002. [CrossRef]
12. Zhang, Z.-Q. Prediction of Maximum Section Flattening of Thin-walled Circular Steel Tube in Continuous Rotary Straightening Process. *J. Iron Steel Res. Int.* **2016**, *23*, 745–755. [CrossRef]
13. Zhang, Z.Q.; Yan, Y.H.; Yang, H.L. A simplified model of maximum cross-section flattening in continuous rotary straightening process of thin-walled circular steel tubes. *J. Mater. Process. Technol.* **2016**, *238*, 305–314. [CrossRef]
14. De Morais, A.B. A thick bondline beam model for the adhesively bonded 3-point bending specimen. *Int. J. Adhes. Adhes.* **2020**, *96*, 102465. [CrossRef]
15. Lu, H.; Zang, Y.; Zhang, X.; Zhang, Y.; Li, L. A General Stroke-Based Model for the Straightening Process of D-Type Shaft. *Processes* **2020**, *8*, 528. [CrossRef]
16. Essa, A.-E.; Nasr, M.; Ahmed, M. Variation of The Residual Stresses and Springback in Sheet Bending from Plane-Strain to Plane-Stress Condition using Finite Element Modelling. In Proceedings of the International Conference on Applied Mechanics and Mechanical Engineering, Military Technical College Kobry El-Kobbah, Cairo, Egypt, 19–21 April 2016; Volume 17.
17. Xiao-Lin, W.; Zhao-Bo, Q. Analyzing an Implemented Mechanism of Intelligent System and Its Work Flow for Straightening Machine of Heavy Beam. In Proceedings of the 2010 International Conference on Intelligent System Design and Engineering Application, Changsha, China, 13–14 October 2010; Volume 1, pp. 323–326. [CrossRef]
18. Zhang, Y.; Lu, H.; Ling, H.; Lian, Y.; Ma, M. Analytical Model of a Multi-Step Straightening Process for Linear Guideways Considering Neutral Axis Deviation. *Symmetry* **2018**, *10*, 316. [CrossRef]
19. Zhang, Y.; Lu, H.; Wang, Y.; Zhang, X.; Zhang, J.; Ling, H. Variable Span Multistep Straightening Process for Long/Extra-Long Linear Guideways. *IEEE Access* **2019**, *7*, 107491–107505. [CrossRef]
20. Jin, L.; Yang, Y.-F.; Li, R.-Z.; Cui, Y.-W.; Jamil, M.; Li, L. Study on Springback Straightening after Bending of the U-Section of TC4 Material under High-Temperature Conditions. *Materials* **2020**, *13*, 1895. [CrossRef]
21. Eggertsen, P.-A.; Mattiasson, K. On the identification of kinematic hardening material parameters for accurate springback predictions. *Int. J. Mater. Form.* **2011**, *4*, 103–120. [CrossRef]
22. Kim, S.C.; Chung, S.C. Synthesis of the multi-step straightness control system for shaft straightening processes. *Mechatronics* **2002**, *12*, 139–156. [CrossRef]
23. Balic, J.; Nastran, M. An on-line predictive system for steel wire straightening using genetic programming. *Eng. Appl. Artif. Intell.* **2002**, *15*, 559–565. [CrossRef]
24. Song, Y.; Yu, Z. Springback prediction in T-section beam bending process using neural networks and finite element method. *Arch. Civ. Mech. Eng.* **2013**, *13*, 229–241. [CrossRef]
25. Ling, H.; Yang, C.; Feng, S.; Lu, H. Predictive model of grinding residual stress for linear guideway considering straightening history. *Int. J. Mech. Sci.* **2020**, *176*, 105536. [CrossRef]
26. Yoshida, F.; Uemori, T. A model of large-strain cyclic plasticity and its application to springback simulation. *Int. J. Mech. Sci.* **2003**, *45*, 1687–1702. [CrossRef]
27. Yoshida, F.; Uemori, T.; Fujiwara, K. Elastic-plastic behaviour of steel sheets under in-plane cyclic tension-compression at large strain. *Int. J. Plast.* **2002**, *18*, 633–659. [CrossRef]
28. Hajbarati, H.; Zajkani, A. A novel analytical model to predict springback of DP780 steel based on modified Yoshida-Uemori two-surface hardening model. *Int. J. Mater. Form.* **2019**, *12*, 441–455. [CrossRef]
29. Zhao, K.M.; Lee, J.K. Finite element analysis of the threepoint bending of sheet metals. *J. Mater. Process. Technol.* **2002**, *122*, 6–11. [CrossRef]
30. Chaboche, J.L.; Lemaitre, J. *Mechanics of Solid Materials*, 1st ed.; Cambridge University Press: Cambridge, UK, 1990.
31. Armstrong, P.J.; Frederick, C.O. *A Mathematical Representation of the Multiaxial Bauschinger Effect, G.E.G.B. Report RD/B/N*; University of Leicester: Leicester, UK, 1996; p. 731.

 materials

Article

Monotonic Tension-Torsion Experiments and FE Modeling on Notched Specimens Produced by SLM Technology from SS316L

Michal Kořínek [1,*], Radim Halama [1], František Fojtík [1], Marek Pagáč [2], Jiří Krček [3], David Krzikalla [1], Radim Kocich [4] and Lenka Kunčická [5]

1 Department of Applied Mechanics, Faculty of Mechanical Engineering, VSB-Technical University of Ostrava, 17. listopadu 2172/15, 708 00 Ostrava, Czech Republic; radim.halama@vsb.cz (R.H.); frantisek.fojtik@vsb.cz (F.F.); david.krzikalla@vsb.cz (D.K.)
2 Department of Machining, Assembly and Engineering Metrology, Faculty of Mechanical Engineering, VSB-Technical University of Ostrava, 17. listopadu 2172/15, 708 00 Ostrava, Czech Republic; marek.pagac@vsb.cz
3 Department of Mathematics and Descriptive Geometry, Faculty of Mechanical Engineering, VSB-Technical University of Ostrava, 17. listopadu 2172/15, 708 00 Ostrava, Czech Republic; jiri.krcek@vsb.cz
4 Department of Materials Forming, Faculty of Materials Science and Technology, VSB-Technical University of Ostrava, 17. listopadu 2172/15, 708 00 Ostrava, Czech Republic; radim.kocich@vsb.cz
5 Institute of Physics of Materials, Academy of Science of Czech Republic, Žižkova 513/22, 616 62 Brno, Czech Republic; kuncicka@ipm.cz
* Correspondence: michal.korinek@vsb.cz

Abstract: The aim of this work was to monitor the mechanical behavior of 316L stainless steel produced by 3D printing in the vertical direction. The material was tested in the "as printed" state. Digital Image Correlation measurements were used for 4 types of notched specimens. The behavior of these specimens under monotonic loading was investigated in two loading paths: tension and torsion. Based on the experimental data, two yield criteria were used in the finite element analyses. Von Mises criterion and Hill criterion were applied, together with the nonlinear isotropic hardening rule of Voce. Subsequently, the load-deformation responses of simulations and experiments were compared. Results of the Hill criterion show better correlation with experimental data. The numerical study shows that taking into account the difference in yield stress in the horizontal direction of printing plays a crucial role for modeling of notched geometries loaded in the vertical direction of printing. Ductility of 3D printed specimens in the "as printed" state is also compared with 3D printed machined specimens and specimens produced by conventional methods. "As printed" specimens have 2/3 lower ductility than specimens produced by a conventional production method. Machining of "as printed" specimens does not affect the yield stress, but a significant reduction of ductility was observed due to microcracks arising from the pores as a microscopic surface study showed.

Keywords: stainless steel 316L; additive manufacturing; multiaxial loading; plasticity; digital image correlation method; hill yield criterion; isotropic hardening; finite element method (FEM)

Citation: Kořínek, M.; Halama, R.; Fojtík, F.; Pagáč, M.; Krček, J.; Krzikalla, D.; Kocich, R.; Kunčická, L. Monotonic Tension-Torsion Experiments and FE Modeling on Notched Specimens Produced by SLM Technology from SS316L. *Materials* **2021**, *14*, 33. https://dx.doi.org/10.3390/ma14010033

Received: 27 November 2020
Accepted: 21 December 2020
Published: 23 December 2020

Publisher's Note: MDPI stays neutral with regard to jurisdictional claims in published maps and institutional affiliations.

Copyright: © 2020 by the authors. Licensee MDPI, Basel, Switzerland. This article is an open access article distributed under the terms and conditions of the Creative Commons Attribution (CC BY) license (https://creativecommons.org/licenses/by/4.0/).

1. Introduction

The austenitic stainless steel AISI 316L is one of the most utilized constructional materials for various parts in the power industry and beyond. It was investigated in the conventional wrought state [1], while it has been loaded in tension, torsion, and even combinations of both. Nevertheless, it is increasingly utilized in the additively manufactured form [2], as it opens new possibilities. This was the motivation for the conference paper [3], which is the forerunner of this further extended paper. Stainless steel 316L (SS316L) may be optimized and applied in an organic shape or can even serve as a custom made part or machine element utilized in the repair or reconstruction of a structure, where commercial products are not available or are hardly producible by conventional manufacturing, such as machining. Various process parameters used during the additive manufacturing of SS316L have been examined [4,5]. One of the important outputs are the mechanical properties [6,7] or porosity [8]. The building direction also plays a vital role [9], and the final surface roughness is of particular interest [10]. It is well known that printing the layer by layer and natural cooling from the bottom directly lead to the formation of residual stresses in the material. The influence of the scanning strategy on the resulting residual stresses was also intensively studied for SS316L in recent years [11–13]. There are many studies available, which are focused on mechanical properties research, including anisotropy induced by the selective laser melting (SLM) process in the literature. However, there is still missing information about the plasticity of SS316L prepared by SLM under various multiaxial stress states. To simulate critical loading states and nonstandard events in technical practice, it is important to realize the necessary monotonic experiments using biaxial testing machines. This is an important motivation for research in the field of multiaxial plasticity.

This paper presents new results for the deformation response obtained during monotonic multiaxial loading of specimens made from SS316L, produced by SLM technology in the "as printed" state. Almost all specimens were printed in the vertical direction to prevent the effect of residual stresses. Due to the character of the specimens that contain notches, the digital image correlation (DIC) method was used. The DIC method is a progressive optical-numerical method suitable for 3D analysis of structural components under uniaxial and multiaxial loading in the full-field [14,15]. Averaged characteristics gained in this experimental study with DIC measurements were used for the validation of a numerical model based on the finite element method (FEM).

2. Materials and Methods

In preprocessing phase, the specimens were created with SOLIDWORKS 2019 (Dassult Systemes SoliDWorks, France, version2019). Subsequently, Powder Bed Fusion 3D printing technology, selective laser melting, was used for the production using a 3D printer Renishaw AM400 (Renishaw, New Mills, UK, 2016), and the material was atomized SS316L powder. This is an additive manufacturing technology, where the laser scans and selectively melts the atomized metal powder particles, bonding them together and building a model layer-by-layer [16,17]. In the beginning of the process, the building chamber was filled with inert gas argon to minimize the oxidation of the metal powder. The layer thickness was set to 50 μm, and the chessboard strategy was used. The strategy translates by 5 mm in the horizontal direction X and Y and rotates for the optimum homogeneous distribution of stress [18].

In general, each specimen always had a different percentage and volume of porosity. The porosity varied with various parameters, such as the number of layers laid and the printing time. The study [19] dealt with the optimization of those parameters, and the production parameters from this study were used to produce the specimens in our study. These production parameters guaranteed a porosity of more than 99.9%. Other 3D printing parameters are shown in Table 1 [19,20] (QuantAM, SW made by company Renishaw) [21]. Building time was 76 h. The part orientation and the position in the chamber, the 3D printing preview, and the chessboard strategy preview in the cross-section are presented in Figure 1. The specimens were separated from the base plate with a band saw. Due to the printing direction, the effect of residual stresses was not expected.

Table 1. 3D printing parameters.

3D Printer:	Renishaw AM400
Powder description:	SS Powder AISI 316L (DIN 1.4404)
Powder Particle Size:	15–45 μm
Layer Thickness:	50 μm
Focus Size:	70 μm
Print Strategies:	Chessboard
Border Power:	110 W
Border Exposure Time:	100 μs
Border Point Distance:	20 μm
Hatches Power:	200 W
Hatches Exposure Time:	80 μs
Hatches Point Distance:	60 μm
Jump speed:	5000 $mm \cdot s^{-1}$
Dosing time:	7 s
Melting range:	1371 to 1399 °C
Concentration of Oxygen:	<0.1% O_2
Inert Gas:	Argon
Purity:	5.0 (99.998%)

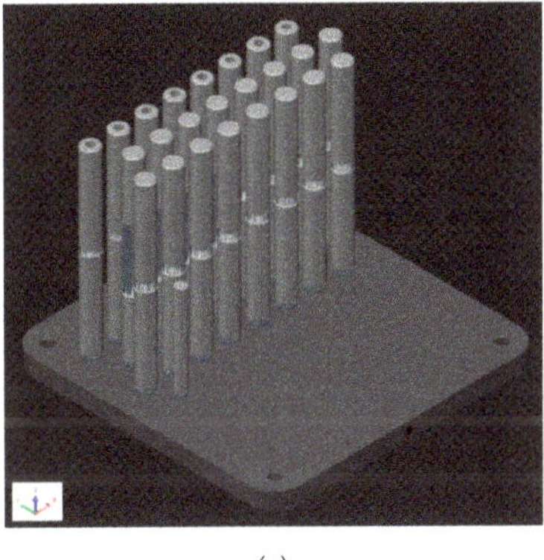

(**a**)

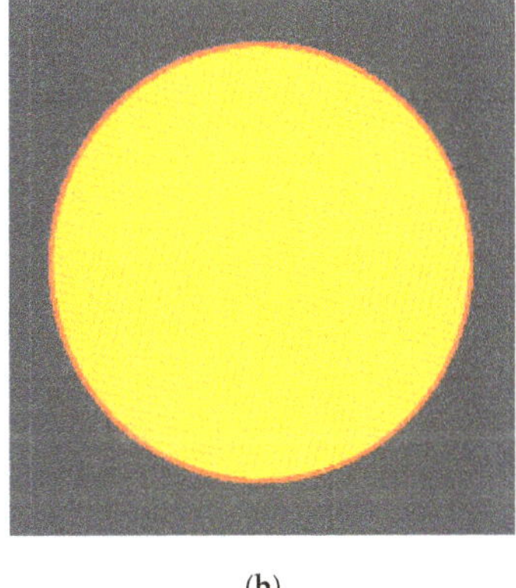

(**b**)

Figure 1. Part orientation and position in chamber: (**a**) 3D printing preview; (**b**) chessboard strategy preview in the cross-section.

The specimens were not further machined (outer surface) or heat treated (in the "as printed" state), therefore had naturally high surface roughness. The geometry of the notches considered in this study is shown in Figure 2. Only one type of specimenswas tubular, other were solid bodies. Tubular (specimens *A*) had to be drilled to the required internal diameter. Each specimen was 160 mm long, and the outer diameter was 15 mm. In addition, the standard tensile test was performed. The solid specimens were loaded only in tension. The tubes were subjected to two different loading modes: pure tension and pure torsion. Each mechanical test with pure axial loading was repeated four times. The torsion test was repeated only two times. The testing machine, LabControl 100 kN/1000 Nm (Opava, Czech Republic), was used. Multiaxial tests were done under deformation control with a 2 mm per min elongation rate for tension and under 0.157 radians per min twist rate for torsion. All tests were conducted at ambient temperature. The results of the tests were evaluated in the form of force (torque) vs. elongation (twist) diagrams.

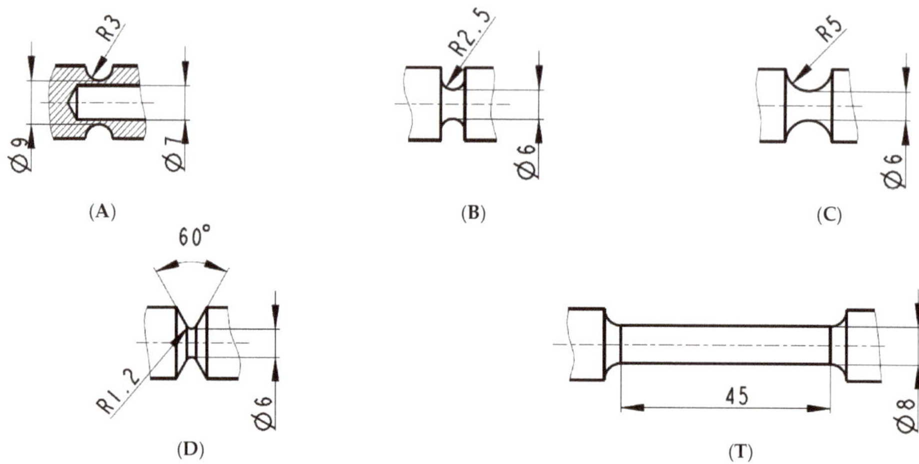

Figure 2. Specimens geometry: The tubular specimen with a notch for multiaxial loading (**A**), solid specimens with a notches for axial loading (**B–D**), and unnotched specimen for the standard tensile test (**T**).

DIC measurements were used to monitor the deformation. This method is characterized by the creation of a light area with dark points, also known as a pattern. Two optical sensors were used for this measurement to get 3D strain data. The sensors were high resolution cameras. The principle of DIC measurement was for two images of the specimen to be compared at different loading states using the appropriate facet size in pixels. Simultaneously, the images from both cameras were correlated in real-time time to get the contours of the specimen's surface. Advantages of this method are the ability to monitor the deformation of very complex shaped areas and the determination and real-time evaluation of the required quantities (displacement, strain, velocity, acceleration, or even stress [14]). Some mechanical properties, such as Young's modulus and Poisson's ratio, are also measured by this non-contact method [22]. The MERCURY RT system, provided by Sobriety company (Kuřim, Czech Republic), was used for all DIC measurements. This software was also useful for configuration and calibration of cameras. The optical probe was virtually created on the specimen before starting the measurements. This probe had to be aligned for both cameras. The optical probe provided an initial length and had to be visible during the whole mechanical test.

The study was supplemented with investigation of surfaces via scanning electron microscopy (SEM, Tescan Orsay Holding a.s., Brno, Czech Republic), and fractography was also performed. Preparations of the samples for the analyses were performed via ultrasound cleaning. The structure investigations were carried out using a Tescan Lyra 3 FIB/SEM microscope. The images were taken with the accelerating voltage of 10 kV.

3. FE Modelling

Since additive manufacturing technology is becoming increasingly popular, it is important to examine if the FEM can sufficiently predict accurate results with respect to the experimental response of materials. A number of analyses were thus performed in this study to validate the finite element model response under several loading conditions. Validation of the FEM was performed in terms of the comparison of results of the FEM with results from experiments.

For the purposes of the FEM, the specimens were modelled as cut-outs of length equal to the initial length measured by the probe during experiments. Various levels of symmetry were utilized to reduce the computation time. For tensile analyses, 1/8 symmetry was used, and for torsion analyses, half symmetry was used. Models were meshed using linear hexagonal elements (SOLID185). Usage of mapped mesh and sizing settings ensured a regular and sufficiently-sized mesh to capture stress and strain gradients accurately (an example of mesh is in Figure 3b). All analyses were prescribed in the form of macros in Ansys Parametric Design Language (APDL) for easy, fast running, and automatic postprocessing of desired results.

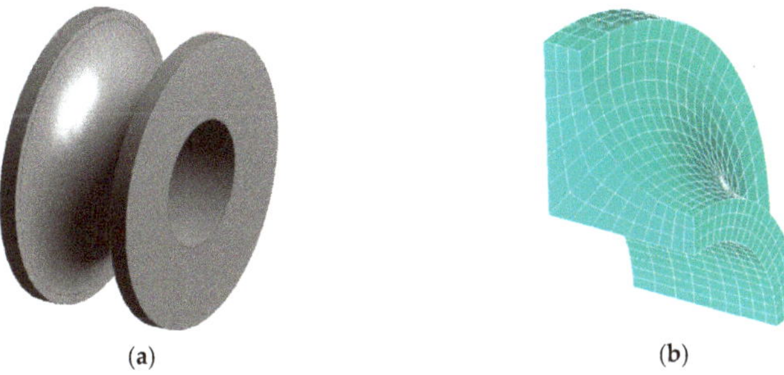

Figure 3. (**a**) Model of full specimen *A* cut-out, and (**b**) meshed specimen *A* utilizing 1/8 symmetry.

Boundary conditions were set in accordance with the experimental loading conditions and symmetry assumptions. For the simulation of the tensile test, nodal displacement and symmetry plane boundary conditions were used. Those were applied on nodes within appropriate faces. See Figure 4a for an example of boundary conditions for tensile test simulations as it is similar for all specimens.

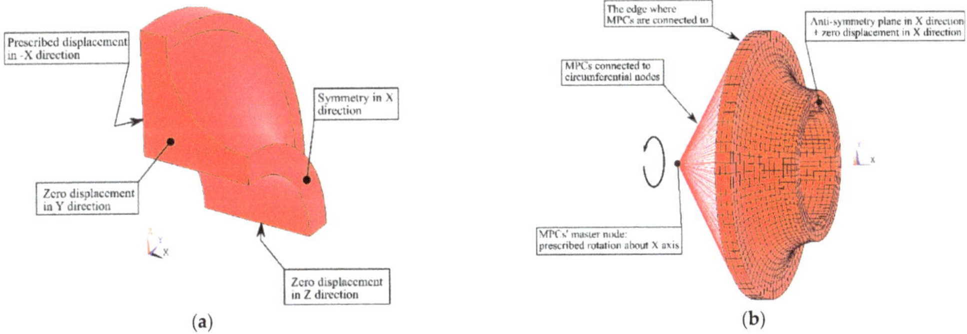

Figure 4. Description of boundary conditions for tensile load simulation (**a**) and for torsion load simulation (**b**) in simulations on specimen *A*.

For the simulation of the torsion test of the specimen *A*, structural multipoint constraints (MPC184) were utilized to load the specimen in torsion. The nodal displacement in "x" direction and antisymmetry plane in "x" direction were used (see Figure 4b). Fix of displacement of nodes in "x" direction was sufficient since Ansys's antisymmetry plane formulation treats, in this case, all other degrees of freedom. Observing the boundary conditions, one can see that all simulations were displacement-controlled.

To simulate the material response of specimens under conditions of tension and torsion, a suitable material model was needed. The chosen material model consists of nonlinear isotropic hardening, together with either Hill yield criterion or von Mises yield criterion. The linear elastic part of the model obeys Hooke's law for three-dimensional problem

$$\begin{gathered}\varepsilon_x = \frac{1}{E}\left[\sigma_x - \mu\left(\sigma_y + \sigma_z\right)\right],\ \varepsilon_y = \frac{1}{E}\left[\sigma_y - \mu\left(\sigma_x + \sigma_z\right)\right],\ \varepsilon_z = \frac{1}{E}\left[\sigma_z - \mu\left(\sigma_y + \sigma_x\right)\right] \\ \gamma_{xy} = \frac{\tau_{xy}}{G},\ \gamma_{yz} = \frac{\tau_{yz}}{G},\ \gamma_{xz} = \frac{\tau_{xz}}{G}\end{gathered} \tag{1}$$

where σ_i are the normal stress components, ε_i are the normal strain components, τ_{ij} are the shear stress components, γ_{ij} are the shear strain components, and G is the shear modulus. Thus, elasticity requires two input parameters of Young's modulus E and Poisson ration μ. Values of the parameters are denoted in Table 2.

Table 2. Elastic parameters.

Parameter	Value	Unit
E	183	GPa
M	0.3	-

Isotropic hardening was suitable for this study since the loading was monotonic. Isotropic hardening during plastic deformation caused a uniform increase of the yield surface. This resulted in increased yield stress. Thus, the yield condition took the form:

$$f(\sigma) - Y = 0 \tag{2}$$

where $f(\sigma)$ is a function of the stress tensor σ and Y is the yield stress defining the current size of the yield surface. For the description of isotropic hardening, the Voce law was used. However, Voce law is a combination of linear and nonlinear isotropic variables and has the form:

$$Y = \sigma_Y + R \tag{3}$$

where σ_Y is the initial yield stress and R is a new internal variable. The evolution of R is done by superposition of two parts:

$$dR = dR_1 + dR_2,\ dR_1 = R_0 dp,\ dR_2 = b(R_\infty - R)_2 dp \tag{4}$$

where R_0 and R_∞ are material parameters and dp is the increment of accumulated plastic strain.

By integration of Equation (4), with zero initial values of p, R_1, and R_2, respectively, and use of Equation (3), the constitutive equation is obtained:

$$Y = \sigma_Y + R_0 p + R_\infty\left(1 - e^{-bp}\right). \tag{5}$$

The meaning of the material parameters is as follows: R_0 is the slope of the saturation stress, R_∞ is the difference between the saturation stress and the initial yield stress, and b is the hardening parameter that governs the rate of saturation of the exponential term. Values of the parameters were optimized by the nonlinear least square method from the tensile test on the specimens printed in the vertical direction and are denoted in Table 3.

Table 3. Voce law parameters.

Parameter	Value	Unit
σ_Y	575	MPa
R_0	950	MPa
R_∞	60	MPa
b	125	-

Hill yield criterion and von Mises criterion were used in this material model to compare their accuracy with respect to real additive manufactured specimen responses. Hill criterion was anisotropic, independent of hydrostatic pressure, and depended on the orientation of the stress relative to the axis of anisotropy, thus suitable for materials in which the microstructure influences the macroscopic behavior of the material, which is the case for additive manufactured steels [23,24]. Hill yield criterion was, in this study, utilized for the modelling of yield strength anisotropy based on build direction [25–28]. Hill yield criterion's stress function for an ideally plastic material has the form:

$$f(\sigma) \equiv F(\sigma_{22}-\sigma_{33})^2 + G(\sigma_{33}-\sigma_{11})^2 + H(\sigma_{11}-\sigma_{22})^2 + 2L\sigma_{23}^2 + 2M\sigma_{31}^2 + 2N\sigma_{12}^2 = \sigma_y^2 \quad (6)$$

where F, G, H, L, M, and N are auxiliary coefficients, which are functions of the ratio of the scalar yield stress parameter σ_Y and the yield stress in each of the six stress components. The relationships between coefficients and ratios are as follows [29–31]

$$\begin{gathered} F = \tfrac{1}{2}\left(\tfrac{1}{R_{22}^2} + \tfrac{1}{R_{33}^2} - \tfrac{1}{R_{11}^2}\right),\ G = \tfrac{1}{2}\left(\tfrac{1}{R_{33}^2} + \tfrac{1}{R_{11}^2} - \tfrac{1}{R_{22}^2}\right),\ H = \tfrac{1}{2}\left(\tfrac{1}{R_{11}^2} + \tfrac{1}{R_{22}^2} - \tfrac{1}{R_{33}^2}\right) \\ L = \tfrac{3}{2}\left(\tfrac{1}{R_{23}^2}\right),\ M = \tfrac{3}{2}\left(\tfrac{1}{R_{13}^2}\right),\ N = \tfrac{3}{2}\left(\tfrac{1}{R_{12}^2}\right), \\ R_{11} = \tfrac{\sigma_{11}^y}{\sigma_y}, R_{22} = \tfrac{\sigma_{22}^y}{\sigma_y}, R_{33} = \tfrac{\sigma_{33}^y}{\sigma_y}, \\ R_{12} = \sqrt{3}\tfrac{\sigma_{12}^y}{\sigma_y},\ R_{23} = \sqrt{3}\tfrac{\sigma_{23}^y}{\sigma_y},\ R_{13} = \sqrt{3}\tfrac{\sigma_{13}^y}{\sigma_y} \end{gathered} \quad (7)$$

where the directional yield stress ratios R_{ii} and R_{ij} are related to the isotropic yield stress parameter σ_y, and σ_{ij}^y is the yield stress in the direction given by the value of subscripts i and j. Almost all directional yield stress ratios for uniaxial and torsional loading were equal to 1. The only different directional yield stress ratio was the directional ratio R_{33}, which was equal to 0.87. The ratio R_{33} corresponded to the axial direction of all vertically printed specimens, thus introducing the effect of building direction into the material model.

The Von Mises yield criterion is isotropic, independent of hydrostatic pressure, and commonly used for metals, polymers, etc. In this study, accuracy of the von Mises yield criterion was examined by comparing with the real material response. The Von Mises stress function takes the form:

$$f(\sigma) = \sqrt{\frac{(\sigma_1-\sigma_2)^2 + (\sigma_2-\sigma_3)^2 + (\sigma_1-\sigma_3)^2}{2}} \quad (8)$$

where σ_i are principal stresses.

4. Results

All specimens were subjected to loading, as described above. Each test was deformation-controlled. Values of applied force or torque were recorded by the testing machine, and the values of deformation were recorded by the DIC system. Because the experimentally measured data embodies natural oscillations, the presented force vs. elongation diagrams were smoothed using functions of the Curve Fitting Toolbox in Matlab; see Figure 5 showing an example from monotonic tensile tests. The combination of moving average and smoothing splines led to sufficient results. Optimal smoothing parameters were chosen with respect to the size and character of the data sets.

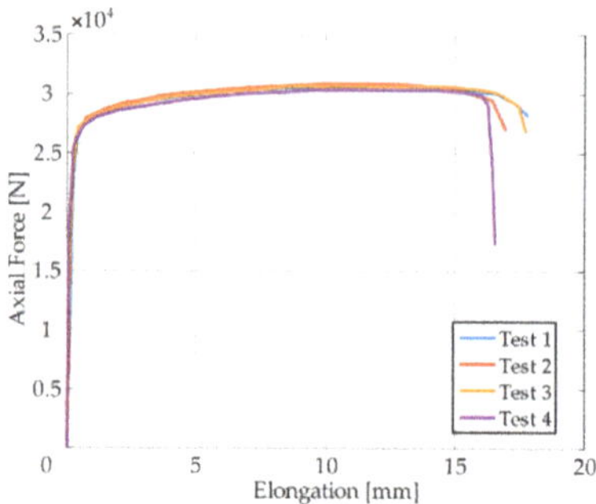

Figure 5. Dependency of force on elongation from tensile test of vertically printed specimens *T*.

Presentation of the simulation results and their comparison with respect to the experimental response of the "as printed" specimens follow next. The comparison was performed in the form of plots comprised of experimental smoothed responses in dashed lines and simulation results in full lines (see Figures 6–10). Contours obtained by DIC measurements and FEM simulations were also compared for each specimen. A comparison of DIC and FEM contours served as the retrospective control of the simulation results with the calibrated material model.

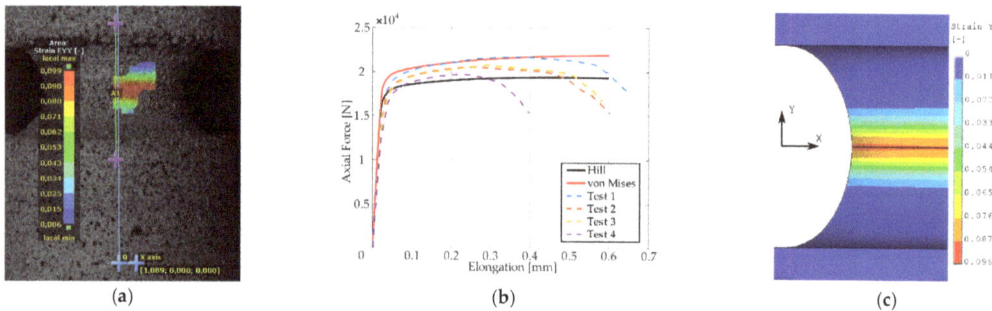

Figure 6. Longitudinal strain distribution from the digital image correlation (DIC) measurement (**a**), dependency of force on elongation compared to experimental and simulation results (**b**), and finite element method (FEM) result with Hill's criterion material model (**c**) for specimen *A*.

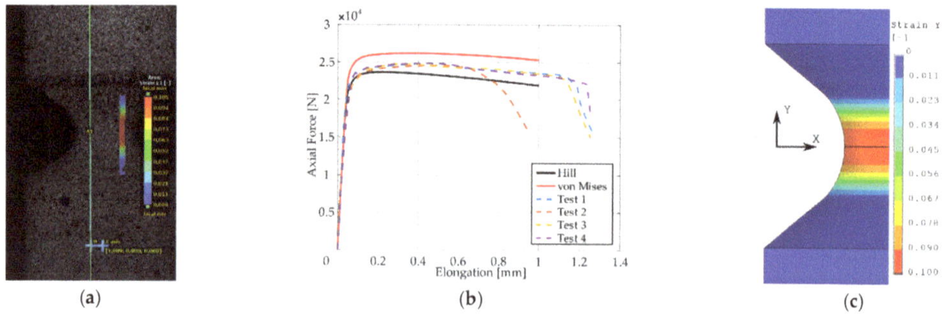

Figure 7. Maximal principal strain distribution (**a**), dependency of force on elongation compared to experimental and simulation results (**b**), and FEM result with Hill's criterion material model (**c**) for specimen *B*.

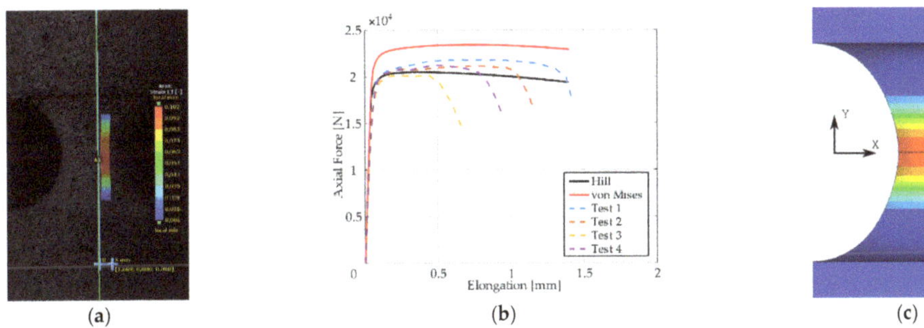

Figure 8. Maximal principal strain distribution (**a**), dependency of force on elongation compared to experimental and simulation results (**b**), and FEM result with Hill's criterion material model (**c**) for specimen *C*.

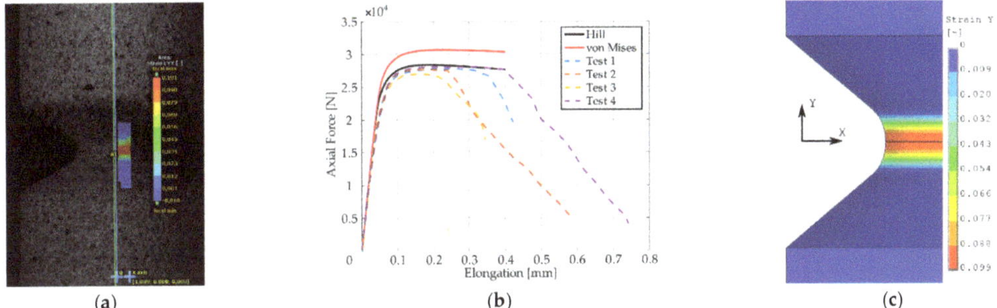

Figure 9. Longitudinal strain distribution (**a**), dependency of force on elongation compared to experimental and simulation results (**b**), and FEM result with Hill's criterion material model (**c**) for specimen *D*.

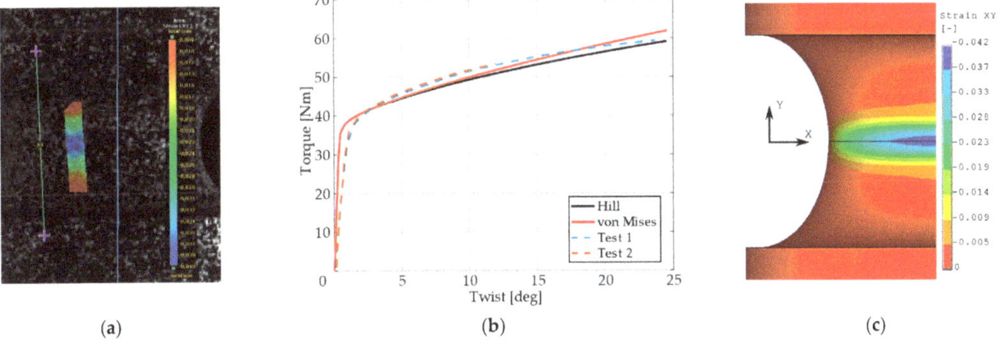

Figure 10. Shear strain distribution for torsional test (**a**), dependency of torque on twist compared to experimental and simulation results (**b**), and FEM result with Hill's criterion material model (**c**) for specimen *A*.

Ductility was calculated from the tensile test results for the specimen without a notch of T-type, considering

$$\delta = \frac{L_u - L_0}{L_0} \times 100\% \tag{9}$$

where L_u is the final length of the specimen after the test and L_0 is the initial gauge length of the testing part of the specimen. Initial gauge length for each specimen was

40 mm, and the elongations are presented in Figure 5. Specimens printed in the vertical direction without the machined outer surface had ductility from 42% to 45%. "As printed" specimens showed approximately 2/3 higher ductility than machined specimens printed in the vertical direction. The machined specimens had ductility 13–15%. The cutting velocity for the machined specimens was 60 m·min^{-1}, the feed was 0.25 mm rev^{-1}, the depth of cut was 1 mm, and they had roughness of Ra 0.8. This difference is also evident in Figure 11a, where the true stress–true strain curves for the three types of specimens are compared. The printed SS316L revealed surprisingly good ductility even when printed in the vertical direction (43% in comparison with 60% of the conventional SS316L), but just for the "as printed" variant.

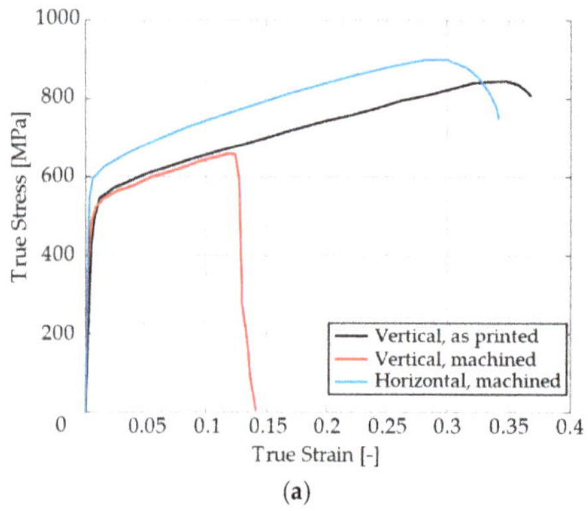

(a)

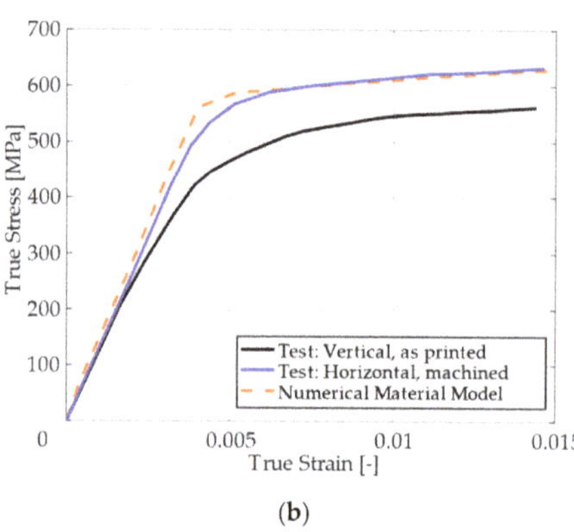

(b)

Figure 11. Three characteristic true stress–true strain curves for stainless steel 316L specimens produced by selective laser melting technology (**a**), and a detail of the material model response and experiment for the specimen "as printed" in the vertical direction and machined in the horizontal direction (**b**).

The dependency of force on elongation was compared with machined specimens, which were printed in the vertical and horizontal directions. Vertically printed specimens had approximately the same yield stress, regardless of whether they were machined or "as printed", as can be seen in Figure 11a. Thus, the machining used did not influence the microstructure.

However, the ductility differed significantly, as described above. It is also clear from Figure 11a that the vertically printed specimens had a lower yield stress than the horizontally printed specimens. This difference was approximately 75 MPa. This difference is evident from Figure 11b, where a detail of a vertical "as printed" and horizontal machined specimen is shown. The material model was determined for a vertical "as printed" specimen, and after adding just 75 MPa, the material model corresponded to the machined specimen printed in the horizontal direction. It is clearly shown that the machining, which was set in a manner to not affect stress-strain behavior of the material, can still lead to the reduction of material ductility.

The experimental work also included detailed microscopic observations of both the "as printed" and machined tensile test specimens. The analyses were performed on the surfaces of both the tensile specimens of type *T*, as well as at the locations of fractures. The fractured "as printed" and machined tensile specimens are depicted in Figure 12a,b, respectively. The figures clearly show that the surfaces of both specimens exhibited macroscopic differences, as the "as printed" specimen had a more or less silky smooth surface, whereas the machined specimen exhibited a shiny metallic surface. Figure 13a,b depict detailed images of the surfaces of both the "as printed" and machined specimens, respectively (acquired from the working part of the specimens featuring the smaller diameter). As can be seen, the "as printed" specimen featured a compact surface but with visible bumps (Figure 13a). Subsequent machining disturbed the surface and uncovered the voids beyond the bumps (Figure 13b). These features are supposed to originate from the 3D printing process. A detail of the original vertically-printed material, from which the specimens were fabricated, is depicted in Figure 13c.

(a) (b)

Figure 12. Fractured tensile test specimens: "as printed" (**a**); machined (**b**).

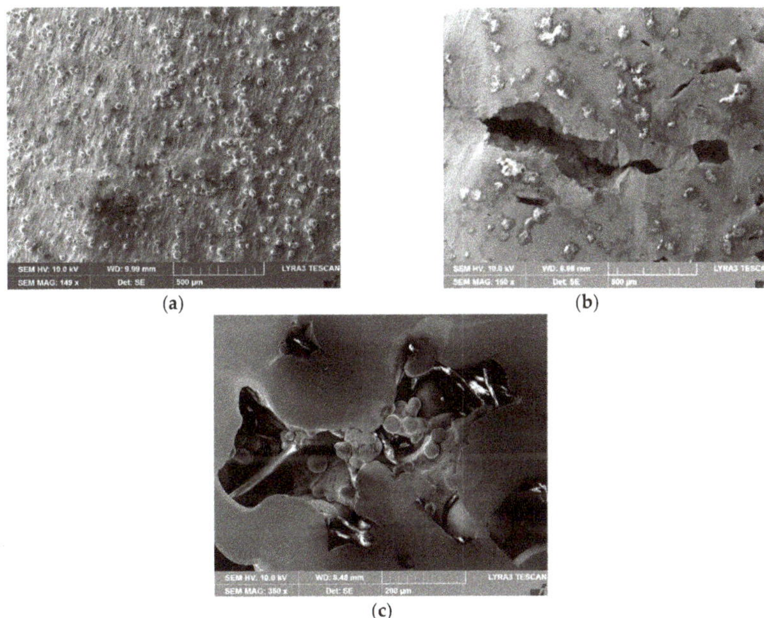

(a) (b) (c)

Figure 13. Surfaces of tensile test specimens: "as printed" (**a**), machined (**b**), and detail of surface of vertically printed material, i.e., original material for tensile test specimens (**c**).

Detailed images of the fractures of the "as printed" and machined specimens are depicted in Figure 14a–d. Mutual comparison of both fracture surfaces reveals that the characters of both fractures were different. The "as printed" specimen exhibited the character of ductile fracture (Figure 14a) but also the presence of local unmelted powder particles (Figure 14b). The machined specimen exhibited only the local occurrence of ductile fracture and prevailing features of brittle fracture (Figure 14c); the cracking seemed to have occurred along the boundaries of the powder particles, i.e., inter-granular fracture occurred (Figure 14d). The specimen also featured a frequent occurrence of unmelted powder particles, especially in the dimples, which originated from tensile loading during mechanical testing. Together with the surface cracks introduced by previous machining, these phenomena most probably contributed to the decreased strength and ductility of this specimen.

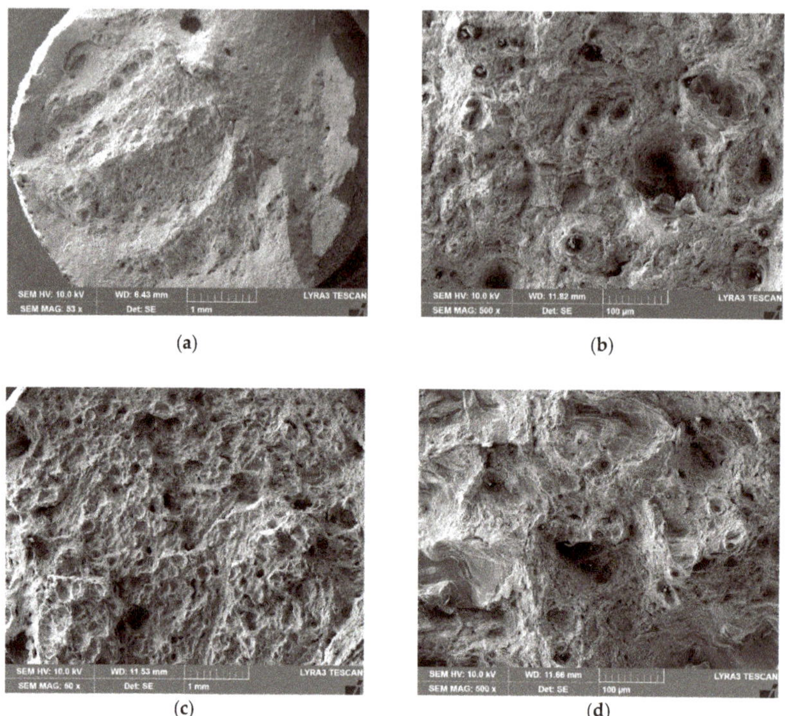

(**a**) (**b**) (**c**) (**d**)

Figure 14. Details of fracture surfaces: "as printed": 53× magnified (**a**), "as printed": 500× magnified (**b**); machined: 50× magnified (**c**), and machined: 500× magnified (**d**).

5. Discussion

During the test, when specimens with different notches were subjected to monotonic loading, it was evident that all specimens were stable up to the ultimate strength and showed very similar dependence of force (torque) on elongation (twist) for each notch type. Significant differences were observed in material fractures. This effect was attributed to the fact that the specimens were left in the "as printed" state. During production by the SLM method, inhomogeneity and porosity arose especially in the notches. This inhomogeneity and porosity can act as an indicator of material fracture. Porosity can affect the mechanical properties because these are cavities that lead to the fracture of the material. Comparison of the experimental and FEM results revealed the following: observing the responses under tension for specimen *A* (Figure 6), one could see that the numerical model using the Hill's yield criterion (further FEM-Hill) described the experiment with the lowest ductility and axial force better. On the contrary, the numerical model using the von Mises yield criterion (further FEM-Mises) described the experiment with the highest ductility and axial force better. Thus, for this specimen, it cannot be unequivocally said which material model describes the experiment better. Next, observing the responses under tension for specimens *B* and *C* (Figures 7 and 8), one could see that the FEM-Hill described the yield region better than FEM-Mises. Nevertheless, after reaching the axial force 'plateau', the experimental responses were generally in between the two numerical model responses, with FEM-Hill rather underestimating and FEM-Mises overestimating the axial force. To conclude the insights, it can be said that FEM-Hill described the experimental responses better in this case, since it accurately captured the beginning of the nonlinear response (which corresponded to the yield region) compared to FEM-Mises. Next, observing the responses under tension for specimen *D* (Figure 9), one could see that the FEM-Hill described the experimental response generally better during the whole examined elongation range. In this case, the FEM-Mises clearly overestimated the axial force. Finally, observing the responses under torsion for specimen *A* (Figure 10), one could see that the FEM-Hill and FEM-Mises responses were identical up to the twist angle of about 10 degrees and then a minor difference could be observed. Both numerical models then described the experimental response well, mainly in the plastic region.

The experiments also showed a significant difference in ductility between specimens that were left in the "as printed" state and specimens that were machined. Ductility was evaluated on specimens without notches. The machined specimens had a prescribed roughness of Ra 0.8. It should be noted that both types of specimens were printed in the vertical direction. The primary idea of why the machined specimens had a 2/3 lower ductility was that the layers that were laid during 3D printing (and the pores were primary located in them) were parallel to the machine knife during machining. As a result of machining, the partial segments broke out more easily and microscopic cracks appeared on the surface of the material with the occurrence of pores (stress concentration). This lead to an earlier fracture. This effect could also be observed during the microscopic analysis of both types of specimens in this study. On the specimens that were produced by printing in the horizontal direction, the machine knife then acted perpendicular to the laid layer during machining and therefore the microscopic segments did not have to be so easily broken out.

6. Conclusions

In this study, four types of notched specimens were investigated under monotonic testing. Loading path was pure tension and pure torsion. In addition, the specimen without notches was investigated by a tensile test. The specimens were made from stainless steel 316L, produced by selective laser melting technology in the "as printed" state. A DIC measurement was used to gather data during testing and for postprocessing.

The main insights can be listed as:

- Specimens "as printed" prepared in the vertical direction had good ductility; only about 15% lower than specimens produced by conventional methods.
- Machining had a negative effect on ductility if the specimen was loaded in the printing direction. "As printed" specimens showed approximately 2/3 higher ductility than machined specimens.
- Additionally, the fracture was different for "as printed" and machined specimens. While "as printed" specimens had the character of ductile fracture, the machined specimen only had the local occurrence of ductile fracture and the prevailing features of brittle fracture.
- Vertically printed specimens had lower yield stress than horizontally printed specimens, approximately by 75 MPa, regardless of whether they were in the "as printed" state or machined.
- The FEM-Mises generally overestimated the axial force, leading to a stiffer response under tensile loading.
- The FEM-Hill showed the ability to better describe the yield range and accurately capture (or slightly underestimate) the axial force within the force 'plateau' range.
- The results confirmed the suitability of the material model with Hill's yield criterion to adequately describe the SLM produced material response under tension as the FEM-Hill (which included the yield stress reduction in the printing direction) better captured the yield region than FEM-Mises (which behaved isotropically).

Regarding the comparison of experimental and FEM analysis results under torsion, it was found that the FEM-Mises and FEM-Hill behaved basically identically within the whole examined twist range, capturing the experimental response well, mainly in the nonlinear part. This implies that the reduction of tensile yield strength in the printing direction in the FEM-Hill did not have the influence on the material response under torsional loading in this case. Nevertheless, further studies are needed to fully capture, understand, and address the torsional behavior and printing direction influence on the response during the torsional loading. The nonlinear isotropic hardening model with Hill yield condition gives acceptable results under given multiaxial stress states. Under monotonic loading, the nonlinear isotropic hardening model is equivalent to Chaboche kinematic hardening model with two back-stress parts when the second back-stress part is linear. Therefore, a good correlation can also be expected for the Chaboche model (in monotonic loading cases).

It should also be noted that the vertically printed specimens used in this study were not heat treated in any way, such as annealing to remove internal stress. This can also affect the results, of course, especially in the case of horizontally printed specimens. However, to maintain a high yield stress of the material, it is advisable to choose a gentle heat treatment that preserves the fine-grained microstructure of the material. The following study deals with the comparison of experiments on notched specimens in the "as printed" state and is finally modified by machining. Attention is paid to the adherence of the specimen geometry and surface roughness [32]. In the field of numerical modeling, numerical procedures capturing ductile failure on "as printed" specimens and comparison with conventionally produced specimens with the inclusion of small punch tests are used [33].

Author Contributions: Conceptualization, M.K., M.P., and R.H.; methodology, R.H.; software, D.K.; validation, F.F., R.H., and D.K.; formal analysis, D.K.; investigation, M.K., R.H., M.P., L.K., F.F., and D.K.; data curation, R.H. and F.F.; writing—original draft preparation, M.K., L.K., R.K., and D.K.; writing—review and editing, R.H.; visualization, J.K.; supervision, R.H. All authors have read and agreed to the published version of the manuscript.

Funding: This work was supported by The Ministry of Education, Youth and Sports from the Specific Research Project (SP2020/23), by the Czech Science Foundation (GACR), grant No. 19-03282S, by The Technology Agency of the Czech Republic, project No. TN01000024, and has been done in connection with the DMS project reg. no. CZ.02.1.01/0.0/17_049/0008407, financed by Structural Funds of the European Union.

Institutional Review Board Statement: Not applicable.

Informed Consent Statement: Not applicable.

Data Availability Statement: The data presented in this study are available on request from the corresponding author. The data are not publicly available due to: All data are presented in the form of graphs in this article.

Conflicts of Interest: The authors declare no conflict of interest. The funders had no role in the design of the study; in the collection, analyses, or interpretation of data; in the writing of the manuscript, or in the decision to publish the results.

References

1. Šebek, F.; Kubík, P.; Hůlka, J.; Petruška, J. Strain hardening exponent role in phenomenological ductile fracture criteria. *Eur. J. Mech. Solids* **2016**, *57*, 149–164. [CrossRef]
2. Marya, M.; Singh, V.; Marya, S.; Hascoet, J.Y. Microstructural development and technical challenges in laser additive manufacturing: Case study with a 316L industrial part. *Metall. Mater. Trans.* **2015**, *46*, 1654–1665. [CrossRef]
3. Kořínek, M.; Halama, R.; Fojtík, F.; Pagáč, M.; Krček, J.; Šebek, F.; Krzikalla, D. Monotonic Testing of 3D Printed SS316L under Multiaxial Loading. In Proceedings of the Experimental Stress Analysis 2020, Ostrava, Czech Republic, 19–22 October 2020; VSB—Technical University of Ostrava: Ostrava, Czech Republic, 2020; pp. 213–224.
4. Metelkova, J.; Kinds, Y.; Kempen, K.; de Formanoir, C.; Witvrouw, A.; Hooreweder, B.V. On the influence of laser defocusing in selective laser melting of 316L. *Addit. Manuf.* **2018**, *23*, 161–169. [CrossRef]
5. Tapia, G.; Khairallah, S.; Matthews, M.; King, W.E.; Elwany, A. Gaussian process-based surrogate modeling framework for process planning in laser powder-bed fusion additive manufacturing of 316L stainless steel. *Int. J. Adv. Manuf. Technol.* **2018**, *94*, 3591–3603. [CrossRef]
6. Pham, M.S.; Dovgyy, B.; Hooper, P.A. Twinning induced plasticity in austenitic stainless steel 316L made by additive manufacturing. *Mater. Sci. Eng.* **2017**, *704*, 102–111. [CrossRef]
7. Chen, X.; Li, J.; Cheng, X.; He, B.; Wang, H.; Huang, Z. Microstructure and mechanical properties of the austenitic stainless steel 316L fabricated by gas metal arc additive manufacturing. *Mater. Sci. Eng.* **2017**, *703*, 567–577. [CrossRef]
8. Tan, Z.E.E.; Pang, J.H.L.; Kaminski, J.; Pepin, H. Characterisation of porosity, density, and microstructure of directed energy deposited stainless steel AISI 316L. *Addit. Manuf.* **2019**, *25*, 286–296.

9. Leicht, A.; Klement, U.; Hryha, E. Effect of build geometry on the microstructural development of 316L parts produced by additive manufacturing. *Mater. Charact.* **2018**, *143*, 137–143. [CrossRef]
10. Kaynak, Y.; Kitay, O. The effect of post-processing operations on surface characteristics of 316L stainless steel produced by selective laser melting. *Addit. Manuf.* **2019**, *26*, 84–93. [CrossRef]
11. Wang, D.; Wu, S.; Yang, Y.; Dou, W.; Deng, S.; Wang, Z.; Li, S. The Effect of a Scanning Strategy on the Residual Stress of 316L Steel Parts Fabricated by Selective Laser Melting (SLM). *Materials* **2018**, *11*, 1821. [CrossRef]
12. Wu, A.S.; Brown, D.W.; Kumar, M.; Gallegos, G.F.; King, W.E. An Experimental Investigation into Additive Manufacturing-Induced Residual Stresses in 316L Stainless Steel. *Metall. Mater. Trans. A Phys. Metall. Mater. Sci.* **2014**, *45*, 6260–6270. [CrossRef]
13. Bartlett, J.L.; Croom, B.P.; Burdick, J.; Henkel, D.; Li, X. Revealing mechanisms of residual stress development in additive manufacturing via digital image correlation. *Addit. Manuf.* **2018**, *22*, 1–12. [CrossRef]
14. Hagara, M.; Lengvarský, P.; Bocko, J.; Pavelka, P. The Use of Digital Image Correlation Method in Contact Mechanics. *Am. J. Mech. Eng.* **2016**, *4*, 445–449.
15. Paška, Z.; Halama, R.; Fojtík, F.; Gajdoš, P. Strain response determination in notched specimens under multiaxial cyclic loading by DICM. In Proceedings of the EAN 2017—55th Conference on Experimental Stress Analysis 2017, Novy Smokovec, Slovakia, 30 May–1 June 2017; TU–Košise: Košice; Slovakia, 2017; pp. 143–149.
16. Gibson, I.; Rosen, D.; Tucker, B.S. Direct Digital Manufacturing. In *Additive Manufacturing Technologies*; Springer: New York, NY, USA, 2014; pp. 375–397. ISBN 978-1-4939-2112-6. [CrossRef]
17. 3D Hubs. Introduction to Metal 3D Printing | 3D Hubs. Available online: https://www.3dhubs.com/knowledge-base/introduction-metal-3d-printing/ (accessed on 16 June 2020).
18. Hajnyš, J.; Pagáč, M.; Mesíček, J.; Petrů, J.; Krol, M. Influence of scanning strategies parameters on residual stress in SLM process according to bridge curvature method for stainless steel AISI 316L. *Materials* **2020**, *13*, 1659. [CrossRef]
19. Hajnyš, J.; Pagáč, M.; Kotera, O.; Petrů, J.; Scholz, S. Influence of Basic Process Parameters on Mechanical and Internal Properties of 316L Steel in SLM Process for Renishaw AM400. *MM Sci. J.* **2019**, 2790–2794. [CrossRef]
20. Data Sheets—Additive Manufacturing. Object Moved©. Available online: https://www.renishaw.com/en/data-sheets-additive-manufacturing--17862 (accessed on 16 June 2020).
21. Renishaw. *Magics Training: Material Profile Editing. Apply Innovation*; Renishaw: New Mills, UK, 2016; Available online: https://www.renishaw.com (accessed on 16 June 2020).
22. Lecompte, D.; Smits, A.; Bossuyt, S.; Sol, H.; Vantomme, J.; Van Hemelrijck, D.; Habraken, A.M. Quality assessment of speckle patterns for digital image correlation. *Opt. Lasers Eng.* **2006**, *44*, 1132–1145. [CrossRef]
23. Hitzler, L.; Hirsch, J.; Heine, B.; Merkel, M.; Hall, W.; Öchsner, A. On the anisotropic mechanical properties of selective laser-melted stainless steel. *Materials* **2017**, *10*, 1136. [CrossRef]
24. Bian, P.; Shi, J.; Liu, Y.; Xie, Y. Influence of laser power and scanning strategy on residual stress distribution in additively manufactured 316L steel. *Opt. Laser Technol.* **2020**, *132*, 106477. [CrossRef]
25. Zhang, N.; Li, H.; Zhang, X.; Hardacre, D. Review of the fatigue performance of stainless steel 316L parts manufactured by selective laser melting. In Proceedings of the 2nd Conference on Progress in Additive Manufacturing, Singapore, 16–19 May 2016; Chua, C.K., Yeong, W.Y., Tan, M.J., Liu, E., Tor, S.B., Eds.; Research Publishing: Singapore, 2016.
26. Afkhami, S.; Dabiri, M.; Alavi, S.H.; Björk, T.; Salminen, A. Fatigue characteristics of steels manufactured by selective laser melting. *Int. J. Fatigue* **2019**, *122*, 72–83. [CrossRef]
27. Song, X.; Feih, S.; Zhai, W.; Sun, C.N.; Li, F.; Maiti, R.; Wei, J.; Yang, Y.; Oancea, V.; Brandt, L.R.; et al. Advances in additive manufacturing process simulation: Residual stresses and distortion predictions in complex metallic components. *Mater. Des.* **2020**, *193*, 108779. [CrossRef]
28. Ferro, P.; Berto, F.; Romanin, L. Understanding powder bed fusion additive manufacturing phenomena via numerical simulation. *Frattura Integrità Strutt.* **2020**, *14*, 252–284. [CrossRef]
29. Colby, R.B. Equivalent Plastic Strain for the Hill's Yield Criterion under General Three-Dimensional Loading. Ph.D. Thesis, Massachusetts Institute of Technology, Cambridge, MA, USA, 2013.
30. Altan, T.; Tekkaya, A.E. Plastic deformation—State of stress, yield criteria flow rule, and hardening rules. In *Sheet Metal Forming: Fundamentals*, 1st ed.; ASM International: Materials Park, OH, USA, 2012; pp. 53–73.
31. Bertin, M.; Hild, F.; Roux, S. On the identifiability of Hill-1948 plasticity model with a single biaxial test on very thin sheet. *Strain* **2017**, *53*, 12233. [CrossRef]
32. Lichovník, J.; Mizera, O.; Sadílek, M.; Čepová, L.; Zelinka, J.; Čep, R. Influence of Tumbling Bodies on Surface Roughness and Geometric Deviations by Additive SLS technology. *Manuf. Technol.* **2020**, *20*, 342–346. [CrossRef]
33. Kubík, P.; Šebek, F.; Petruška, J.; Hůlka, J.; Park, N.; Huh, H. Comparative investigation of ductile fracture with 316L austenitic stainless steel in small punch tests: Experiments and simulations. *Theor. Appl. Fract. Mech.* **2018**, *98*, 186–198. [CrossRef]

MDPI
St. Alban-Anlage 66
4052 Basel
Switzerland
Tel. +41 61 683 77 34
Fax +41 61 302 89 18
www.mdpi.com

Materials Editorial Office
E-mail: materials@mdpi.com
www.mdpi.com/journal/materials